INDUSTRIAL COMBINED HEAT AND POWER
THE POTENTIAL FOR NEW USERS

CONTENTS

DEPARTMENT OF ENERGY

INDUSTRIAL COMBINED HEAT AND POWER: THE POTENTIAL FOR NEW USERS

The prosp[illegible] [illegible]wer (CHP) plant within those indu[illegible] is presently limited or non-existent.

1984–1[illegible]

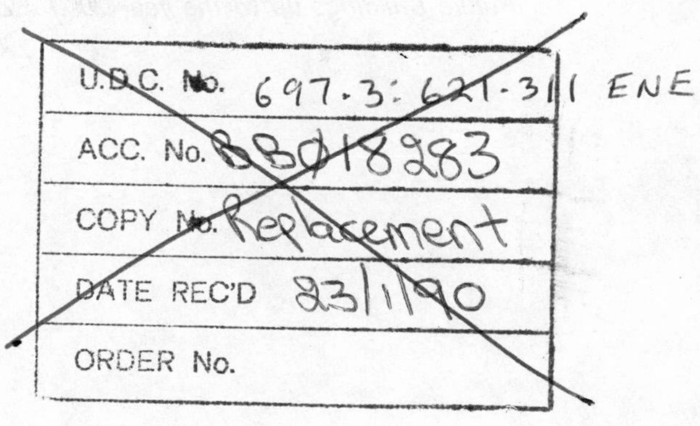

Ove Arup & Partners
13, Fitzroy Street
London
W1P 6BQ
Tel: 071-636-1531

LONDON: HER MAJESTY'S STATIONERY OFFICE

First Published 1988

ISBN 0 11 412944 4

This publication is the seventh in the Energy Efficiency series published by the Energy Efficiency Office. The series is primarily intended to create a wider public understanding and discussion of the efficient use of energy.

The publications in the series do not necessarily represent Government or Departmental policy.

The first six publications in the series:

1. *Energy Efficiency Demonstration Scheme: A Review* (£8.25)

2. *The Pattern of Energy Use in the UK 1980* (£9.80)

3. *Energy use and Energy Efficiency in UK Manufacturing Industry up to the year 2000 (2 volumes)*
   - *Vol 1: Executive Summary and Main Report* (£8.25)
   - Vol 2: Sector Reports containing the Detailed Analyses of the Industries, their Energy Use and Potential Energy Savings (£16.00)

4. *Energy Efficiency in Low Income Households: An Evaluation of Local Insulation Projects* (£5.00)

5. *Combined Heat and power and electricity generation in British Industry 1983–1988* (£13.95)

6. *Energy use and Energy Efficiency in UK Commercial and Public Buildings up to the year 2000* (£26.00)

MANAGEMENT SUMMARY

Overview

1 This report forms one part of a two part study of private industrial electricity generating capacity in the United Kingdom. The first part involved an assessment of existing installations to establish the current level of generating capacity and electrical output and also prospects for the installation of further capacity on those industrial sites which generate electricity already. The second part, reported here, sought to quantify the prospects for investing in combined heat and power (CHP) plant within those industries where CHP capacity is limited or non-existent.

2 In the context of this study, combined heat and power refers to the generation of electricity combined with the utilisation of the heat output to provide, primarily, the requirements of a manufacturing process. For this reason, the study concentrated on quantifying the investment opportunities for CHP plants based on "conventional" technologies in the unit size range 100 kW to 10 MW electrical output. In particular, it did not consider the smaller packaged micro CHP systems based on spark ignition engines.

Scope of Study

3 The work was primarily a "desk" study, based largely on engineering skills and judgement, rather than individual surveys of industrial sites where CHP could be shown to be a cost effective investment opportunity.

4 The major aim was to assess the potential for CHP investment within UK industries where the majority of sites do not operate a CHP plant. Sensitivity analysis was then carried out for individual "typical" sites in order to identify those features which had the greatest influence on CHP economics. An analysis of overall investment potential over a range of fossil fuel and electricity prices was then undertaken.

5 The study also investigated, in a qualitative manner, the role that private or public utility companies could play in the development of industrial CHP investment. It also indicated the improvement in economics which may be achievable by considering group CHP schemes involving several industrial sites. A quantitative assessment of the UK potential for such schemes was <u>not</u> attempted.

6 The derivation of CHP potential involved several major tasks which are outlined in the following paragraphs.

CHP Data Base

7 A CHP data base was compiled which included information regarding the performance, capital cost, operating and maintenance costs of

conventional CHP equipment. The following "prime mover" systems were included:

- Gas Turbines in the size range 0.5 - 10 MWe
- Diesel Engines (0.65 - 10 MWe)
- Dual Fuel Diesel Engines (0.65 - 6.4 MWe)
- Industrial Spark Ignition Engines (0.1 - 2 MWe)
- High Pressure Boilers with Back Pressure Steam Turbines (0.1 - 10 MWe)
- Gas Turbine/Steam Turbine Combined Cycle (4 - 8 MWe)

together with the necessary boiler plant and ancillary items such as natural gas compressors. Allowances were also made in the plant cost estimates for building and other civil work plus engineering design and project management.

Industrial Data Base

8 An industrial data base was compiled which contained information of a statistical nature, derived from government sources, and information relating to "typical" sites within each of the fifty industries selected for assessment. The latter information was obtained in part from published documents and contacts with industrialists. However, it also relied heavily on the more general knowledge of industrial processes and patterns of energy consumption accumulated by the Energy Technology Support Unit, Ove Arup and Partners, Trade associations and others.

Economic Assessment

9 An assessment of CHP economics for each "typical" site was carried out using a computer model. The program assessed over 600 different CHP plant configurations, drawn from the CHP data base. The most cost effective scheme was then selected within particular payback criteria. By scaling up these individual typical site assessments, the "Gross Potential" was obtained for various payback periods. The term "Gross Potential" refers to the total capacity which would be installed if every industrialist decided to implement a CHP scheme, identified by the computer model, which lay within a particular payback period.

Derivation of Investment Potential

10 In practice the Gross Potential represents an upper limit and it was necessary to derive an estimate of the investment potential over the next 5-10 years. A small number of businesses were contacted in sectors of industry which appeared to offer good prospects for CHP.

Their views were sought on the investment criteria employed by their companies for energy efficiency projects with particular reference to CHP. From the data collected a propensity to invest schedule was derived which was used to transform the gross potential into an investment potential. The methodology employed was as follows:

11 Forty individuals were invited to complete a CHP site assessment form. This information was used to assess the technical and economic feasibility of a CHP project on the particular site. Results for their sites were submitted to 22 respondents who were contacted to obtain their views on the prospects for CHP in their company.

12 This exercise served several purposes. It provided supplementary information on sites where a CHP investment might be cost effective and the data was included in the industrial data base. It served to check the model and provided a means of obtaining opinions regarding barriers to investment in CHP, attitudes towards private utility companies etc. Respondents were also asked to judge the chances that a major hypothetical cost saving project would be implemented for payback periods in the range 1 to 5 years. The responses to this question were used as a basis for calculating the "Investment Potential". This refers to the amount of CHP capacity which might actually be installed given the present investment criteria, over the next 5-10 years, within those UK industries where CHP capacity is, at present, limited or non-existent.

Changes in Energy Costs

13 This study commenced in Autumn 1984 and the energy prices used for the initial assessments were based on typical industrial prices prevailing in December 1984. These were:

Interruptible Gas	28.3p/Therm	£2.68/GJ
Heavy Fuel Oil	15.6p/litre	£3.80/GJ
Coal	£62.5/tonne	£2.20/GJ

14 By the end of 1985, the price of heavy fuel oil had declined to a level comparable to the price of interruptible gas. For this reason, both the "Gross Potential" and "Investment Potential" were re-assessed using a heavy fuel oil price of 11p/litre (£2.68/GJ) equivalent to that of interruptible gas.

15 During the first few months of 1986 the crude oil price fell from \$28 per barrel to around \$10-12 per barrel. This resulted in a fall in heavy fuel oil price and a more gradual decrease in the price of interruptible gas. The situation regarding future energy price levels and differentials remained highly uncertain during the time of the

study. Currently uncertainty exists about British Gas pricing policy and the effect of electricity privatisation on CHP operators.

16 The CHP "Investment Potential" is affected by energy price changes in at least two ways. The economics of particular CHP schemes are sensitive to both the ratio of fossil fuel to electricity prices and price differentials between heavy fuel oil, interruptible gas and coal; both these features have changed significantly during the course of this study. Furthermore, industrial investment criteria are dependent on the risks perceived by individual managers. These might include possible changes in product demand, changes to the site operations, changes in fuel price levels and differentials, capital rationing and cost of capital in addition to perceived technical risks associated with a particular investment opportunity. At best a desk study can only take account of these features in a qualitative manner. These uncertainties have been compounded by the recent changes in fuel prices which in due course will presumably be reflected in the general business environment and hence affect the criteria adopted by industrialists for cost saving investments.

CHP Investment Potential

17 In an attempt to estimate investment potential in these changed circumstances sensitivity analysis was applied to a range of fossil fuel and electricity prices. Electricity prices were raised to 1986 levels and capital cost was increased by 5% to allow for general inflation. Revised "Investment Potentials" were then calculated on the basis of equivalent prices for gas and heavy fuel oil in the range 12 to 32p/Therm and for progressively reducing electricity tariffs. The results are presented in Table 1.

18 The data were derived by extrapolating from the results of the original assessments carried out at higher fossil fuel prices. Although the method chosen for these calculations is somewhat unrefined it was felt that more detailed analyses could not be justified in the light of the uncertainties outlined above.

19 Overall it would appear that the total CHP capacity, likely to be installed over the next 5-10 years within those UK industries which have little or no capacity at present, will be, at most, a few hundred MWe. If fossil fuel and electricity prices remain at roughly their current level the indications are that some 200-500 MWe of new CHP capacity will be installed. This corresponds to the installation of 5 to 10 new industrial CHP plants per year.

Opportunities in Industry

20 Estimates of the "Investment Potentials" for individual UK industries are presented in Table 2. These estimates were calculated for the following price scenario:

Heavy Fuel Oil	20p/Therm (7.8p/litre)	£1.89/GJ
Interruptible Gas	20p/Therm	£1.89/GJ
A 5% reduction on 1986 Electricity Prices		

21 Under these circumstances the total estimated "Investment Potential" amounts to some 285 MWe. The data in Table 2 indicate that over 80% of this potential could be achieved within only nine industries; a further 16% could be achieved within 14 industries.

22 Although the level of investment potential in any individual industry is uncertain, it is unlikely that the ranking of industry shown in Table 2 would change significantly with energy prices.

23 To achieve the full Investment Potential of about 285 MW, under this particular price scenario, would require the installation of new CHP systems on about 55 industrial sites. However, about 270 MW could be installed if the 30 larger sites decided to invest. Of these sites, 10 are situated in the chemicals industry, 8 in food and drink, 4 in engineering and 8 elsewhere.

The Role for Utility Companies

24 The comments received from a small number of industrialists revealed that the major barriers to investment in CHP relate to economics rather than to concerns of a technical nature. Similar comments were made by industrialists currently operating CHP plant. In general, companies considering the implementation of energy cost saving projects are seeking payback periods of 2-3 years or less. This in part reflects the perceived uncertainties mentioned earlier.

25 Even with the most favourable energy price scenario the number of CHP projects which could achieve payback periods of this length is relatively small. At the longer payback periods, more typical of CHP schemes, the chances of implementation decrease rapidly.

26 The study showed that, in certain circumstances, the payback period for a "group" CHP scheme comprising several industrial sites can be substantially more attractive than the payback period which could be achieved by any of the individual sites. In particular, smaller sites with individual heat demands below about 10 MW are unlikely to achieve sufficiently attractive economics to warrant investment in 'whole site' CHP plant. When several smaller sites are combined an economically viable CHP project may be possible. In practice, group CHP schemes of this nature are very rare in UK industry. This is perhaps because individual companies tend to view such schemes as simply compounding what is already perceived to be a

financially risky investment by linking its economic success to the continuing profitability of adjacent sites.

27 In principle, such schemes could be financed and operated by separate utility companies; either private or public. Several private utility companies were contacted during the course of the study and in general they were prepared to take a longer term view of CHP investment than most industrialists. The industrialists contacted during the course of this study expressed considerable scepticism regarding the prospects for utility company investment in CHP on their particular sites. However, the recent changes in fossil fuel prices might possibly act as a stimulus to this type of approach to CHP investment.

Conclusions

28 The assessment of the likely level of investment in combined heat and power plant within UK industries, where present capacity is limited or nonexistent, is hampered by several difficulties and uncertainties. Intrinsic problems exist regarding both the assessment of CHP economics on a nationwide scale and the assessment of industrialists' criteria for investing in cost saving schemes. These problems have been compounded by recent changes in energy costs which have resulted in uncertainties regarding future tariffs and the impact these changes may have on future attitudes within UK industry.

29 The results of this study indicate that investment in new CHP plant within the industries assessed is unlikely to amount to more than a few hundred MWe over the next 5-10 years. If fossil fuel and electricity prices remain at roughly their May 1986 level, then some 200-500 MWe CHP capacity may be installed. The bulk of this investment is likely to occur on around 30 sites within some 9 industries.

30 The main barriers to CHP investment seem to be economic. For a variety of reasons, companies tend to view CHP as a financially risky investment. Even with an advantageous energy price scenario, only a relatively small proportion of technically feasible schemes can achieve payback periods of less than 2-3 years. The chance that an industrialist will invest decreases rapidly as the payback period increases.

31 A limited investigation of group CHP schemes showed that where several small sites share one CHP installation, the financial viability can be substantially better than for smaller installations on individual sites.

32 In principle, utility companies, both public and private, could be well placed to take advantage of this type of opportunity. However, some industrialists appeared sceptical of the role for utility companies in this area.

Table 1
Overall Investment Potential (MWe)

Electricity Price	Prices for Gas and Heavy Fuel Oil [p/therm]					
	12p	16p	20p	24p	28p	32p
May 1986 Prices	708	497	367	247	161	125
May 1986 Prices Less 5%	603	432	285	195	122	86
May 1986 Prices Less 10%	497	340	209	139	82	42
May 1986 Prices Less 15%	412	254	153	104	53	23
May 1986 Prices Less 20%	261	165	107	57	27	8

Table 2
Investment Potential In Individual Industries

Gas and Heavy Fuel Oil Price - 20p/therm
Electricity Price - May 1986 Less 5%

INDUSTRY	Investment Potential (MWe installed)	
Chemical		
Fertilisers	41	
Pharmaceuticals	28	
Synthetic Resins	21	
Synthetic Rubber	12	
Soap & Detergents	5	
Photographic Materials	3	
Organic Chemicals	1	
Adhesives	1	
Others	2	114
Food Drink and Tobacco		
Distilling	14	
Brewing	16	
Edible Oils	5	
Miscellaneous Foods	2	
Confectionery	4	
Milk Products	4	
Fruit and Veg. Processing	3	
Tobacco	3	
Fish Processing	2	
Petfood	1	
Others	1	55
Paper		
Paper and Board Manufacture	31	
Paper Conversion	5	36
Textiles		
General Textiles	5	5
Plastic & Rubber Processing		
Rubber	19	
Plastics	4	23
General Engineering		
Vehicles, ships and aircraft	50	
Mechanical Engineering	1	51
GRAND TOTAL		285

INDUSTRIAL COMBINED HEAT AND POWER
THE POTENTIAL FOR NEW USERS

Technical Report

CONTENTS

1.0 INTRODUCTION AND KEY FINDINGS

1.1 Industrial Combined Heat and Power

1 Combined heat and power is the on-site generation of electricity combined with the utilisation of heat recovered from the generating plant. In the industrial context, covered by this study, heat recovered from combined heat and power plant (CHP) is used primarily to fulfil the heat requirements of a manufacturing process.

2 CHP plants have been used in some industries for many years but there have been changes which have affected the environment for CHP and now only some 150 companies operate CHP plants. The relatively high fuel prices experienced between 1973 and 1985 encouraged energy intensive industries to invest in equipment to reduce site heat demands, either by conserving energy or changing to a less energy intensive process or product. This can result in existing CHP plants, especially those based on steam turbines which constitute the majority of the systems, becoming uneconomic to operate with the reduced heat demands. Increased fossil fuel costs, especially oil and gas, relative to electricity have also adversely affected the economics of operating existing CHP plant.

3 [Conversely, the Energy Act (1983) was introduced to encourage the private generation of electricity. The Act requires the electricity boards to offer to purchase electricity supplied to them by private generators at realistic prices.]

4 In view of these changes the Energy Efficiency Office felt that it was timely to re-assess the situation regarding the private generation of electricity within U.K. industry and assess the medium term scope for investment in new CHP systems on sites where electricity is generated at present and on industrial sites which currently have no generation facilities.

1.2 Objectives of this Work

5 The work was split into two parts:

- A survey of existing private generators. This survey was carried out by the Department of Energy. A questionnaire was sent to 150 private generators requesting information relating to 1983. These operators were believed to represent 99% of the electricity generated in industry. The survey showed that between 1977 and 1983 the installed capacity of the private generators fell by 21.7% while the electrical output was reduced by 26.6%. Of the installed capacity in 1983, of 3.6 GW, about 2.1 GWe was CHP plant; the majority of which comprised back pressure steam turbine systems. The survey also identified that although there would be modest investment in new CHP this would be more than offset by likely reductions in generating plant among existing private generators resulting in a net reduction of 245 MWe between 1983 and 1988. However, the recent changes in energy prices plus the availability of more attractive electricity tariffs, as a result of the Energy Act (1983), mean that the prospects for CHP on existing sites may be better than perceived in 1984. This survey was completed during 1985 and has been published by the Energy Efficiency Office (Combined Heat and Power Generation in British Industry 1983-1988, Energy Efficiency Series, No. 5, HMSO 1986).

- A study to determine the likely level of investment in technically proven CHP in industries which currently have little or no generating capacity. This study was intended to provide an initial assessment of the potential for industrial CHP and as such took the form of a "desk" study based on engineering skills and judgement rather than a detailed survey of sites where CHP could be a cost effective investment. The study was carried out, through the Energy Technology Support Unit (ETSU), by Ove Arup and Partners and under the guidance of a Steering Group comprising officials from the Department of Energy, ETSU, the Electricity Council, British Gas, British Coal and Ove Arup and Partners.

1.3 Key Findings

6 The Ove Arup study utilised a computer model to assess the economics of CHP installations on "typical" sites within 50 UK industrial sectors. This was based on information regarding the performance and costs of the various prime mover systems combined with relevant data relating to "typical" sites within each industry. The results were "scaled-up" on the basis of statistical information obtained from a government survey of industrial purchases relating to 1979. Further details are presented in Appendices 2 and 3 and these aspects of the study are summarised in Chapters 3 and 4.

Gross Potential

7 This computerised assessment yielded estimates of the "Gross Potential" for CHP investments within those 50 industries where the present capacity is limited or non-existent. The "Gross Potential" was calculated for various ranges of payback period. the results are displayed graphically in Figures 7.1 and 7.2. Throughout this report, "Gross Potential" refers to the amount of CHP capacity, in terms of MW of electricity generation, which would be installed if all the UK sites which could obtain a payback period in the range indicated decided to implement the

particular opportunity produced by computer analysis. The "Gross Potentials" presented in this report were based on gas and heavy fuel oil prices of 28p/th and a typical electricity tariff prevalent during 1984/85. For instance, if the 81 sites identified by the computer analysis as having the potential for a CHP investment of 5 years or less decided to implement these schemes the total installed capacity would amount to 730 MWe. Details of the calculation of "Gross Potential" and the assumptions involved are presented in Appendix 1 and summarised in Chapter 7.

Likely Investment in CHP

8 In practice only a small proportion of the sites, which contribute to the "Gross Potential" will actually implement their particular opportunities. On an individual site there are a variety of factors which affect such an investment decision. Investment in industrial CHP is often perceived as a relatively "high risk" cost saving investment from the financial and, to a lesser extent, the technical point of view. The risks arise in part from uncertainties regarding future energy prices but also from uncertainties relating to the future level of site output and product mix. These have a direct effect on the site demands for heat and power which influence the economics of CHP. On the technical side, there are general concerns regarding the reliability and the cost of maintaining CHP plants. These risks are compounded by the relatively high capital cost of a CHP investment. Capital is often considered to be expensive or in short supply and a CHP investment may be competing with other opportunities which are related more directly to the main stream business activity. Industry tends to offset these problems and uncertainties by specifying a relatively short payback period for investment in CHP plant. The barriers to investment in CHP are discussed more fully in Chapter 15.

Investment Potential

9 In an attempt to estimate the CHP capacity which might actually be installed in the 50 industries encompassed by this study, a limited number of selected individuals in industrial organisations were contacted. Their subjective views were sought on several aspects of CHP including the chances that their company would invest in a hypothetical cost saving project, of a particular payback period. The replies were used to estimate the proportion of the "Gross Potential" which might be installed over a 10 year period. This quantity is referred to as the 'Investment Potential'. The results are presented in Table 10.1 at a variety of fossil fuel and electricity prices. If prices remain at roughly their present day level, in real terms, the likely level of CHP investment could amount to some 200-500 MWe

over a 10 year period. For details see Chapters 9 to 11 and Appendix 1.

10 In addition to these "site-wide" CHP systems several smaller, retro-fit schemes were also evaluated. These involved the installation of small spark-ignition reciprocating engines driving directly a device such as an air or refrigeration plant compressor, or a small back pressure steam turbine either in a "direct drive" configuration or generating electricity. This type of scheme could result in the installation of a further 60 MW capacity over the next 10 years. Details are presented in Chapter 14 and Appendix 4.

Reducing Capital Cost

If the capital cost were to decrease in real terms this
11 If the capital cost were to decrease in real terms this
would result in enhanced rates of return and would lead to a
would result in enhanced rates of return and would lead to a **disproportionate increase in the level of investment.** If, for example, capital and installation costs were to reduce by 25% the capacity installed over the next 10 years could double to around 500-900 MWe. Two other developments were identified whereby the level of investment in industrial CHP could be increased, namely group and utility schemes. Further details are presented in Chapter 12.

Group CHP Schemes

12 In certain circumstances CHP economics can be improved significantly by combining several neighbouring sites into a "Group CHP Scheme". This is particularly the case where several relatively small sites are combined. Quantification of the potential scope for such arrangements is difficult and not attempted by this study. It will depend on the geographical location of suitable sites and the improved economics will be offset, at least to some extent, by an increased in perceived risk and the contractual complexity of such arrangements. This is discussed in Chapter 16.

Utility CHP

13 An alternative way forward would involve utility company investment in industrial CHP; both private and public. In contrast to the situation in the USA such arrangements are rare in the UK. Several private utility companies were contacted during 1985. Despite the encouragement given to this type of arrangement by the Energy Act 1983, utility companies seemed to view industrial CHP as a fringe activity. They tended to concentrate on heat-only systems especially those which involved a conversion from gas or oil to coal firing. The industrialists contacted were also sceptical regarding the economic benefits which might be obtainable from such an association. On the other hand, utility companies did appear to take a longer term view of

investments in CHP or boiler plant than most industrialists. They may also possess the necessary expertise to finance, operate and manage CHP plant on single sites or as in a "group scheme". This aspect is discussed further in Chapter 17.

Conclusion

14 To summarise, the level of investment in industrial CHP capacity could amount to 200-500 MWe over the next decade or so within those industries where present capacity is limited or non-existent. This estimate is based on the assumptions that energy prices and capital costs will remain at roughly their present levels (in real terms), that industrial investment criteria also remain unchanged and that the relevant sites become generally aware of the benefits that can be obtained by investing in a suitable CHP plant. Unless plant capital and installation costs can be reduced the best way of encouraging more investment in industrial CHP would appear to be via private or public utility companies. When this study was carried out, during 1985, the prospects in this area did not appear encouraging. The more recent changes in energy prices should improve the prospects but further work would be required to quantify the potential.

2.0 METHODOLOGY

15 The study sought to provide an initial "broad-brush" assessment of the potential for installing technically proven cost effective CHP plants within those sectors of U.K. industry where CHP capacity is limited or non-existent.

16 The method adopted took the form of a desk study to determine the most cost effective CHP systems on typical industrial sites and to sum the results over the whole of industry. The work involved several distinct stages:

- Compilation of a data base of CHP plant, mainly from information supplied by manufacturers.
- Compilation of a data base of energy consumption patterns for industrial sites. Statistical data and information from a wide variety of other sources was used, including the in-house knowledge of Ove Arup and Partners and ETSU.
- System assessments to establish the most cost effective CHP installations on typical sites and scaling up these results to represent the whole of industry.
- Contacts with industry to ascertain current attitudes to investment and hence determine the likely installed CHP capacity over the next 10 years.

17 The scope for new CHP installations across industry was quantified in two ways:

- Gross Potential - The total CHP plant capacity that it would be technically feasible to install within certain economic criteria.
- Investment Potential - The total CHP plant capacity that is likely to be installed given current investment trends in industry.

A diagram illustrating the main stages of the methodology is shown in Figure 2.1. Details of the methodology are described in Appendix 1.

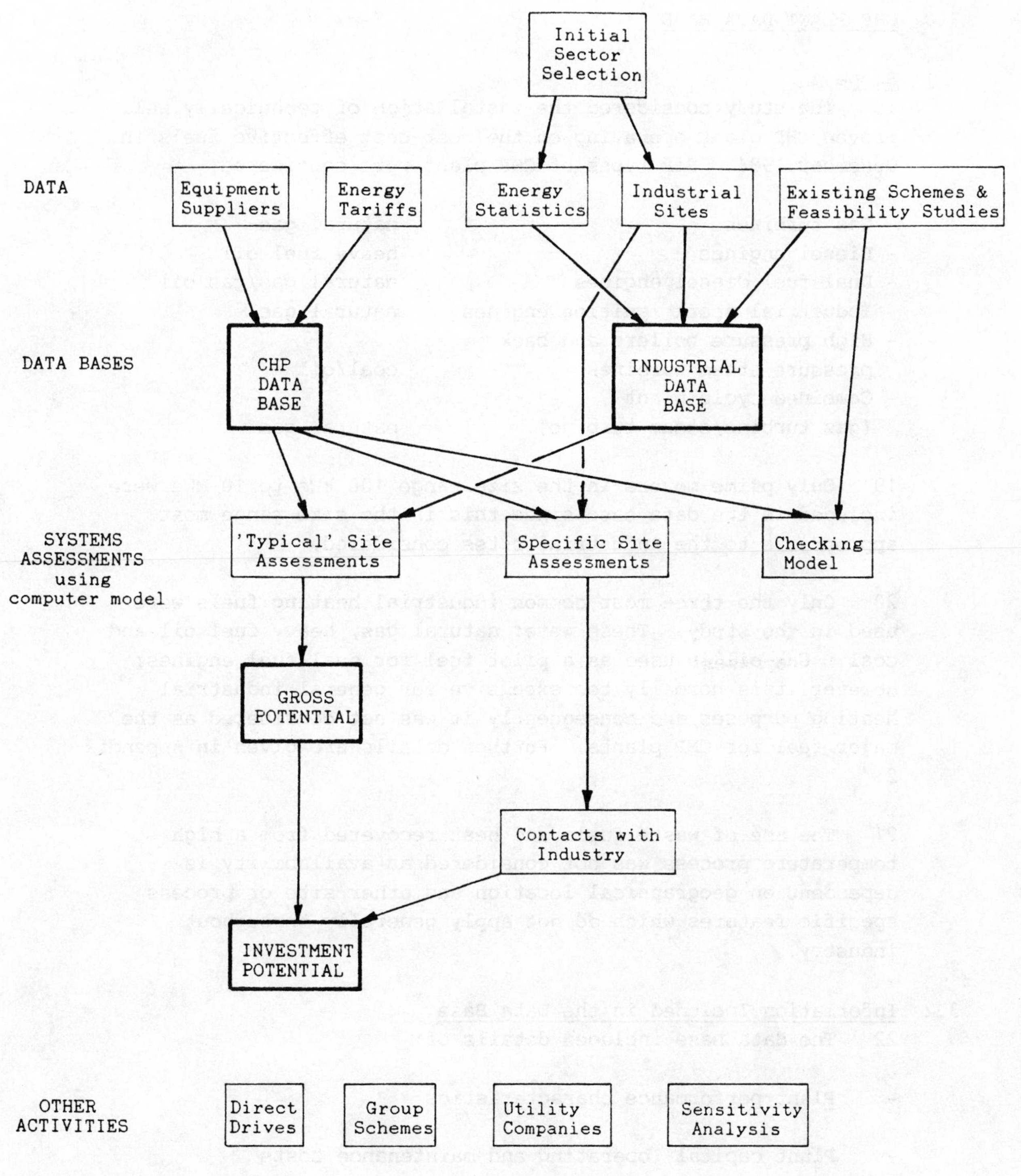

Initial Sector Selection
DATA
Equipment Suppliers
Energy Tariffs
Energy Statistics
Industrial Sites
Existing Schemes & Feasibility Studies
DATA BASES
CHP DATA BASE
INDUSTRIAL DATA BASE
SYSTEMS ASSESSMENTS using computer model
'Typical' Site Assessments
Specific Site Assessments
Checking Model
GROSS POTENTIAL
Contacts with Industry
INVESTMENT POTENTIAL
OTHER ACTIVITIES
Direct Drives
Group Schemes
Utility Companies
Sensitivity Analysis

METHODOLOGY FOR CHP STUDY **Fig. 2.1**

3.0 CHP PLANT DATA BASE

3.1 Scope

18 The study considered the installation of technically well proven CHP plant operating on the most cost effective fuels in December 1984. Six types of CHP plant were considered:

- Gas turbines	natural gas
- Diesel engines	heavy fuel oil
- Dual fuel diesel engines	natural gas/gas oil
- Industrial spark ignition engines	natural gas
- High pressure boilers and back pressure steam turbines	coal/oil
- Combined cycle plant (gas turbine/steam turbine)	natural gas

19 Only prime movers in the size range 100 kWe to 10 MWe were included in the data base since this is the size range most appropriate to the industrial sites considered.

20 Only the three most common industrial heating fuels were used in the study. These were: natural gas, heavy fuel oil and coal. Gas oil is used as a pilot fuel for dual-fuel engines; however it is normally too expensive for general industrial heating purposes and consequently it was not considered as the major fuel for CHP plants. Further details are given in Appendix 2.

21 The use of waste fuels and heat recovered from a high temperature process was not considered as availability is dependent on geographical location and other site or process specific features which do not apply generally throughout industry.

3.2 Information Included in the Data Base

22 The data base included details of:

- Plant performance characteristics

- Plant capital, operating and maintenance costs

A summary of the main performance details which characterise each type of CHP plant and size range available is given in Table 3.1. The plant efficiencies allow for the additional auxiliaries required to operate a CHP plant, as opposed to conventional boilers, which in the case of gas turbines include a gas compressor.

CHP Plant - Key Performance Data

Plant Type	Size Range	Net Electrical Efficiency # of Prime Mover	Overall CHP System Efficiency #		Heat/Power Ratio	
			Unfired	After-fired	Unfired	After-fired
Gas Turbine	< 2MW 2 - 10MW	* 10 - 15% * 15 - 25%	65% 65 - 70%	70 - 73% 70 - 73%	3 - 5 1.5-2.5	6 - 10 3 - 5
Heavy Fuel Oil Diesel Engines	0.65 - 10MW	35 - 37%	65%	75%	0.8	3
Dual-Fuel Diesel Engines	0.65 - 6.4MW	30 - 33%	60 - 70%	71 - 76%	1.1	3
Spark Ignition Engines	100 - 2000kW	28%	72 - 78%	-	1.5-1.7	-
Backpressure Steam Turbines	0.1 - 10MW	2 - 12%	76 - 78%	-	>5	-
Combined Cycle Plant	4 - 8MW	19 - 25%	-	73%	-	2 - 3

* After deduction of typical intake and exhaust losses and auxiliaries including gas compression.

\# Based on Higher Calorific Value.

Table 3.1

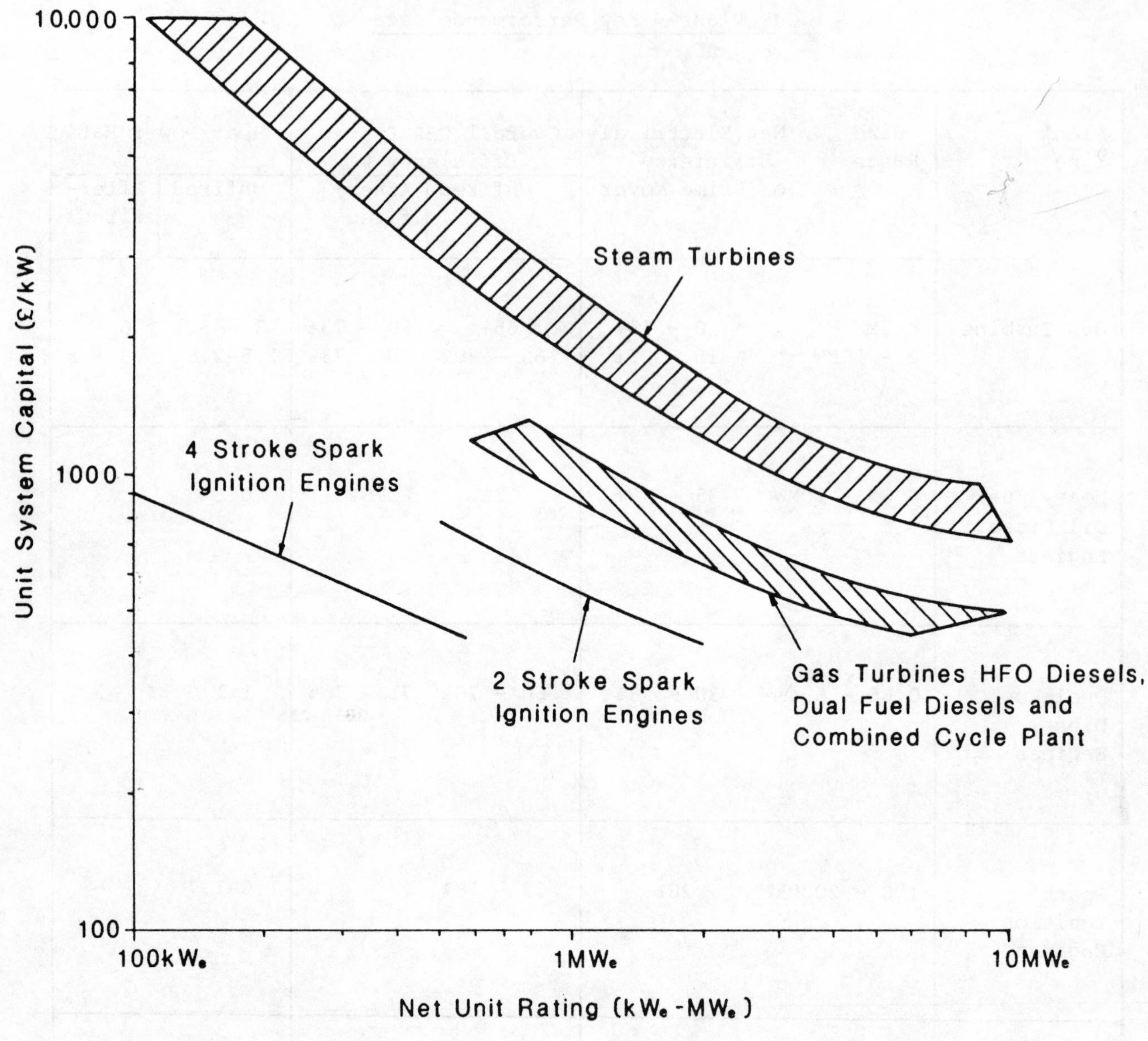

CHP PLANT INSTALLED CAPITAL COSTS **Fig. 3.1**

23 Capital costs were compiled for complete installations. These included not only hardware and installation costs, for all of the main and auxiliary plant, but also the provision of services, civil engineering, design and project management fees and contingencies that would be typical of installations on existing industrial sites. A summary of these costs is shown in Figure 3.1.

24 The plant performance and capital cost data were obtained from selected plant suppliers who are listed in the acknowledgement at the end of this report. The data included in the data base represented conservative plant design to reflect realistic as opposed to best case installations. Full details of the data and assumptions made are given in Appendix 2.

3.3 Comparison of CHP Plant Types

25 Gas turbines are characterised by relatively low heat to power ratios and a high temperature heat output. Relatively high total system efficiencies can be achieved especially if after-firing of the exhaust gases is employed. A disadvantage is the requirement for natural gas at high pressure. Achieving this is costly if high pressure gas is not available from British Gas. For the purposes of this study it is assumed that gas is only available at 26" W.G. (0.065 bar gauge). However, British Gas may frequently be able to supply at a higher pressure which would improve the cost effectiveness of a gas turbine CHP plant.

26 Diesel engines have the highest electrical generation efficiency but a significant proportion of the heat is only available at a low temperature from the engine cooling water. If this cannot be utilised, the total system efficiency may be too low for economic operation. The total system efficiency may be improved however by after-firing the exhaust gases.

27 Spark ignition engines also produce a significant proportion of the heat output as low temperature hot water. As the exhaust gases cannot be after-fired it is important to utilise the engine cooling water in order to achieve a high total system efficiency. Efficient cost effective spark ignition engines are available in smaller unit sizes than other types of plant.

28 Back pressure steam turbines have in the past been the most common type of CHP plant. However the heat to power ratios are high and in general they only appear to be cost effective for some larger applications.

29 The capital costs per unit electrical output of gas turbines, combined cycle and diesel engine CHP plant are in the same order, as shown in Figure 3.1. Spark ignition engines are cheaper at the smaller sizes. Coal fired back pressure steam turbine systems however are considerably more expensive; the high heat to power ratios resulting in the need for large high pressure boilers and the associated auxiliary equipment.

3.4 Retro-fitted Steam Turbines

30 The high capital cost associated with steam turbine CHP plants is associated largely with the cost of a large high pressure boiler to generate the steam. There will be situations however where small back pressure turbines could be retro-fitted onto existing boilers for a much reduced capital cost. This situation is dependent upon the particular conditions prevailing at a site, and cannot be applied generally to the whole of industry. Consequently, retro-fitted steam turbines have been considered separately, as described in Section 14.

4.0 INDUSTRIAL DATA BASE

31 The purpose of the industrial data base was to provide details of the energy consumption patterns in those industries which were likely to have the greatest economic potential for CHP. These consumption patterns were then used as the basis for assessing the economics of operating industrial CHP plants. The data was obtained from two main sources: statistical data of gross energy consumption patterns throughout the whole of industry and; data on individual sites to provide 'hour by hour' profiles of the typical energy usage. The data was used to define 'typical' industrial sites. CHP assessments were carried out on each of these and the results were scaled up to represent the whole of industry.

4.1 Selection of Industries for Assessment

32 Using statistical information from a Department of Energy analysis of the Business Statistics Office 1979 Energy Purchases Enquiry on the energy consumption of industrial sectors, an initial selection exercise was carried out to identify those industries in which CHP could be cost effective. A lower limit of annual energy consumption of 250,000 therms/year was selected. Smaller sites were considered to be unlikely candidates for an industrial CHP installation. Industries where CHP is already well established and those which require high temperature process heat were also excluded.

33 Fifty industries were identified (listed in Appendix 3) in the following broad sectors:

- food, drink and tobacco
- chemicals and allied industries
- rubber and plastics processing
- paper and board
- textiles
- land vehicles, ships and aircraft industries
- mechanical, electrical, electronic and instrument engineering.

34 Each of the 50 industries were divided into as many as three size bands to represent sites of different energy consumption. Each 'typical' site therefore represented a number of actual sites in an industry. These actual sites would use different fuels for their boiler plant and the number of sites were split between gas, heavy fuel oil and coal to represent this diversity. The data in each size band was modified to take account of changes in site heat to power ratio since 1979. And the splits between the fossil fuels burned were adjusted to take account of known conversions to coal since 1979. No attempts were made to

allow for conversions from oil to gas. Also no allowances were made for site closures since 1979.

35 The resulting statistical data for each size band in each industry was then used to:

- Establish the overall energy consumption of a 'typical' site in each size band.
- To provide a method for 'scaling-up' the results of the CHP assessments on each 'typical site' to represent the whole of industry.

Fuller details of this selection process are described in Appendix 3.

4.2 Energy Consumption Data for 'Typical' Sites

36 The overall annual energy consumptions for each 'typical' site were derived from the adjusted statistical data described above. More detailed information was then assembled on the energy demand patterns of each typical site. This included details of the split of consumption between the various consumers of heat, the quality of the heat required, and the daily demand profiles for heat and electricity and the variations throughout the year, based upon the common practices in each industry. Figure 4.1 shows the data assembled to define each 'typical' site. The full industrial data base is presented in Appendix 3.

4.3 Sources of Information

37 The information used to define each 'typical' site was obtained from a wide variety of sources. The overall energy consumptions in the key industries were taken primarily from government statistics and publications concerning national industrial energy use.

38 The details of the demand patterns of each typical site were obtained from:

- Departments of Energy and Trade and Industry Publications, including the Industrial Energy Thrift Scheme Reports and the Energy Audit Series.
- Publications concerned with the use of energy in industry.
- Information provided by industrial federations and research associations.

- Details of site energy consumption, taken from CHP assessment forms, that were provided by 2 industrial establishments.

- In house knowledge of ETSU and Ove Arup & Partners.

A full list of the sources of information is provided in Appendix 3.

39 The information assembled in the industrial data base contains the data necessary to assess the cost effectiveness of CHP plants on 'typical' sites in 50 major industries. This data was not previously available. A wide variety of sources were tapped and a considerable amount of judgement and experience, on the part of Ove Arup and Partners and ETSU, was brought to bear in establishing the data base. Several assumptions were made, concerning changes in industry since the 1979 Energy Purchases Figure 4.1 Enquiry and also in defining the energy demand patterns of individual sites. In particular:

- that the changes in heat to power ratio identified since 1979 are due entirely to changes in heat demand;

- that the shortfall in the coverage of industry of the 1979 purchasers enquiry was similar to the number of site closures since 1979.

40 The effect of some of these assumptions on the estimates of potential for CHP are considered later. Nevertheless the quality of the information in the industrial data base is believed to be sufficient to fulfil the objectives of the study, namely, to provide an initial "broad brush" estimate of the potential for industrial CHP within those sectors of industry where CHP is not yet established.

ENERGY DEMANDS	CONSUMPTION DETAILS	DEMAND PROFILES
		S = SUMMER DAY W = WINTER DAY N = NON WORKING DAY
ELECTRICITY	(i) Total annual consumption	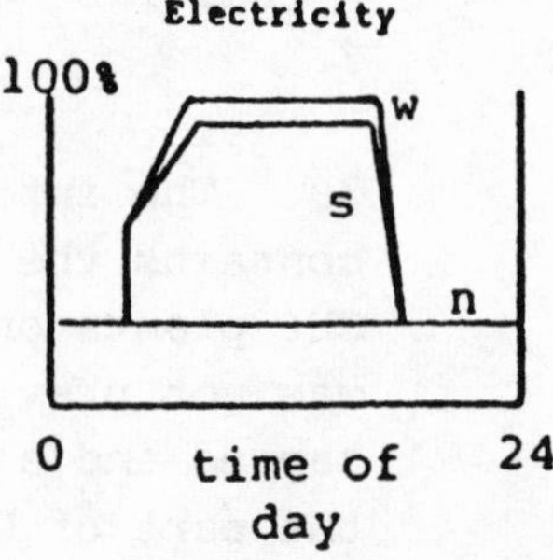
HEAT (Fossil fuel for existing boilers)	(i) Total annual consumption (ii) Steam/Hot water condition (iii) Fraction of demand which can be met by water at less than 85° C (iv) Percentage steam condensate return (v) Efficiency of existing boilers (based on HCV)	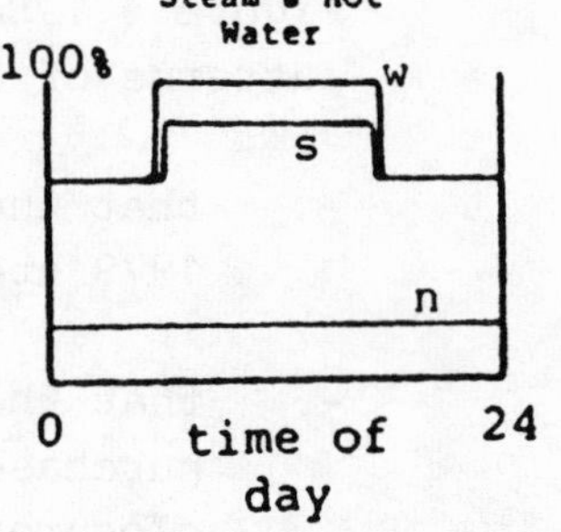
HEAT (Fuel for directly fired plant suitable for gas turbine exhaust)	(i) Total annual consumption (ii) Temperature required	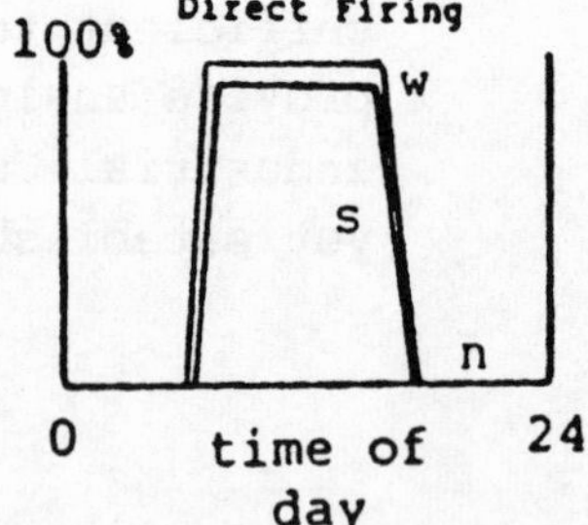

DATA DEFINING AN INDUSTRIAL SITE **Fig. 4.1**

5.0 ASSESSMENT OF CHP SCHEMES

5.1 Computer Model

41 Assessments of the economics of operating CHP plant on industrial sites were undertaken using a computer model. The model used data from 'typical' sites in the industrial data base and plant from the CHP plant data base and undertook an hour by hour economic analysis to identify the most cost effective mode of operating the plant. The hourly costs were then summed over a year to establish the annual operating cost. The program also identified the cost of operating the site without CHP and hence the savings resulting from CHP were assessed. The costs taken into account by the program are indicated in Figure 5.1.

42 The computer model was able to work through the different CHP plants contained within the CHP plant data base and identify the economics of operating many different configurations of plant. For this the most cost effective plant was easily identified. Typically about 600 different CHP plant configurations would be assessed in the analysis of an individual 'typical' site.

43 Checks were carried out to validate the computer model using data from CHP feasibility studies on actual sites. Close agreement was found between the results from the model and the feasibility studies.

44 The computer model was used in two ways:

- To assess the most cost effective CHP plant configuration for each of the 'typical' sites in the industrial data base, and hence derive the economic potential for industrial CHP.
- To assess the most cost effective CHP plant configuration for selected industrial sites as part of the exercise to seek the views of industrialists on investments in CHP.

The computer model is described in more detail in Appendix 2.

CHP PLANT

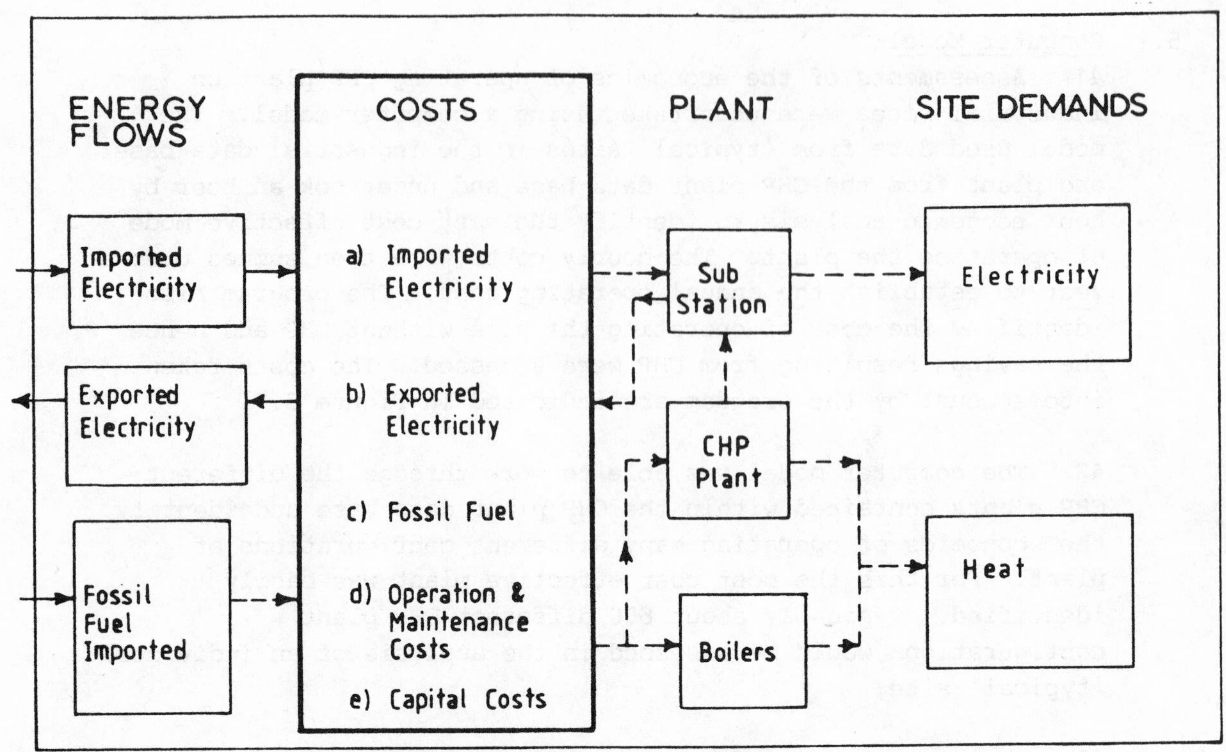

CONVENTIONAL PLANT

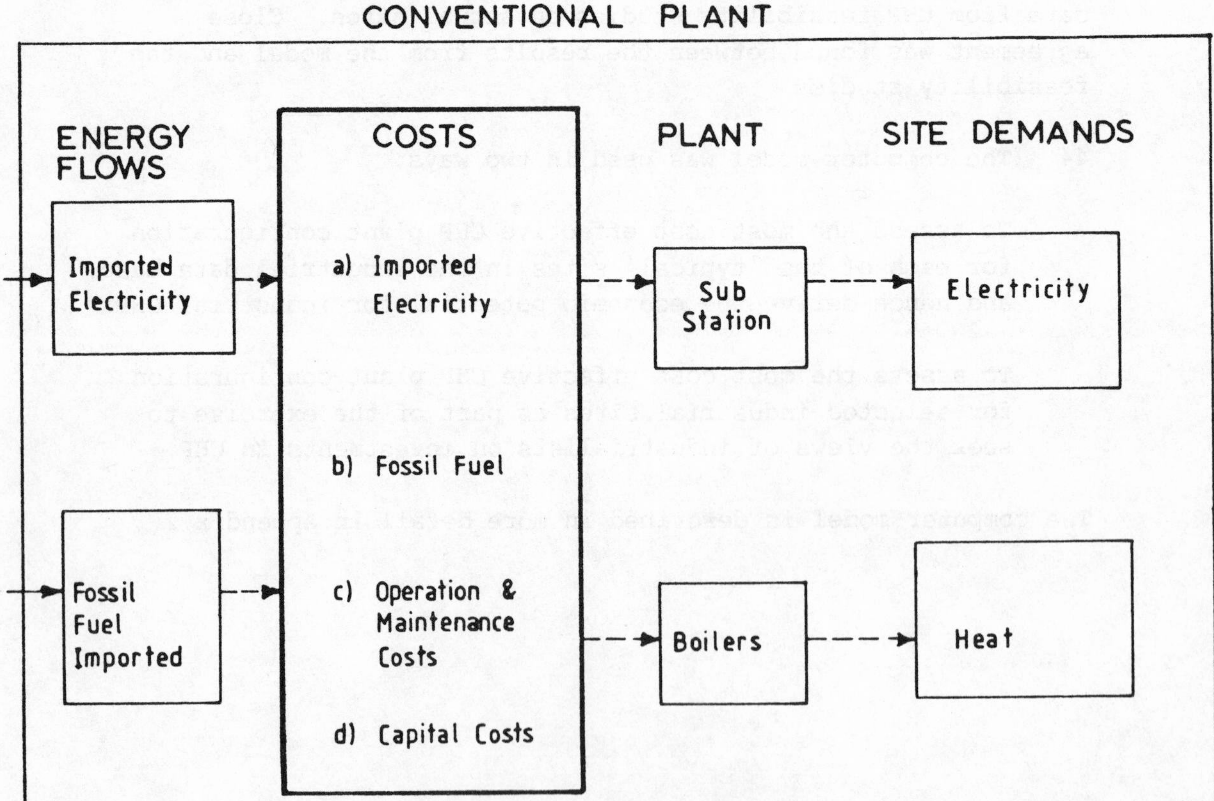

COMPARISON OF COSTS ON A CHP AND A CONVENTIONAL SITE **Fig. 5.1**

5.2 CHP Preliminary Assessment Charts

45 The results of the assessments of each of the 'typical' sites have been summarised in the form of "CHP Preliminary Assessment Charts". These enable an initial CHP assessment to be made from five simple parameters which characterise the site.

46 These charts have been produced for four types of CHP plant: gas turbines; heavy fuel oil diesels; spark ignition engines and steam turbines. They are shown in Figures 5.2 to 5.5. Details of how to use the charts are provided in Appendix 1.

47 The charts estimate the payback period for the most cost effective CHP plant and it has been found that these paybacks are normally within ±20% of those calculated by the computer program.

48 Although general features of industrial sites are contained inherently within the charts, site details, which could affect the choice of CHP plant type and size, are not allowed for. Nevertheless the charts do provide a simple rough guide to an industrialist wishing to make a preliminary CHP assessment, although it is important to stress that such assessments are only valid for industrial sites which fall within the ranges of the charts.

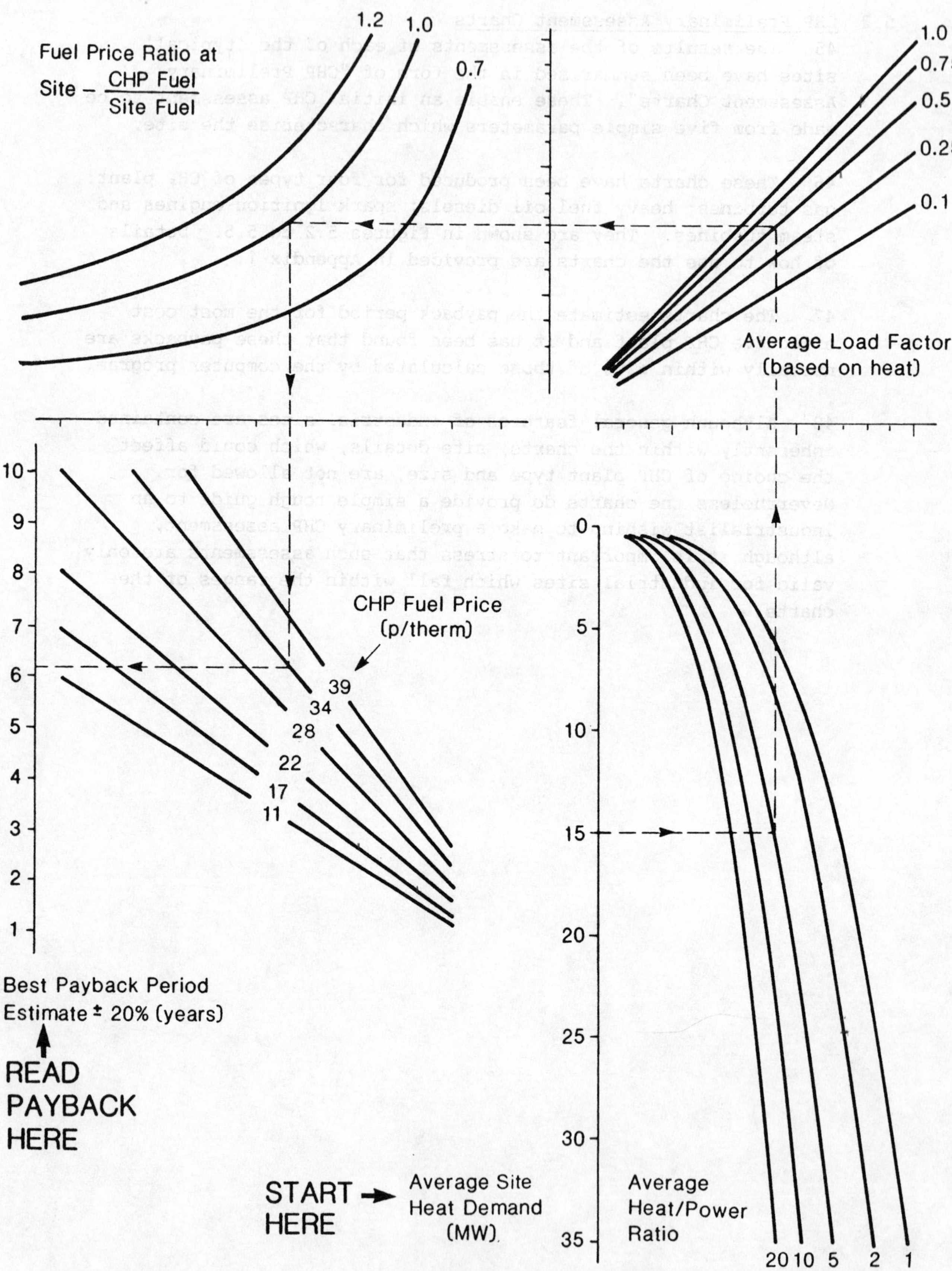

PRELIMINARY CHP ASSESSMENT CHART - GAS TURBINE PLANT (Based on 1985 capital costs)

Fig. 5.2

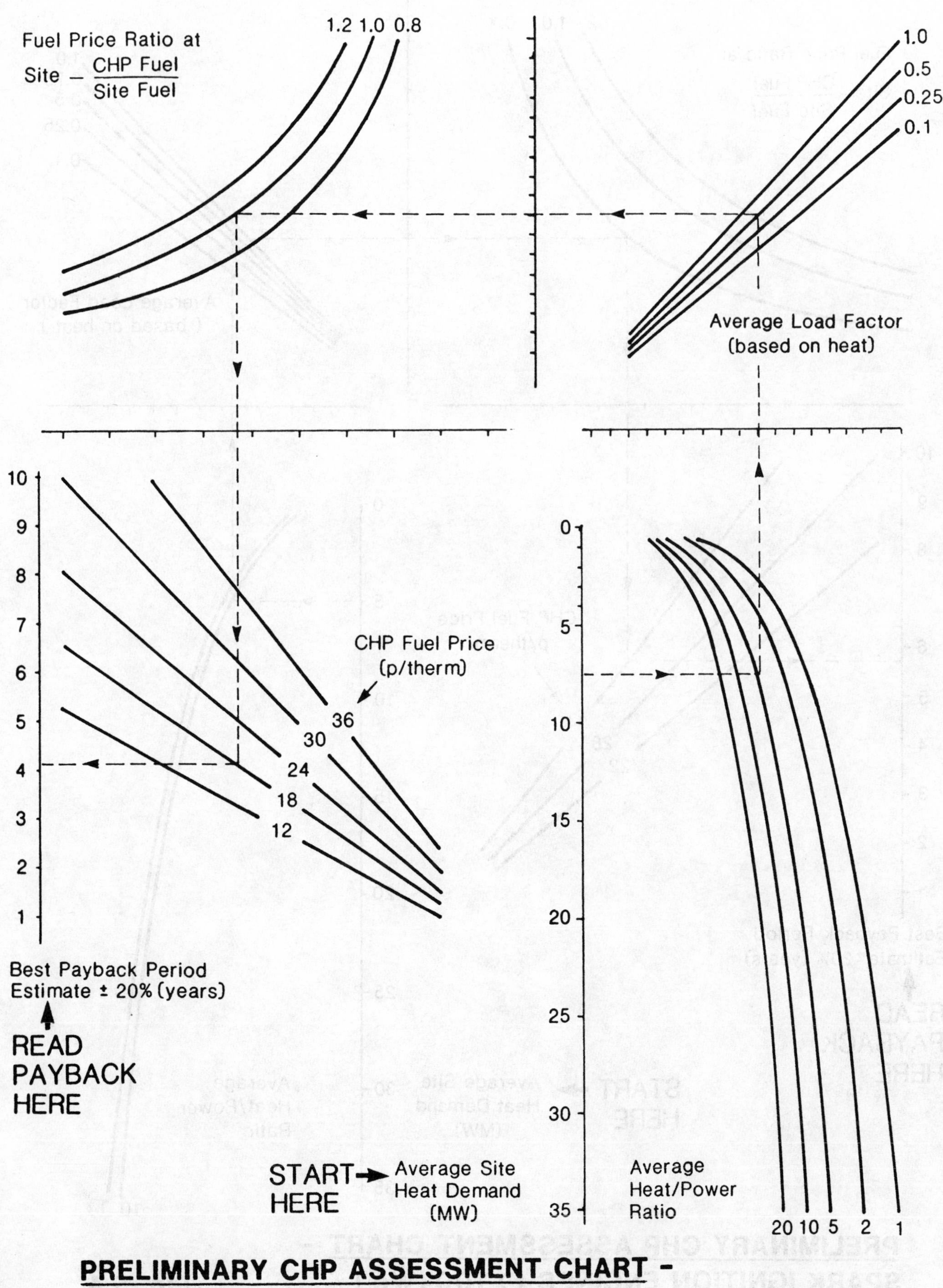

PRELIMINARY CHP ASSESSMENT CHART – HEAVY FUEL OIL DIESELS

(Based on 1985 capital costs)

Fig. 5.3

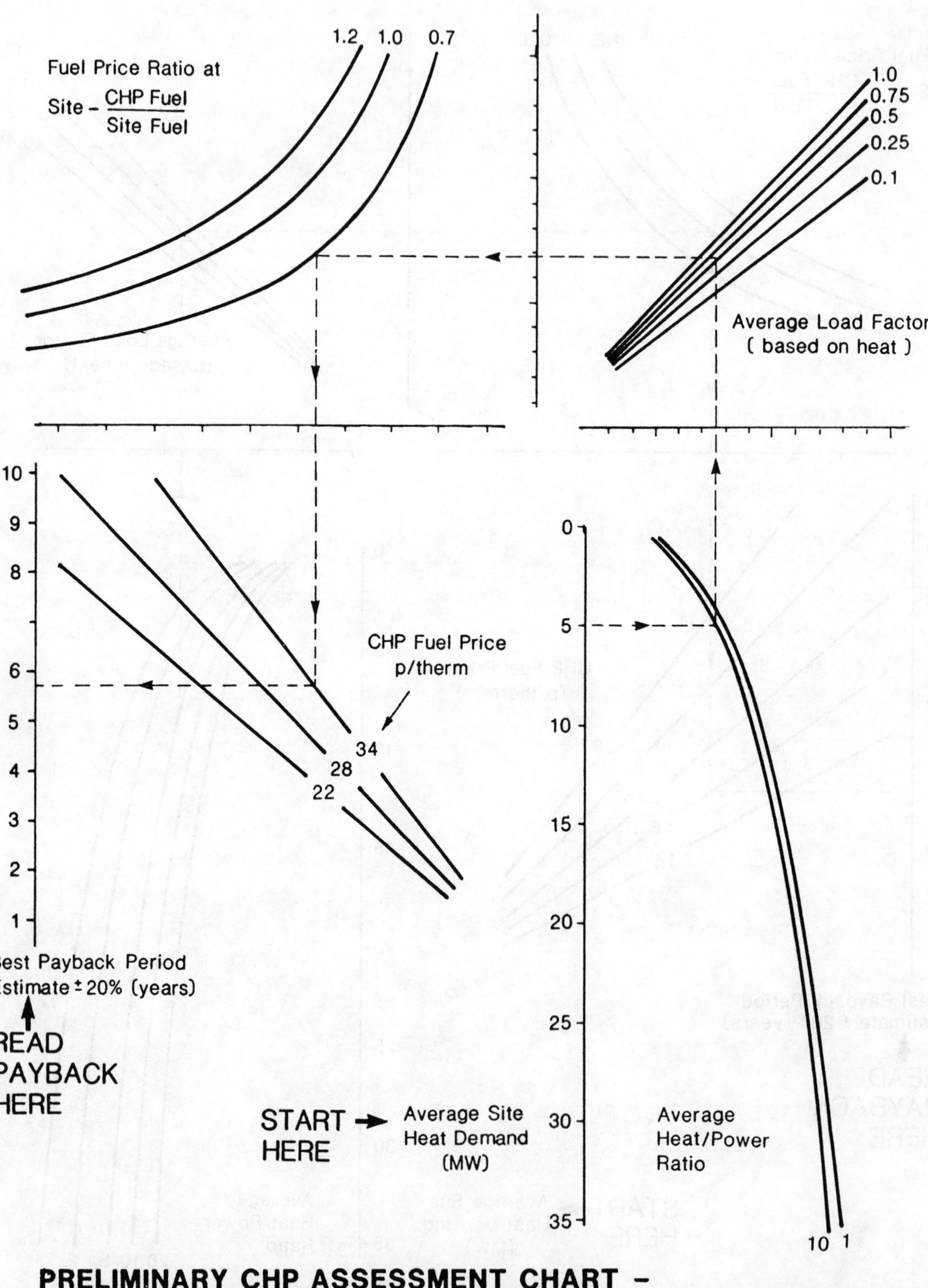

PRELIMINARY CHP ASSESSMENT CHART – SPARK IGNITION ENGINES (>100kWe.)

Industrial Sites (Based on 1985 capital costs)

Fig. 5.4

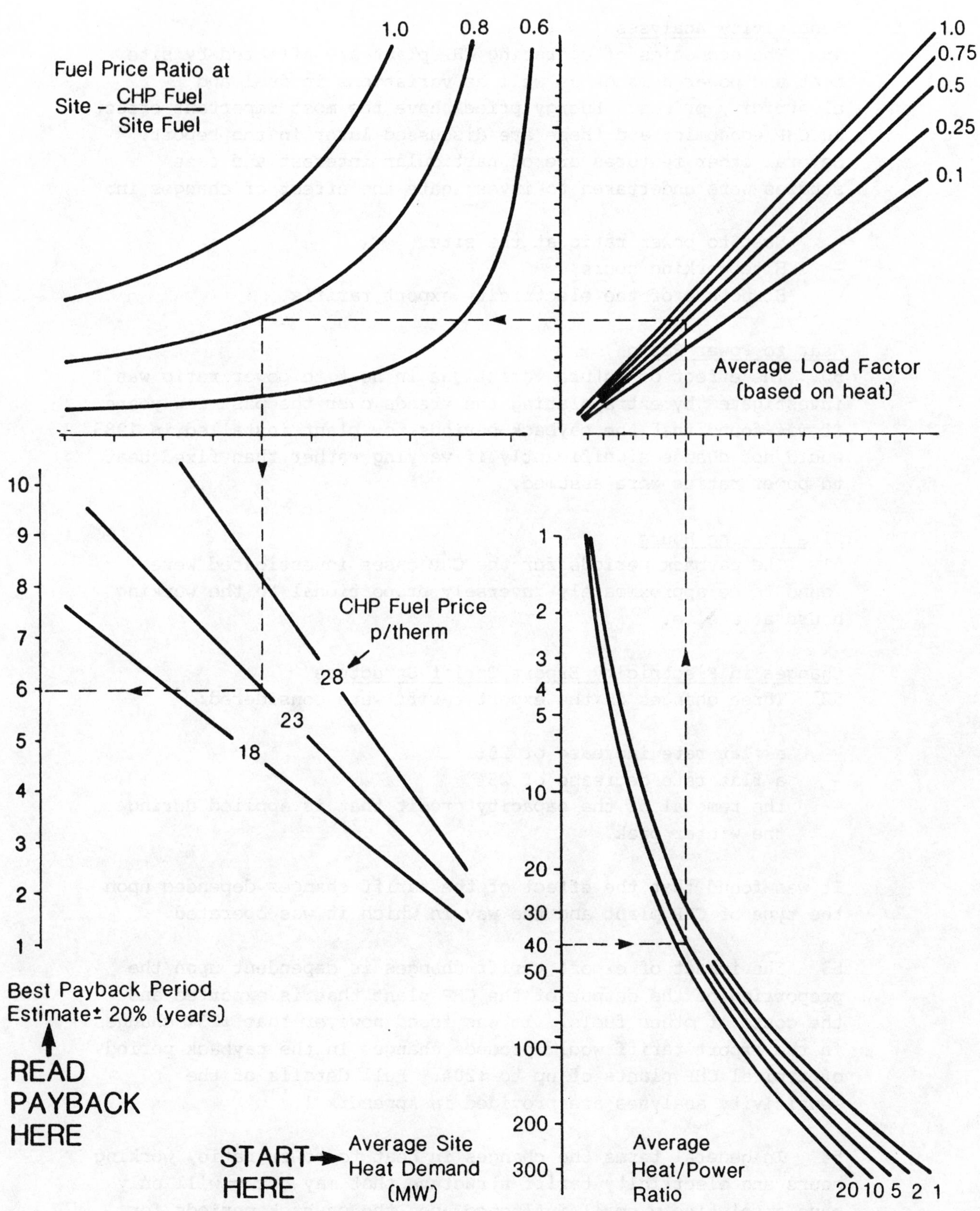

PRELIMINARY CHP ASSESSMENT CHART - COAL FIRED BOILERS + STEAM TURBINE PLANT (Based on 1985 capital costs) **Fig. 5.5**

5.3 Sensitivity Analyses

49 The economics of operating CHP plant are affected by site heat and power demands as well as variations in fuel and electricity prices. Energy prices have the most important effect on CHP economics and these are discussed later in the report. Several other features are of particular interest and case studies were undertaken to investigate the effect of changes in:

- Heat to power ratio at the site
- Site working hours
- Structure of the electricity export tariffs.

Heat to Power Ratio

50 The effect of future variations in heat to power ratio was investigated by extrapolating the trends over the past few years. It was found that the payback periods for plant installed in 1985 would not change significantly if varying rather than fixed heat to power ratios were assumed.

Site Working Hours

51 The payback periods for the CHP cases investigated were found to be approximately inversely proportional to the working hours at a site.

Changes in Electricity Export Tariff Structure

52 Three changes in the export tariff were considered:

- a flat rate increase of 25%
- a flat rate decrease of 25%
- the removal of the capacity credit that is applied during the winter peak.

It was found that the effect of the tariff changes depended upon the type of CHP plant and the way in which it was operated.

53 The impact of export tariff changes is dependent upon the proportion of the output of the CHP plant that is exported and the costs of other fuels. It was found however that ±25% changes in the export tariff would produce changes in the payback period of typical CHP plants of up to ±20%. Full details of the sensitivity analyses are provided in Appendix 1.

54 In general terms the changes in heat to power ratio, working hours and electricity tariff structure that may occur will only have a relatively small influence upon the payback periods for CHP plant. The potential for CHP will be much more dependent upon fossil fuel price changes.

6.0 ENERGY COSTS

6.1 Fossil Fuels

55 During the course of the study there have been profound changes in prices of the fossil fuels. See Figure 6.1. These resulted from the effects of the industrial dispute in the coal industry followed by the dramatic fall in world oil prices starting in January 1986.

56 The original assessments of the potential for CHP were carried out in the spring of 1985 using fuel prices as of December 1984. During 1985 the price of heavy fuel oil fell and the potential for CHP was re-assessed using a price of heavy fuel oil equivalent to that of interruptible natural gas. In the first few months of 1986 the crude oil price fell by more than 50%. The investment potential for CHP was thus recalculated over a range of fuel prices and electricity tariffs.

57 The original December 1984 fuel prices used in the assessment of the potential for CHP were:

Natural Gas	28.3p/therm	£2.68/GJ
Coal	£62.5/tonne	£2.20/GJ
Heavy fuel oil	15.6p/litre	£3.80/GJ

58 After the initial fall in the price of oil the potential for CHP was re-assessed using the following prices:

Natural Gas	28.3p/therm	£2.68/GJ
Coal	£62.5/tonne	£2.20/GJ
Heavy fuel oil	11.0p/litre	£2.68/GJ

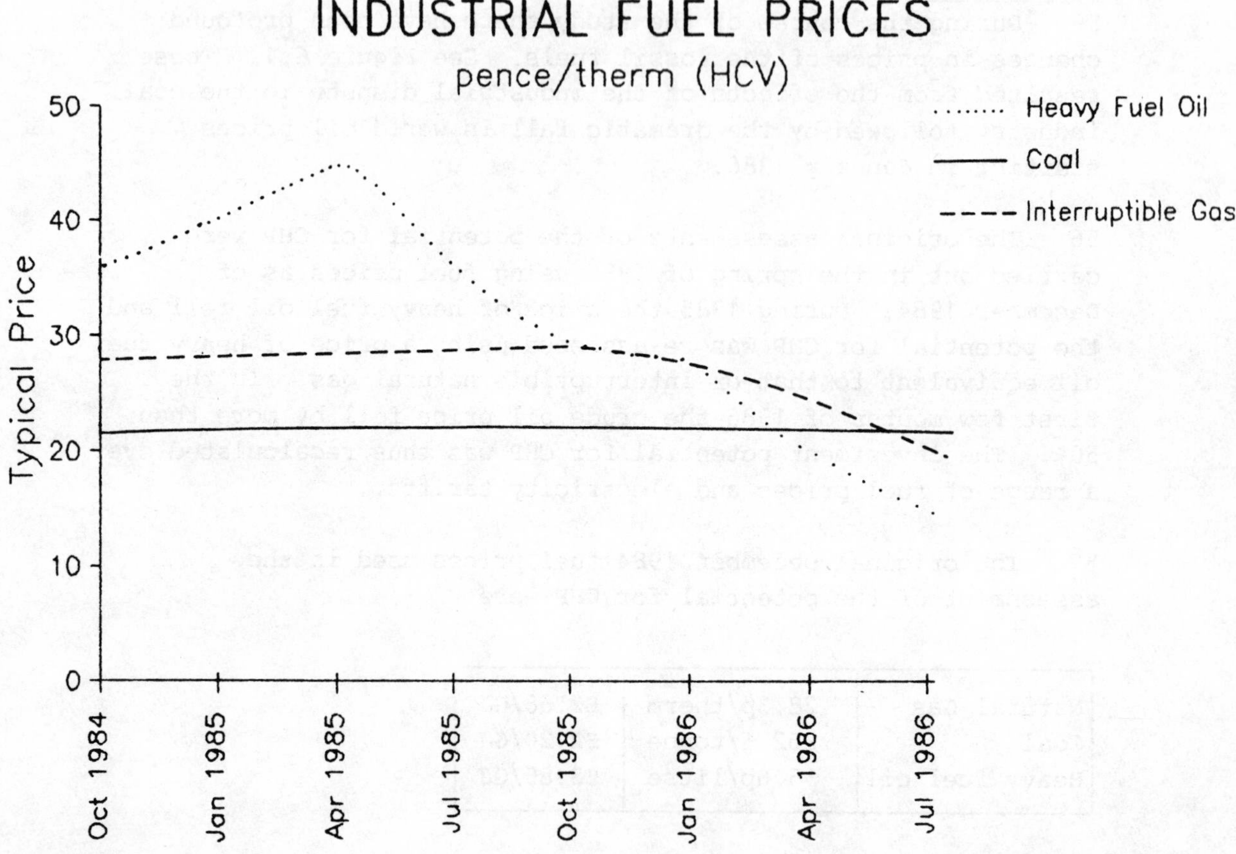

Fuel Price Changes (Oct. 1984 - Jul. 1986)

Fig. 6.1

6.2 Electricity Tariff

59 The Energy Act (1983) encourages the private generation of electricity. Tariffs for the purchase of electricity from private generators are published by the electricity boards and these are discussed in Appendix 1.

60 Tariffs based on maximum demand are used in the study as they are most widely used in industry. The London Electricity Board 1984/85 tariff was selected for the study as it was close to the national average maximum demand tariff. A summary is shown in Table 6.1 indicating the tariff for a non-CHP site as well as the site with private generation.

61 When an industrial consumer installs a CHP plant, the tariff will normally change from a conventional maximum demand tariff to a private generation tariff and this is assumed throughout the study. Some companies may not purchase electricity under a maximum demand tariff and this will affect the cost effectiveness of a particular CHP installation.

Electricity Tariff Summary

Based on London Electricity Board Tariff: Ref 1984/85		Non - CHP Maximum Demand Tariff	Maximum Demand Tariff for Private Generators
Monthly Standing charge		£51.00	£66.50
Monthly capacity charge		£0.75/kVA	£0.75/kVA
Monthly Maximum demand charge	November and March	£3.85	£3.85
	December to February	£5.45	£5.45
Unit Charges for imported electricity	Midnight to 07.00hrs	1.54p/kWh	1.54p/kWh
	07.00hrs to Midnight	3.21p/kWh	3.21p/kWh
Unit Payments for exported electricity	00.30 to 7.30hrs	-	1.37p/kWh
	07.30 to 20.00hrs Mondays to Fridays in November and March	-	3.46p/kWh
	07.30 to 20.00hrs Mondays to Fridays in December to February	-	6.05p/kWh
	At all other times	-	2.43p/kWh
Fuel cost adjustment, based on a fuel cost of £51.50/Tonne	Imported	+ 0.286p/kWh	+ 0.286p/kWh
	Exported electricity	-	+ 0.26p/kWh

Table 6.1

7.0 DERIVATION OF GROSS POTENTIAL

7.1 Methodology

62 The gross potential, is defined as the total CHP plant capacity that it is technically feasible to install within certain economic criteria. This was calculated for each of the 50 selected industries over a number of payback period ranges. The payback period ranges chosen were <1, <2, <3, <4, <5, <6, <8, Table 6.1 and <10 years and were selected to give coverage of the following groups:

- Payback periods which might be acceptable to a manufacturing company; probably 2 to 4 years or less. Equivalent to an internal rate of return greater than 20%.

- Payback periods which might be acceptable to a CHP utility company; probably less than 4 to 6 years. Equivalent to an internal rate of return of 10-20% or more.

- Long term investments which would generate an internal rate of return of 5% or more (equivalent to payback periods of less than approximately 8 years).

63 Internal rate of return is an alternative economic criterion to simple payback period that can be used for comparing alternative investments. This quantity also takes into account the useful life of the installation. It was assumed that a typical industrial CHP installation will have a notional life of 10 years and that the annual savings (after correcting for inflation) remain constant. A simple relationship was then derived between payback period and internal rate of return and this is shown in Table 7.1. As used in this study, the calculation of internal rate of return assumes constant energy prices and a fixed output from the CHP plant, i.e. the site heat and power demands do not change. In some cases these assumptions may be inappropriate and the internal rates of return should be calculated separately.

64 The assessment of gross potential involved identifying the most cost effective CHP plant configuration, using the computer model described in Section 5.1, for each payback period range for each 'typical' site and for a number of different existing boiler fuels at each site. The gross potential for each industry was then calculated by multiplying the installed CHP capacity for each 'typical' site and existing boiler fuel combination by the number actual sites represented by this combination. The results for each industry were then summed to provide a gross potential for all of the industries considered. Details of the assessment of gross potential are given in Appendix 1.

Comparison of Simple Payback Period with Internal Rate of Return

Simple Payback Period (Years)	Internal Rate of Return* (%)
1	100
2	49
3	31
4	21
5	15
6	11
7	7.0
8	4.3
9	2.0
10	0

* A notional plant life of 10 years is assumed.

Table 7.1

7.2 Gross Potential in Industry

65 The gross potential was originally assessed using the fossil fuel prices prevailing in December 1984. By December 1985 the price of heavy fuel oil had fallen to a level comparable with the price of interruptible gas and the gross potential was therefore re-assessed using a price of 11p/litre for heavy fuel oil equivalent to 28.3p/therm for interruptible natural gas.

66 The results of the re-assessment of gross potential split by industry, are shown in Table 7.2. Details of the heat and power demands in each industry are provided in Appendix 2.

67 This assessment of gross potential has certain inherent assumptions, in particular that:

- Each of the 50 key industries can be represented by several 'typical' sites and that CHP assessments carried out on each typical site can be scaled up to represent the situation for the whole of each industry.

- The allowances made in updating the information from the 1979 Energy Purchases Enquiry represent the real changes in industry, although the study did not seek to take any account of site closures over the period 1979 to 1985.

- The relative levels of the fuel and electricity tariffs do not change significantly during the life of the plant.

- The site output, pressures and product mix remain constant and hence the heat and power demand patterns of the site do not change significantly over the life of the project.

- That all sites could, in principle, obtain supplies of all three fossil fuels considered.

- No consideration was given to schemes using heat recovered from a process or schemes using waste materials as fuel.

- Many of the technical assumptions in the CHP data base were of a conservative nature.

68 Alternative energy conserving investments which may be more attractive than CHP are not taken into consideration in the assessment of gross potential although they will affect the likely level of investment.

Gross Potential for CHP

	Gross Potential for each Range of Payback Period *									
	<4 years		<5 years		<6 years		<8 years		<10 years	
	MWe	No. of Sites	MWe	No. of Sites	MWe	No. of Sites	MWe	No.of Sites	MWe	No. of Sites
Chemicals	268	24	374	33	429	40	524	132	633	189
Food/Drink	20	4	73	23	282	60	529	135	769	327
Engineering	58	3	77	4	131	103	850	162	1340	167
Paper	39	2	122	17	122	17	398	51	493	54
Textiles	-	-	-	-	11	20	136	54	367	244
Rubber, Plastics	-	-	86	4	99	4	267	46	356	189
TOTAL	385	33	732	81	1071	244	2704	580	3958	1170

* As each payback period band is cumulative the potentials cannot be added.

Table 7.2

69 Some of these assumptions have been investigated by the sensitivity analyses described in Section 5.3. The effects of changes in fuel price are described later.

70 On balance it is considered that the inaccuracies involved in making these assumptions will tend to cancel out and that the "true" gross potential is unlikely to be significantly smaller than indicated in Table 7.2.

7.3 Types of CHP Plant in the Gross Potential

71 A breakdown of the types of CHP plant that make up the gross potential is shown in Table 7.3. It can be seen that most of the potential is for gas turbine based or spark ignition engine CHP plant.

72 The type of CHP plant most likely to be cost effective on a site depends upon a number of features of the site and the relative levels of the fuel and electricity prices. Some of these are discussed below.

73 Two important features of a CHP plant enhance cost effective operation - a high total system efficiency and a high electrical generation efficiency. Back pressure steam turbine CHP plants have a low electrical generation efficiency, combined with a relatively high capital cost, this makes other types of plant more cost effective. On the other hand steam turbines benefit from being able to use the cheapest fuels but they are only likely to be cost effective for very large sites, or where a cheap fuel, such as a combustible waste material, is available. The latter option was not considered within the study for the reasons stated earlier. Internal combustion engines, and diesel engines in particular, have the highest electrical generation efficiencies. However much of the heat output is in the form of low temperature hot water. Unless this heat can be utilised gas turbines will often be a more cost effective option. Gas turbines however require a high pressure gas supply and when this is not available a significant proportion of the electrical output (up to 10%) may be required for gas compression.

Gross Potential for CHP
(Type of CHP plant)

Type of Plant	Gross Potential for each Range of Payback Period [MWe]				
	<4	<5	<6	<8	<10
Gas Turbine	270	322	516	1571	2244
Spark Ignition Engine	-	12	23	218	496
Combined Cycle	115	398	532	915	1217
TOTAL	385	732	1071	2704	3958

Table 7.3

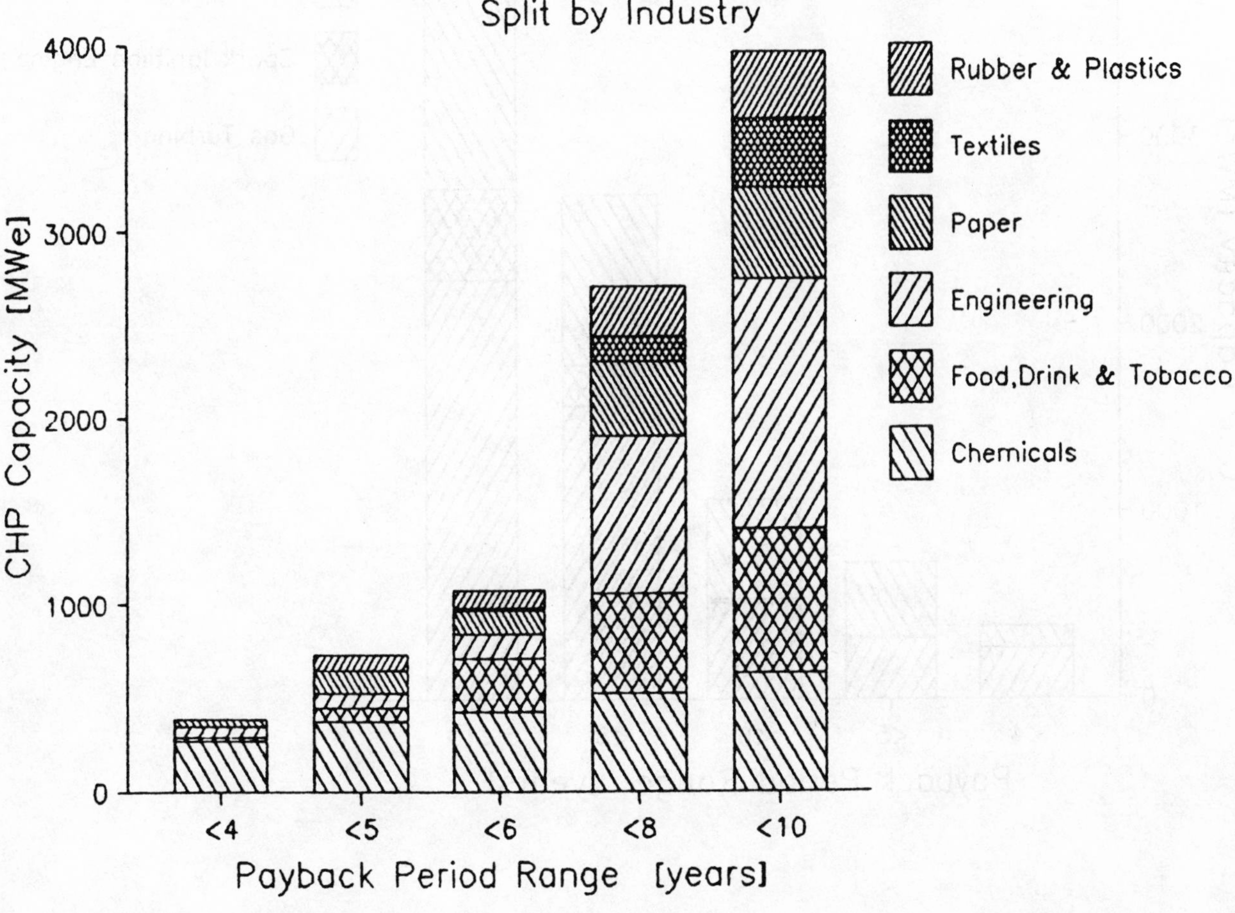

GROSS POTENTIAL IN INDUSTRY **Fig. 7.1**

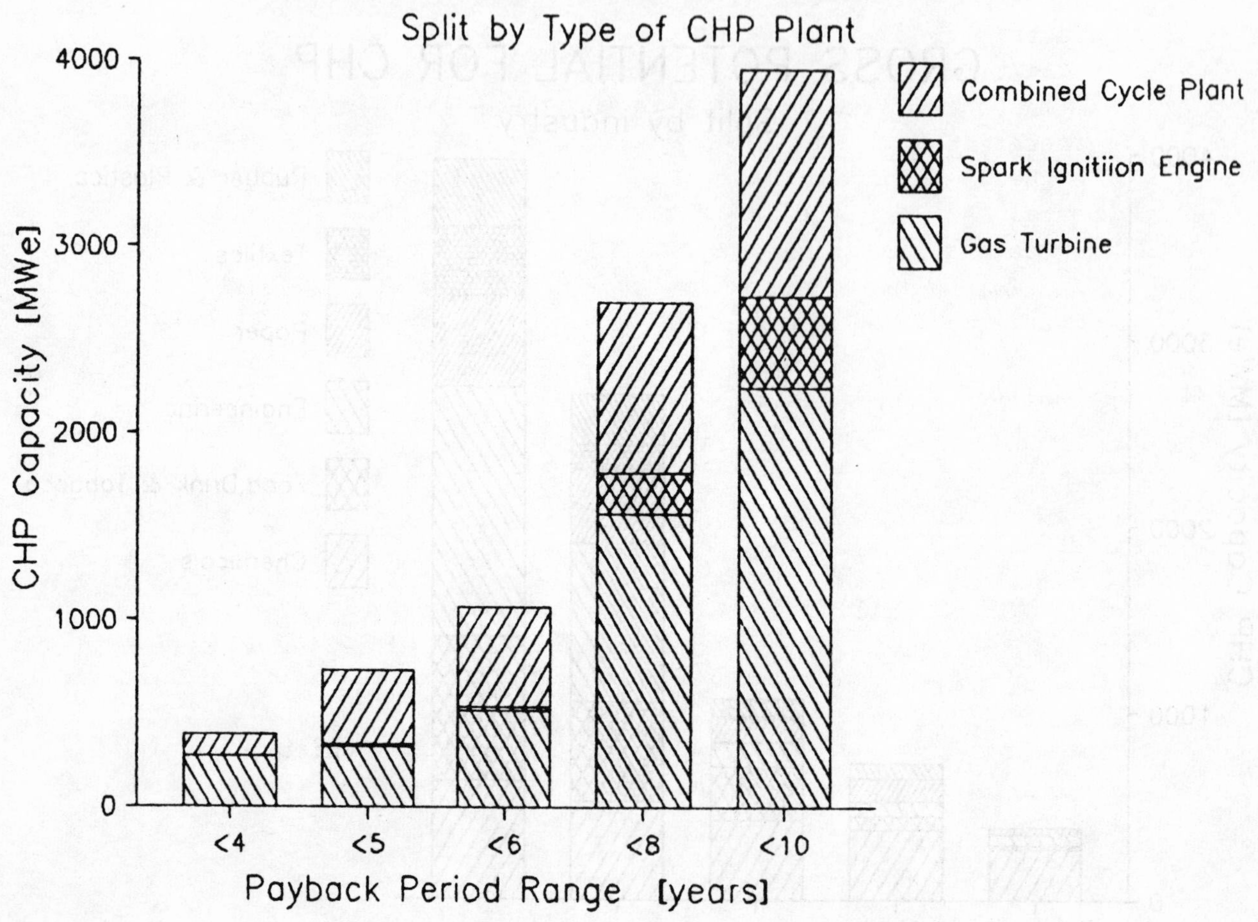

GROSS POTENTIAL IN INDUSTRY **Fig. 7.2**

74 Over the past year the prices of heavy fuel oil and interruptible natural gas have fallen significantly relative to electricity. The effect of this is to make electricity relatively more valuable and this improves the cost effectiveness of those CHP plants (diesel engines) which have a high electrical generation efficiency. Thus as fossil fuel prices fall relative to the price of electricity the gross potential is likely to contain a greater proportion of diesel engine CHP plant and a reduced proportion of gas turbine plant.

75 The fall in the price of heavy fuel oil and gas has also adversely affected the potential for back pressure steam turbine systems based on coal fired boilers. Coal is no longer significantly cheaper than oil or gas and the high capital cost of these steam turbine systems makes them less attractive than gas turbine or internal combustion engine CHP plant.

7.4 Assessing the Likely Investment

76 The actual investment in CHP will only be a proportion of the gross potential depending on the propensity to invest in CHP plant. A number of people in industry were contacted to assess the economic criteria that would govern investment in CHP plant and an 'investment potential' was estimated. This 'investment potential' is defined as the total CHP capacity which might actually be installed within those industries that do not have CHP plants already given the present investment criteria in industry.

8.0 CONTACTS WITH INDUSTRY

77 To obtain the views of people within industry and their comments regarding CHP 40, site assessment forms were sent to selected individuals within industrial organisations. CHP assessments were carried out using the computer model on each of the 25 completed forms. The results were returned to the individuals and 22 of them were contacted with the aims of obtaining feedback on likely attitudes towards CHP implementation and information on the criteria employed for judging cost reduction investment proposals as a basis of estimating likely market penetration.

78 This information was obtained by seeking information on the following topics:

- past involvement with CHP and coal conversion studies;
- reactions to the assessment of CHP economics at the site;
- prospects for implementing a CHP scheme at the site;
- subjective probability of implementing a hypothetical cost saving investment as a function of payback and capital cost;
- expectations of energy price movements and attitudes to energy costs;
- barriers to investment in CHP plant;
- attitude towards utility companies.

Details of the questions asked and comments received are described in Appendix 1.

79 The sample of industrialists selected was not random nor was it intended to be a statistically significant sample in terms of business activity, size distribution or pattern of energy consumption. They were however known to be interested in energy related matters.

80 Some had already expressed an interest in CHP and most were within companies where CHP might be an economically viable investment. The individuals were therefore aware of the benefits of energy conservation and CHP and in a better position to give informed opinions about attitudes to investment in CHP. This was considered to be important for this particular study.

9.0 DERIVATION OF INVESTMENT POTENTIAL

81 The gross potential represents the total of CHP capacity which would be installed if all of the CHP opportunities within a particular payback period range were implemented. In practice the proportion of the gross potential that is likely to be installed will depend on the selection criteria used by industrialists for cost saving investments. These criteria will be based largely upon the return on investment, the capital cost of a project and the degree of technical and economic risk perceived.

82 The contacts with industrialists were used as a means of estimating these investment criteria.

83 The 22 individuals contacted were asked to judge the probability of implementing a hypothetical cost saving investment with a particular capital cost and payback period. The results of the exercise are expressed graphically in Figure 9.1. The number of individuals contacted in this exercise was limited and in view of this, and non-random method of selection, the graph shown in Figure 9.1 cannot be considered to be statistically robust.

84 The propensity to invest, shown in Figure 9.1, was applied to the gross potential, shown in Table 7.2, to estimate the investment potential which is shown in Table 9.1. The investment potential is an estimate of the CHP capacity which may actually be installed within the next 10 years or so assuming that industry is made aware of the opportunities. This investment potential reflects the following price scenario, which prevailed in the latter half of 1985:

Interruptible Natural Gas	28.3p/therm
Heavy fuel oil	28.3p/therm
Coal	23.2p/therm

Electricity tariff - LEB Maximum Demand 1984/85.

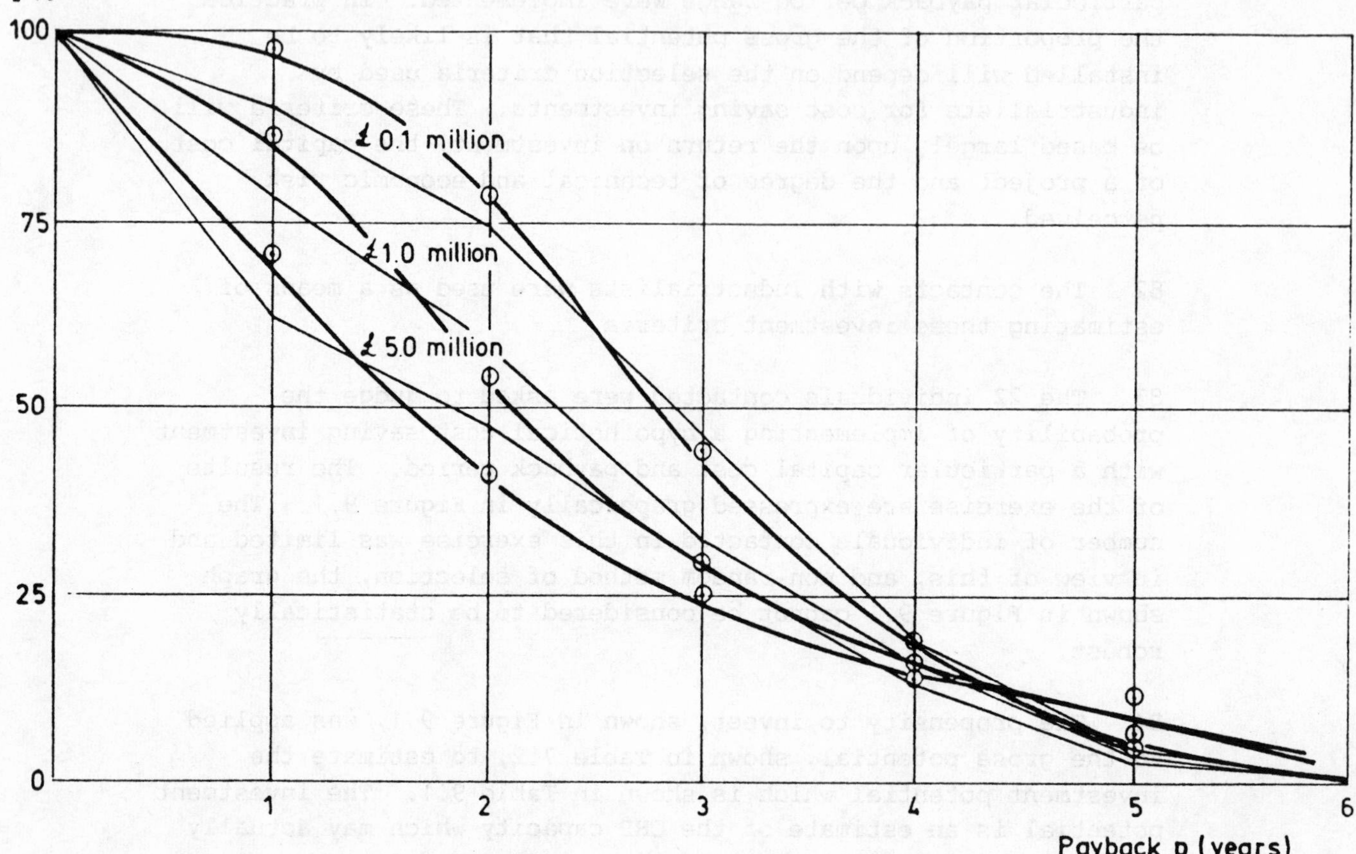

Function used:

$$I = \frac{100}{1 + \left(\frac{p}{E}\right)^{a}}$$

where E and a are coefficients determined by the least squares method

MODELLING OF PROPENSITY TO INVEST VERSUS PAYBACK PERIOD

Fig. 9.1

Comparison of Propensity to Invest

Source	Capital Cost	Payback Period (years) 2	3	4	5	6
Dept of Trade and Industry *	<£100,000	95%	42%	16%	9%	2%
Present Study	<£100,000	79%	57%	20%	9%	5%
Present Study	£1 million	53%	29%	17%	10%	7%
Present Study	£5 million	40%	24%	16%	11%	8%

* Heat Recovery Consultancy Scheme

Table 9.1

85 The total investment potential for those industries, which currently have little or no CHP capacity, is thus 125 MW and is likely to result from installations on 21 sites. The estimated capital cost for the investment would be £60 million and annual primary energy savings of 200,000 TCE are likely.

86 Table 9.2 indicates that the potential lies mainly in the chemical industry; an industry where many of the very large sites have traditionally operated CHP plants. However there is also potential in other industries where there are major process heat demands with high load factors.

Investment Potential for CHP

Industry	Investment Potential MWe	No. of Sites	Capital Cost £ million	Primary Energy Savings ktce
Chemicals	64	6	32	102
Food/Drink	18	5	9	29
Engineering	16	7	8	26
Paper	17	2	8	27
Textiles	-	-	-	-
Rubber, Plastics etc.	10	1	5	16
TOTAL	125	21	£63 million	200

Table 9.2

10.0 SENSITIVITY OF 'INVESTMENT POTENTIAL' TO ENERGY PRICE CHANGES

87 At the beginning of 1986 the price of crude oil fell from $28 per barrel to around $10-12 per barrel and this resulted in a steadily decreasing price for heavy fuel oil, Figure 6.1. As a consequence, both the levels and differentials of the fuel and electricity prices have changed considerably; heavy fuel oil which was once the most expensive, became the cheapest and this has prompted reductions in the prices of interruptible gas, coal and electricity. The situation in the immediate future now remains uncertain. This could have a significant effect on the level of CHP investment in the UK for two reasons:

- The economic viability of CHP plants is very sensitive to energy price levels and differentials;

- Attitudes to CHP investment in industry will also change. Falling energy prices will improve the profitability of many site operations and the general attitudes regarding future business prospects. The economics of CHP will improve while the viability of most other energy efficiency measures will worsen. This will affect the way in which the risks of CHP investment are perceived. These changes in the business environment will take place over a period of time as fuel price patterns are established and perceptions of future trends are developed. At present no clear changes in attitude are evident as there is much uncertainty regarding fuel price trends.

88 These changes have added extra uncertainties to those intrinsic in a study of this nature. However an attempt to accommodate them has been made by estimating the likely investment potential over a range of fossil fuel and electricity prices. This was done by assessing the affect of energy price changes on a number of individual sites and then extrapolating the results for the whole of industry. The analysis, although crude, was felt to be adequate for the purposes of estimating the investment potential, bearing in mind other major uncertainties, such as changes in the propensity to invest in CHP, which cannot be quantified in any way at present.

Investment Potential for CHP
(for a range of energy prices)

Electricity Price	Prices for Gas and Heavy Fuel Oil [Pence/therm]					
	12	16	20	24	28	32
May 1986 Prices	708	497	367	247	161	125
May 1986 Prices Less 5%	603	432	285	195	122	86
May 1986 Prices Less 10%	497	340	209	139	82	42
May 1986 Prices Less 15%	412	254	153	104	53	23
May 1986 Prices Less 20%	261	165	107	57	27	8

Table 10.1

89 The investment potential, quantified in Section 9, at a heavy fuel oil and interruptible gas price of 28p/therm was reassessed, initially, by increasing capital costs by 5% and using 1986 electricity tariffs. This brought it up to date by allowing for inflation. The investment potential was then calculated over a range of prices for heavy fuel oil and interruptible gas from 12 to 32p/therm and a range of progressively reducing unit charges for electricity. The results are shown in Table 10.1. In this analysis no attempt was made to update the propensity to invest curves, Figure 9.1, to take account of any changes in investment attitudes.

90 Table 10.1 shows that the investment potential ranges from zero, at the 'worst' price scenario to 700 MWe at the 'best'.

91 Fuel prices in the summer of 1986 were somewhat below 20p/therm and a conservative estimate of the investment potential was made using a fuel price of 20p/therm and an electricity price 5% below the spring 1986 level. From Table 10.1 this gives an investment potential of 285 MWe.

92 It is possible that, taking into account all of the uncertainties and assumptions, investment in CHP within those industries which have little or no capacity at present could be in the region of 200-500 MWe over the next 10 years, assuming that energy prices remain at their summer 1986 levels, in real terms, and that industry becomes aware of the opportunities for and benefits of CHP investments.

11.0 INVESTMENT POTENTIAL - BREAKDOWN

11.1 Breakdown by Industry

93 The investment potential of 285 MWe, assessed at a fuel price of 20p/therm and a 5% reduction on the 1986 electricity prices, is shown broken down by industry in Table 11.1. Of this, over 80% of the potential is achieved within 9 industries and requires installations on some 30 sites. The average size of each CHP plant is thus 7.5 MWe.

These 9 industries are generally those where there are large sites which have large steady demands for process heat throughout the year.

95 As fuel prices have fallen and the investment potential, shown in Table 10.1, rises, then more industries will have potential for CHP as it becomes more attractive on sites with smaller heat demands and lower load factors.

11.2 Choice of CHP Plant Technology

96 The investment potential for CHP assessed at a price for heavy fuel oil and interruptible gas of 28p/therm, comprises mainly gas turbine based CHP plant. The method used for estimating the investment potential at lower fuel prices, Table 10.1, cannot provide a realistic estimate of the proportions of each type of CHP plant. However, for the reasons described below, it is likely that there will be a much greater proportion of diesel engine CHP plant in the investment potential shown in Table 11.1. The factors which affect the choice of CHP plant are described as follows:

Effect of energy prices on CHP plant economics

97 Two price differentials are important in controlling the cost effectiveness of operating CHP plants: the price differential between the fuel for the CHP plant and electricity purchased from the public supply and; the price differential between fuel for the CHP plant and the alternative fuel for producing heat from heat only boilers, although often these fuels are the same. Thus, the more expensive the price of electricity relative to the CHP plant fuel the more cost effective is the CHP plant, and the more expensive the alternative boiler fuel the more cost effective is the CHP plant. The present situation, in which fossil fuel prices have decreased in price significantly, relative to electricity, has resulted in CHP schemes becoming increasingly attractive, as shown by the effect on investment potential shown in Table 10.1.

INVESTMENT POTENTIAL FOR CHP

Prices for Gas and Heavy Fuel Oil - 20p/therm

Electricity Price - May 1986 Less 5%

INDUSTRY	MWe	No. of Sites
Chemical		
- Fertilisers	41	
- Pharmaceuticals	28	
- Synthetic Resins	21	
- Synthetic Rubber	12	
- Soap & Detergents	5	
- Photographic Materials	3	
- Organic Chemicals	1	
- Adhesives	1	
- Others	2	
TOTAL	114	13
Food Drink and Tobacco		
- Distilling	14	
- Brewing	16	
- Edible Oils	5	
- Miscellaneous Foods	2	
- Confectionery	4	
- Milk Products	4	
- Fruit and Veg. Processing	3	
- Tobacco	3	
- Fish Processing	2	
- Petfood	1	
- Others	1	
TOTAL	55	13
Paper		
- Paper and Board Manufacture	31	
- Paper Conversion	5	
TOTAL	36	5
Textiles		
- General Textiles	5	
TOTAL	5	4
Plastic & Rubber Processing		
- Rubber	19	
- Plastics	4	
TOTAL	23	2
General Engineering		
- Vehicles, ships and aircraft	50	
- Mechanical Engineering	1	
TOTAL	51	18
GRAND TOTAL	285	55

Table 11.1

Site specific features

98 CHP plants are most cost effective when a high proportion of the heat output from the prime mover is utilised. Diesel engine CHP plants are at a disadvantage in this respect as much of the heat is in the form of low temperature hot water, for which there may not be a demand. Similarly, as the exhaust gas temperatures from diesel engines are relatively low this can limit the efficiency of waste heat recovery boilers. Conversely, diesel engines have low heat to power ratios and competitively priced units are available in smaller sizes then gas turbines. Consequently on sites with low heat demands diesel engine CHP plant can be more cost effective than gas turbine plant if there is a demand for low temperature heat.

99 As fuel prices decrease it is likely therefore that the proportion of internal combustion engine CHP plant (diesel engines in particular) in the investment potential will increase as diesel engine CHP becomes more cost effective on smaller sites. The ultimate choice of plant on any site will be dependent however on specific features of the site which affect the total system efficiency of a CHP plant. In the case of diesel engine CHP this will often centre on the demand for the low temperature hot water from the engine cooling system. This can be used to meet domestic hot water demands or for preheating boiler feedwater, the requirements for which will depend on the proportion and temperature of the condensate return at the site.

12.0 SENSITIVITY TO CAPITAL COSTS

100 Payback periods are inversely proportional to capital cost and therefore a reduction in capital cost would increase the investment potential. The investment potential with general 25% reduction in capital cost is shown in Table 12.1. Comparison with Table 10.1 gives an indication of the scale of increase in potential. At lower fossil fuel prices and higher electricity prices the increase in potential is about 70-90%, whereas at lower electricity prices and higher fossil fuel prices the increase in potential rises to over 200%.

101 The relatively large changes in investment potential for a small change in the capital cost are attributable to the combined effect of the sharp increase in gross potential and steep decrease in propensity to invest as the payback periods increase from 3 to 6 years. Many CHP schemes have payback periods which fall just outside the criteria for investment required in industry, a 25% reduction in capital cost would have the effect of bringing many more of these schemes within industry's investment criteria.

102 Considering a fuel price scenario, of about 20p/therm for fossil fuels and a 5% reduction in the 1986/87 electricity tariff, a 25% reduction in capital cost increases the investment potential from 285 to 658 MWe. This would represent a total capital investment by industry of some £250 million spread over the next 10 years.

Investment Potential for CHP with 25% Reduction in Capital Costs
(for a range of energy prices)

Electricity Price	Prices for Gas and Heavy Fuel Oil [Pence/therm]					
	12	16	20	24	28	32
May 1986 Prices	1187	881	739	593	404	281
May 1986 Prices Less 5%	1044	820	659	495	339	206
May 1986 Prices Less 10%	878	712	516	328	213	127
May 1986 Prices Less 15%	783	607	407	245	159	83
May 1986 Prices Less 20%	647	396	252	160	80	46

Table 12.1

13.0 ENERGY SAVINGS

103 Although the overall system efficiency of a CHP plant is often lower than an industrial boiler, the marginal efficiency for electricity generation is in the region of 55-75%, depending on the plant type. This is significantly higher than the generation and distribution efficiency of the public supply of about 30%. Thus although the fossil fuel consumption of a site with a CHP plant will be larger than that of a site operating heat only boilers, primary energy savings will be made on a national scale.

104 The likely investment in CHP over the next 10 years, assuming that energy prices remain roughly at their present level, is in the range 200-500 MWe and it is estimated that the resulting primary energy savings would be in the range 300-700 ktce/year. This investment potential comprises gas and oil fired plant.

14.0 RETRO-FIT CHP

105 The six main types of CHP scheme used in the assessment of investment potential all involve the installation of completely new equipment. Three types of retrofit CHP plant have also been investigated; two direct drive plants and a 'retrofit' steam turbine generating set:

a) Spark ignition engines directly driving air compressors with heat recovery from the compressor and engine.

b) Steam turbines utilising steam from existing boilers directly driving items of plant.

c) Steam turbine generating sets utilising steam from existing boilers.

106 Direct drive CHP involves the use of a prime mover to drive an item of plant (such as an air compressor, or refrigeration plant) directly. Not only can the waste heat from the prime mover be recovered but the losses in the generator and electric motor, which would occur if conventional CHP and electric drive were used, are avoided. Retrofit steam turbine plants could be fitted to existing steam boiler plant where the boiler is capable of raising steam at a higher pressure than the site distribution pressure; the turbine is used in place of a pressure reducing valve.

107 A site suitable for a direct drive CHP plant should have large single items of plant, such as compressors, pumps or refrigeration plants, which operate continuously for a large proportion of the year. Suitable drives would be in the range 100 kW to 1 MW fulfilling a base load duty.

108 In the case of steam turbine prime movers the site should also have a large base load low pressure steam demand. The retrofit steam turbines described will have relatively high heat to power ratios, typically 20:1, thus a steady steam demand of 10 MW would be required for a 500 kW steam turbine installation of this type. Details of the retrofit systems analysed are given in Appendix 4.

109 It has not been possible to quantify the potential for retrofit CHP plant in any detail as the information regarding suitable drives and boilers is not readily available. Consequently only a very rough estimate of the investment potential has been made.

110 Details of the method used are given in Appendix 4. The fuel prices used are:

Natural Gas	28.3p/therm
Heavy Fuel Oil	28.3p/therm
Coal	23.2p/therm

Electricity tariff - LEB Maximum Demand 1984/85.

111 It is estimated that the investment potential for the retrofit plants described is 60 MW. This would be in addition to the investment potential described in Section 9 and represents a significant additional CHP capacity.

112 The capital cost of this retrofit plant is somewhat lower than the other main types of new plant considered in the study as it is retrofitted onto existing plant. It therefore represents an attractive CHP investment wherever suitable applications occur.

15.0 BARRIERS TO INVESTMENT

113 The opportunity provided by the industrial contacts was used to obtain some guidance as to the most likely barriers to the adoption of industrial CHP. The details are described in Appendix 1. It was found that the views expressed were broadly similar to those identified in the survey of existing users.

Payback and Capital Costs

114 The principal barrier to the implementation of CHP schemes is without doubt the relatively poor payback period obtained at the majority of sites, and the relatively high capital costs of such schemes.

115 It was found that many companies tend to allocate available funds to investments which impinge on the market side of the company's business (products, production, distribution, customer service, promotion), rather than on the cost side of the business. This is reflected both by the fact that several companies contacted employed different criteria for judging the merits of 'main stream' and 'cost saving' investments, as well as in the diminishing willingness to invest as the capital costs increase, as shown by the propensity to invest curves, Figure 9.1.

116 Capital costs have a dual influence on the willingness to invest: they affect both the payback and the propensity to invest. This is compounded by the fact that CHP plant capital costs are typically an order of magnitude greater than the costs usually associated with heat generating plant or other energy cost saving investment.

117 It is concluded that the level of investment in CHP plant is likely to be highly sensitive to influences affecting capital costs.

Uncertain Site Conditions

118 The second most important barrier to the wider adoption of CHP plant concepts is uncertainty regarding the future of the site in question. Several concerns were identified:

- uncertain future for the site as a whole;

- uncertainties regarding specific products made at the site;

- uncertainties regarding future energy prices, and particularly whether to opt for coal, oil or gas based CHP against the backdrop of the dispute in the coal industry followed by falling oil prices.

Other Influences

119 In comparison with payback, capital costs, and site uncertainties, other influences on the willingness to invest in CHP are comparatively minor. They include:

- various site specific problems (existence of heat surplus, seasonality, etc);

- technical worries or uncertainties (lack of skills, high cost of technical support).

120 At the time this exercise was undertaken the price of oil had not yet fallen dramatically and there was surprising little concern over medium to long term energy price movements as they might affect attitudes to CHP. With the fall in the oil price there are now uncertainties as to how long energy prices will remain low.

121 Overall, these various concerns lead many people in industry to view a CHP plant as a "high risk" investment. Hence, the general unwillingness to accept payback periods of much more than 2-4 years.

Removing the Barriers

122 If more CHP capacity is to be installed these barriers must be overcome and several approaches can be envisaged:

- The payback periods may be improved somewhat by careful design to ensure that the CHP plant operates at the optimum system efficiency. Technical improvements in plant could help to increase efficiencies and reduce capital costs, especially for the smaller sizes of gas turbine which would improve the payback periods on smaller sites. Group CHP schemes could also help to reduce payback periods.

- The acceptable payback periods may be increased if utility companies, possibly with the backing of financial institutions, become involved in CHP contracts to provide heat and power to industrial sites.

16.0 GROUP CHP SCHEMES

123 There are circumstances where the payback period for CHP at individual small industrial sites is considered to be too lengthy. The payback periods can be reduced if several sites are combined to form a group CHP scheme.

124 An assessment of the national potential for group CHP schemes is beyond the scope of this study but a method of estimating the best payback period for a group CHP scheme has been established. This method is described in detail in Appendix 5.

125 Examples showing the improvement in payback period that can be achieved from group schemes are presented in Appendix 5 and several general conclusions can be drawn from the analysis:

The marginal benefits of group schemes are much greater when small sites rather than larger sites are combined. Larger individual sites can offer attractive payback periods for CHP plant individually. Smaller sites with average heat demands below about 10 MW are rather unattractive individually and combination of several sites can produce a much more attractive group scheme.

Where a large site is combined with several smaller sites the resulting group scheme payback will not be significantly shorter than that achieved by the large site alone.

126 Group CHP schemes are rare at present. This may be because industrialists do not wish to become involved in schemes which depend upon other companies remaining in business. This is seen as merely compounding the risks of what is often seen as an already risky investment. Furthermore they are contractually complex and individual companies may not wish to undertake the lengthy negotiations needed to implement a scheme which is peripheral to their main business activities.

127 Utility companies, who specialise in the supply of energy, should be very well suited to the installation and operation of group CHP schemes. They will have the financial and management resources to initiate a group CHP scheme which may be too large an undertaking for any one site in a group.

17.0 THE ROLE FOR UTILITY COMPANIES

128 The main barrier to CHP investment is the relatively low rate of return, required by industry, for cost saving schemes. On the whole, utility companies both private and public are prepared to take a longer term view and hence accept investment opportunities which yield a lower rate of return. Nine utility companies were contacted and asked for their views on CHP contracts. Generally they were prepared to invest in schemes with payback periods longer than 5 years corresponding to rates of return of less than 15%.

129 The utility companies are in a good position to install and operate CHP plants as the provision of energy services is their main function. They may also be better equipped to provide the technical expertise necessary to install and operate CHP plants in an effective manner. Industrial companies often have to take commercial risks regarding the production of chemicals, paper, etc. They may not wish to compound this situation by investing in what they may see as 'risky' steam raising systems.

130 The utility companies are thus in a position to be able to take up some of the gross potential for CHP with a payback period which is considered too long by industry. It is possible therefore that if these companies were to take advantage of this opportunity the investment potential for CHP could be well in excess of that estimated above. They are also ideally suited to become involved in group CHP schemes comprising several sites where any one site may be reluctant to taken responsibility for the whole scheme. The utility companies do not appear to have been actively involved in seeking out and implementing CHP schemes.

131 In other countries utility companies are already involved with CHP and are at the forefront of the investment in Holland and parts of the USA where considerable CHP investment is taking place.

132 Some UK utility companies have tended to concentrate on coal conversion schemes. At the present fuel prices these are less attractive and this might have an effect on utility company attitudes. But there is also some reluctance on the part of industry to become involved with private utilities.

18.0 ADVICE TO INDUSTRY

133 For potential industrial CHP users who are interested in investigating the prospects for CHP at their site it is recommended that an initial rough assessment is made using the preliminary assessment charts described in Section 5. The charts are only accurate to within about ±20% and do not indicate the optimum size of the plant.

134 If the payback period for CHP from the charts is roughly comparable to that which is considered acceptable, then it is worth taking the investigations further. The following procedure may be adopted:

- First, survey the energy usage at the site and compile a set of data which defines the heat and power loads on a daily and a seasonal basis. This may require the installation of some monitoring equipment as suitable heat and power demands are not often recorded.

- Second, ensure that there are no other cost effective ways of reducing the energy consumption at the site by heat recovery, insulation measures, better control or production scheduling or by investing in more modern equipment. On a complex site a process integration study is a good way of assessing these possibilities.

- Finally, undertake a detailed CHP feasibility study, which will identify the prospects for CHP at a particular site. This will involve a comparison of several types of CHP plant and possible fuels, discussions with fuel suppliers and the Electricity Board to identify any special requirements for the systems, budget costs and savings resulting from the operation of CHP systems and recommendations of the best option.

135 Energy consultants can be called in to report on any of the three aspects described above. Two principal measures of qualification can be used in choosing a consultant:

- whether the consultant has the necessary expertise.

- whether the consultant is a member of an appropriate professional or Chartered Institution.

19.0 CONCLUSIONS

136 The main aim of the study has been to quantify the potential for CHP investment within the UK industries where CHP is not widespread at present. The potential has been determined in terms of an investment potential, defined as the quantity of CHP that is likely to be installed in industry over the next 10 years given current investment criteria but assuming that industry at large is aware of the possibilities.

19.1 Investment Potential

137 There are many inherent difficulties in undertaking a 'desk' study of this nature that uses predominantly statistical information to establish industrial energy consumption patterns. In particular site specific features which affect CHP economics can only be taken into account on a global basis for each industry. There are also considerable problems in establishing the criteria used by industry in assessing whether to invest in CHP, particularly as the fuel prices are currently so unstable. In view of these uncertainties the best estimate of investment potential would indicate that several hundred megawatts of CHP are likely to be installed over the next 10 years. If energy prices remain roughly at their current level then the new capacity is likely to be in the range of 200-500 MWe.

138 In addition there is the potential for a further 60 MWe or so of 'retrofit' CHP capacity.

139 Of the 50 industries investigated only 9 appear to have a significant investment potential, based on current prices and investment trends, and the CHP installations are likely to be made on some 50 sites within these industries. This represents a rate of investment in the order of 20-50 MWe per year. The capital costs of these installations is in the range of £400-500/kWe

19.2 Barriers to CHP

140 It is technically feasible to install considerably more CHP capacity than the investment potential indicates, in the order of ten times the investment potential within the payback period of less than 10 years. The reason for such a low rate of implementation is mainly that the payback periods and capital costs for CHP plant are mostly larger than industry is prepared to accept for this type of non-production-related, cost reducing investment. Many companies do not wish to commit themselves to a long term investment, albeit with a good return, when they see the future holding many uncertainties.

19.3 <u>Group CHP Schemes</u>

141 There is the prospect of improving the economics of CHP on some smaller sites if group CHP installations are considered. These could in principle increase the investment potential. Group schemes may be seen by industry as being an even more risky investment as their success depends upon the continued profitability of a number of sites.

19.4 <u>Role for Utility Companies</u>

142 Private utility companies are prepared to accept payback periods for investment in CHP which are longer than those required by industry. They are in the business of supplying energy, and operating CHP plants for customers is therefore part of their business. At present however they have little involvement in this country. This may in part be attributable to the doubts that some industrial companies have about the merits of utility companies and it is evident that there are some 'social' barriers to be overcome before the utilities become more involved.

Acknowledgements

We acknowledge the help and assistance of the following companies who provided technical information.

Gas Turbines

A.E.G. Kanis
Centrax Ltd.
GEC Ruston Gas Turbines
Kongsberg
Sulzer
Dale Electric
Ingersoll Rand
Noel Penny Turbines

Steam Turbines

N.E.I. APE W.H. Allen
KKK Ltd.
Turbodyne Dresser
A.E.G. Kanis
R.G. Joliffe & Co. Ltd.
Peter Brotherhood

Internal Combustion Engines

N.E.I. APE Ltd. - Crossley Engines
N.E.I. APE Ltd. - W.H. Allen
Sulzer
Harland and Wolff Ltd.
Stork Diesel
Dale Electric
Deutz Engines
Caterpillar
R.A. Lister
Mirlees Blackstone
Applied Energy Systems
Petbow Ltd.
Rolls Royce Perkins Engines
Clark Kincaid
Jenbacher
GEC Dorman Diesels
MAN-GHH

Boiler Plant

Foster Wheeler Power Products
Senior Green
Babcock Power Ltd.
APV Spiro Gills
Stone-Platt Crawley
Wanson Co. Ltd.

We also acknowledge the help and assistance of the following organisations and companies who provided heat and power demand details of industrial sites.

Beecham Pharmaceuticals
British Cocoa Mills
British Paper and Board Industry Federation
BIP Chemicals
Clayton Aniline
Control Data Corporation
Cray Valley Products
Croda Oil Refining
Eli Lilly
Ford Motor Company
Hickson and Welch
Ind Coupe
IMI Muntz
St. Ivel
Kodak Ltd.
J. Lyons
Metal Box plc
Nestle
Proctor and Gamble
Rockware Glass
Rowntree Mackintosh
Scottish Agricultural Inds.
Scottish Grain Distillers
Shirley Institute
Weetabix Ltd.

We are also grateful for the help and assistance given by the Department of Energy, Economics and Statistics Division and the Energy Technology Support Unit.

NOMENCLATURE

Abbreviations

CHP	Combined Heat and Power
DEn	Department of Energy
ESI	Electricity Supply Industry
ETSU	Energy Technology Support Unit
FP	Fuel Price Ratio, as defined in Appendix 4
HP	Heat to Power Ratio, as defined in Appendix 4
HT	Heat Demand (MW), as defined in Appendix 4
IRR	Internal Rate of Return
ISO	International Standards Organisation
LF	Load factor, as defined in Appendix 4
MCR	Maximum Continuous Rating

Units

GJ	Gigajoules
kW	Kilowatts
kWe	Kilowatts of electrical power
kWh	Kilowatt - hour
MW	Megawatts
MWe	Megawatts of electrical power
MWh	Megawatt - hour
kTherms	Thousand therms
TCE	Tonnes of Coal equivalent (Unit of Primary Energy)
ktce	Thousand tonnes of coal equivalent
mtce	Million tonnes of coal equivalent

Energy Conversion Factors

A=B x Factor		B				
		kWh	MWh	Therms	GJ	TCE
	kWh	1	1000	29.3	277.8	7220
	MWh	0.001	1	0.0293	0.2778	7.22
A	Therms	0.03412	34.12	1	9.479	246.5
	GJ	0.0036	3.6	0.1055	1	26
	TCE	0.0001385	0.1385	0.00406	0.0385	1

To convert a quantity B in particular units, to a quantity A in different units, multiply the quantity B by the factor from the table

Glossary

After-firing	Burning more fuel in the exhaust gases from an internal combustion engine or gas turbine.
Back pressure turbine	A steam turbine in which the exhaust steam is utilised in a heating process rather than being condensed.
Combined cycle plant	Plant where the hot exhaust gases from a gas turbine are used to raise steam to power a steam turbine.
Combined heat and power	The combined production of electricity and usable heat from one plane.
Condensate	The condensed steam returned from a heating application to the power station.
Degree day	The daily difference in °C between a base temperature of 15.5°C (or any specified base) and the 24 hour mean outside temperature (when it falls below the base temperature).
Existing boilers	The steam or hot water raising boilers that already exist on a site before the installation of a CHP plant.
Gross potential for CHP	The quantity of CHP that it is technically feasible to install with in specified economic criteria.

High pressure steam	Steam at a pressure greater than 250 psig.
Internal rate of return	The minimum rate of interest that could be paid for the capital employed over the life of the project without making a loss.
Interruptible gas	Gas supplied under a contract which allows for the temporary cessation of supplies at short notice.
Load Factor	The ratio of the consumption within a specified period to the consumption that would result from continuous use of the maximum or other specified demand over the same period.
Low pressure steam	Steam at a pressure of 250 psig or less.
Payback period	The time for the cumulative savings resulting from a project to equal the capital investment.
Prime mover	Engine which can be used to produce mechanical energy from a fuel or heat source.
Primary energy	Energy that has not been subjected to any conversion or transformation process.
Typical site	A site of average energy consumption and with energy demand profiles typical of actual industrial sites within a sector (or sectors) of industry and within a band of annual energy consumption.

INDUSTRIAL COMBINED HEAT AND POWER
THE POTENTIAL FOR NEW USERS

APPENDIX 1
METHODOLOGY

CONTENTS

1.0 METHODOLOGY CONCEPT

1 The aim of this study is to estimate the scope for installing cost effective CHP plants within those sectors of UK industry where CHP capacity is limited or non-existent. The method adopted was to identify those industries where CHP was likely to be cost effective and then to assess the cost effectiveness of CHP in each case. This was achieved by defining several hypothetical sites in each industry which had average energy consumption and typical heat and power demand profiles representing the most common practice in the industry. On each of these 'typical' sites the most cost effective CHP plants were identified using a computer model to assess the economics of operating numerous CHP plant configurations at the site. For each 'typical' site the most cost effective CHP plants were identified for a range of economic criteria. These plants were then scaled-up to give an estimate of the capacity that could be installed in each industry. The results were then summed over all of the industries considered to produce a total for the whole of industry. This 'desk' exercise indicated the quantity of CHP which was technically feasible for a particular range of payback period and was termed the Gross Potential.

2 In order to estimate the CHP plant capacity which may actually be installed, contacts were made with industry to determine the economic criteria that might be used in making investments in CHP. These criteria were applied to the Gross Potential to estimate an Investment Potential - the quantity of CHP plant that might actually be installed in the industries considered over the next 10 years or so.

3 A diagram showing the main stages of the methodology is shown in Figure 1.1. The first stage of the work involved preparing data bases: the industrial data base contained details of each 'typical' site; the CHP plant data base contained performance and cost details of all of the different CHP plant configurations that were appropriate to the study. These two data bases were used by the computer model to assess the economics of CHP at each of the 'typical' sites. This in turn led to the derivation of the Gross Potential. Assessments of the economics of CHP at a number of specific sites were also carried out and used as case studies when seeking comments from industry. The economic criteria for investing in CHP, identified from these contacts with industry, were then used to derive the Investment Potential.

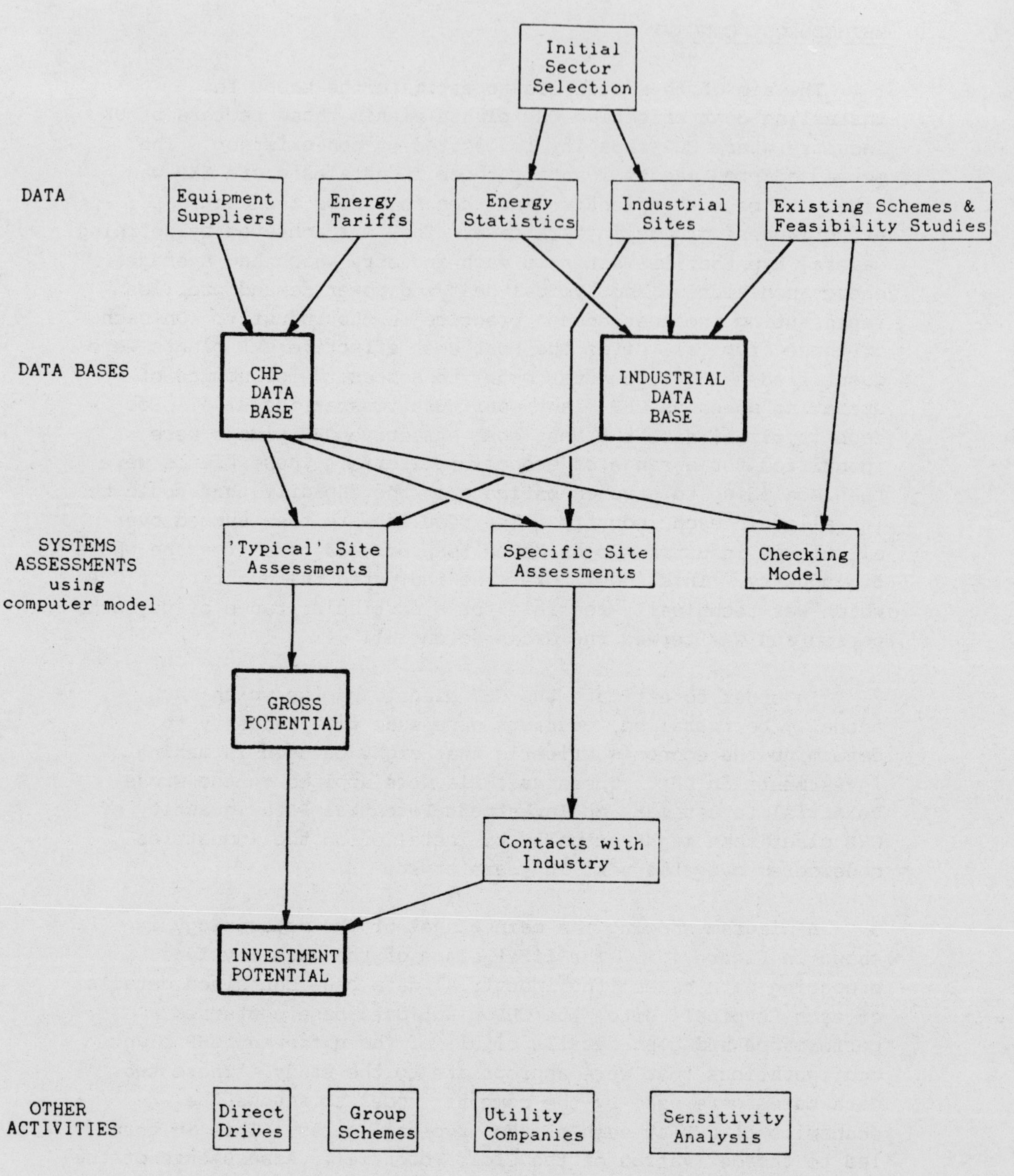
Initial
Sector
Selection
DATA
Equipment
Suppliers
Energy
Tariffs
Energy
Statistics
Industrial
Sites
Existing Schemes &
Feasibility Studies
DATA BASES
CHP
DATA
BASE
INDUSTRIAL
DATA
BASE
SYSTEMS
ASSESSMENTS
using
computer model
'Typical' Site
Assessments
Specific Site
Assessments
Checking
Model
GROSS
POTENTIAL
Contacts with
Industry
INVESTMENT
POTENTIAL
OTHER
ACTIVITIES
Direct
Drives
Group
Schemes
Utility
Companies
Sensitivity
Analysis

METHODOLOGY FOR CHP STUDY **Fig. 1**

2.0 PREPARATION OF DATA BASES

4 Details of the preparation of the CHP Plant Data Base and Industrial Data Base are given in Appendices 2 and 3 respectively.

3.0 SYSTEM ASSESSMENTS

3.1 Method

5 The factors affecting the economic viability of a CHP installation are complex. A CHP system can be assessed most realistically by considering the operation of the plant under all conditions of heat and electricity demand that are likely to occur on a site. And by taking into account fuel costs, maintenance costs and the variation in the electrical tariff throughout the year. The most effective way of undertaking this analysis is by the use of a computer model to simulate the operation of a CHP plant.

Computer Model for Assessment of CHP Systems

6 A full description of the computer model and the assumptions it contains is given in Appendix 2. The CHP computer model includes a method for establishing the optimum mode of operating a CHP plant at an industrial site to obtain the greatest cost saving, and a method of calculating the quantity of those savings.

7 The optimum mode of operating the CHP plant is established for each hour of each day of the year by considering the CHP plant operating in an electricity matching or heat matching mode, with import or export of electricity. The hourly running costs of operating in either mode are compared with the running cost with the plant switched off. The mode with the lowest running cost is considered to be the optimum. Having established the optimum mode the running costs are summated over a whole year for both a CHP plant and a 'heat only' plant at the site. The annual running cost savings can be estimated and hence a payback period calculated for the capital cost. The analysis takes into consideration the costs shown below:

a) Fuel cost for the CHP plant.

b) Fuel cost for existing boiler plant (to top up the CHP plant).

c) Maintenance cost of CHP plant (including allowances for plant breakdown and maximum demand).

d) Cost of imported electricity.

e) Cost benefit of exporting electricity.

8 It is necessary to evaluate the performance for each hour of each day to take into consideration the hour by hour changes in

the demands for heat and power and the changes in electricity tariff throughout the day and year.

Use of the Computer Model

9 The computer model was used in three ways during the study.

1. Industrial Sector Assessments

The model was used in the assessment of the potential for CHP across industry. Each of the typical sites in the industrial data base (as described in Appendix 3) was analysed using the model. The most appropriate CHP plant, of the 600 different configurations tested by the model, was then selected using the criteria described in Section 4.0.

2. Site Assessments for Individual Industrial Sites

A number of actual industrial sites were analysed using the CHP model to identify the potentially most attractive CHP schemes. The results of the analyses were circulated to the industrial sites concerned for their comments. In all, 25 industrial sites were investigated.

3. Checking the Model

In order to check the results predicted by the computer model a number of test runs were made. In these, data on actual CHP sites was compiled from the results of detailed feasibility studies that had been undertaken. The computer model was used with this data and the results compared with the results of the feasibility studies.

10 Details of some of these comparisons are given in Appendix 2 where checks on the following types of plant and site are described.

i) Gas turbine plant at a papermill.
ii) Coal fired boiler and steam turbine at a food factory.
iii) Heavy fuel oil diesel engine at a rubber products site.
iv) Gas turbine plant at a site where the exhaust gas was used directly for drying.

11 It was found that the computer model produced results closely similar to those produced by the feasibility studies.

3.2 A Method for Estimating the Most Cost Effective Payback Period for CHP at an Industrial Site

12 During the course of the study, the assessment of CHP options at many industrial sites produced a wealth of results. These have been used to produce a simple method of estimating the payback period for the most cost effective CHP plant at an industrial site. The assessment charts were derived using a multiple linear regression technique. This is described in more detail in Section 3.5.

13 Assessment charts were produced for each of the following four different types of CHP plant: gas turbine plant; heavy fuel oil diesel plant; spark ignition engine plant and coal fired boiler with back pressure steam turbine plant. Each assessment chart enables the payback period for the most cost effective CHP plant to be estimated from five simple characteristics of any industrial site.

3.3 ASSESSMENT CHARTS

14 The assessment charts are shown in Figures 2 to 5 for each of the four different types of CHP plant. The payback period is determined from five parameters which define the application. These are:

a) Annual average heat demand (MW).
b) Annual average heat to power ratio (dimensionless).
c) Annual average load factor of the heat demand (dimensionless).
d) Price ratio - fuel price for CHP plant/fuel price for existing boilers at the site.
e) Current CHP plant fuel price (p/therm).

Each of these parameters is defined as follows:

a) The annual average heat demand (HT), expressed in MW, defined as:

$$HT = \frac{\text{Total Annual Boiler Fuel Consumption (GJ)} \times 1000 \times \text{Boiler Efficiency (\%)}}{8760 \times 3600 \times 100}$$

b) The heat to power ratio (HP), a dimensionless quantity, defined as:

$$HP = \frac{\text{Annual Boiler Fuel Consumption (GJ)} \times \text{Boiler Efficiency (\%)}}{\text{Annual Electricity Consumption (GJ)} \times 100}$$

c) The load factor (LF), also a dimensionless quantity, based on the heat demand. It is defined as:

$$LF = \frac{\text{Annual Average Heat Demand (MW) (as defined above in (a)}}{\text{Typical Maximum Heat Demand (MW)}}$$

The typical maximum steady heat demand is the steady heat demand during a typical day at the time of year when the heat demand is highest. This is most likely to occur during the winter period on an average day in December, January or February. Short peaks such as the warm-up period should not be included. The typical maximum steady heat demand will always be lower than the installed boiler capacity of a plant and often in the region of 50% - 80% of the peak heat demand occurring during the warm-up period or on an exceptionally cold day. An example of a typical maximum steady heat demand is shown in Fig. 6.

d) This price ratio is non-dimensional and any units can be used, provided that they are consistent.

e) The current CHP fuel price should be expressed in p/therm (based on HCV).

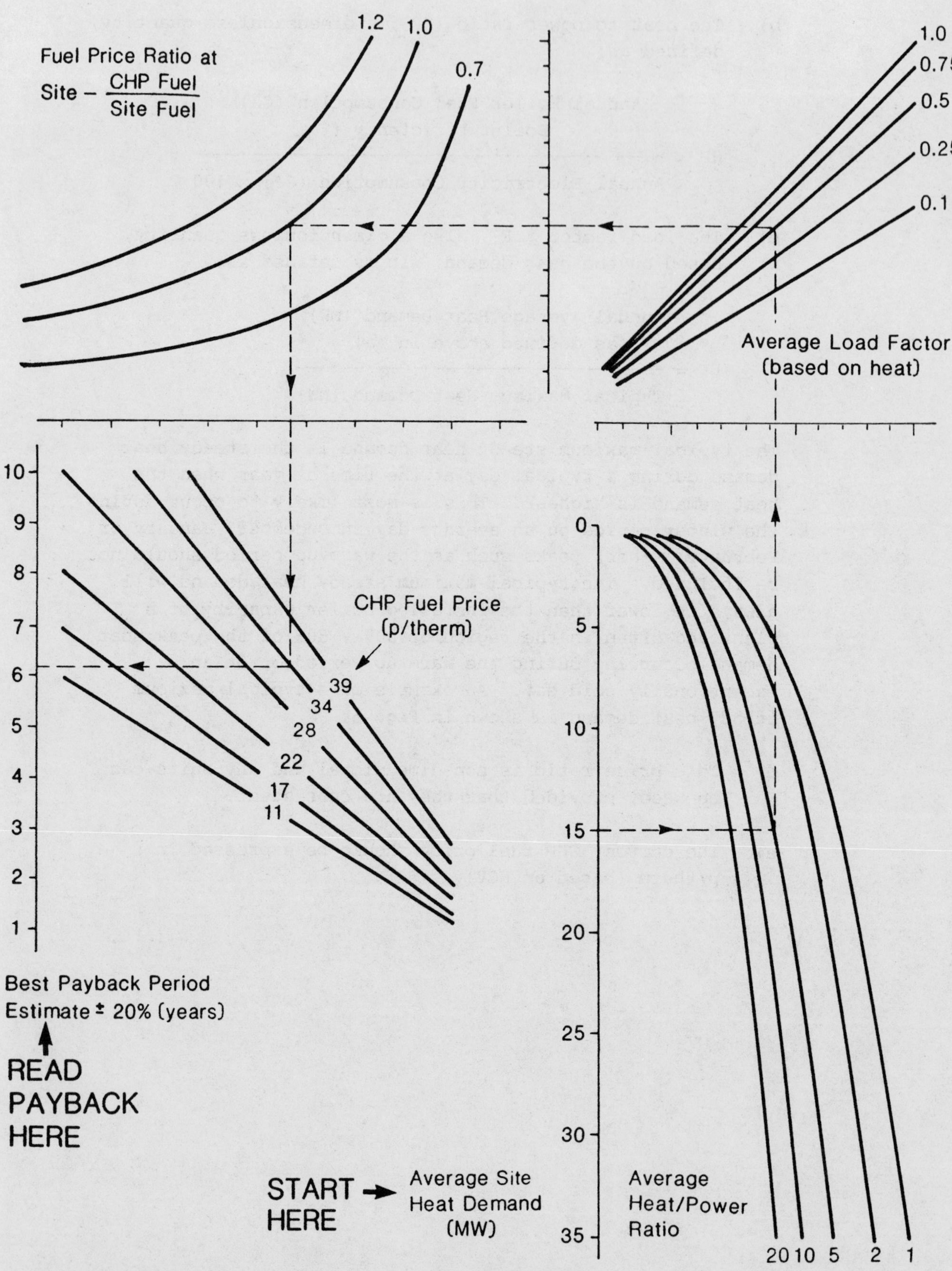

PRELIMINARY CHP ASSESSMENT CHART – GAS TURBINE PLANT
(Based on 1985 capital costs)

Fig. 2

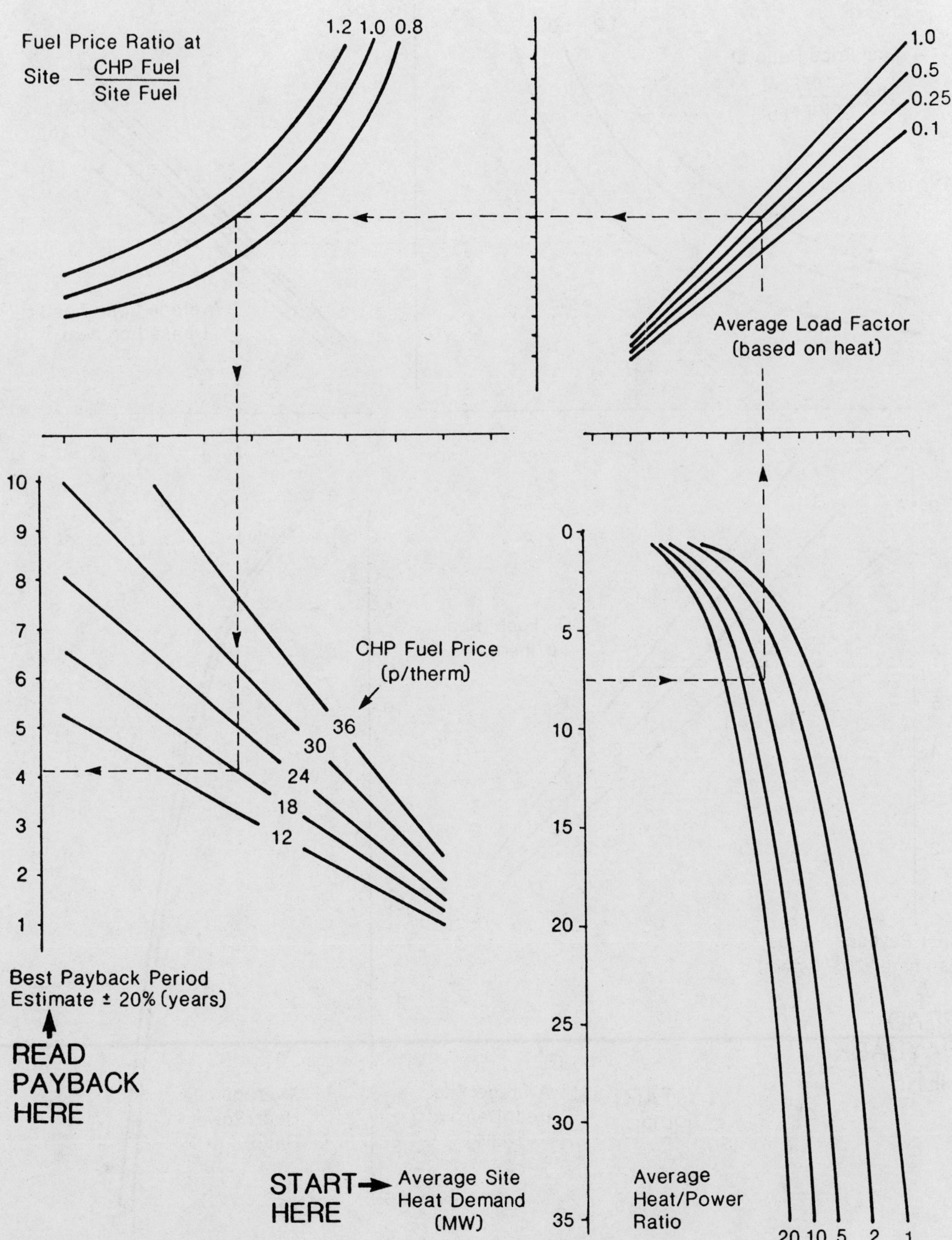

PRELIMINARY CHP ASSESSMENT CHART – HEAVY FUEL OIL DIESELS

(Based on 1985 capital costs)

Fig. 3

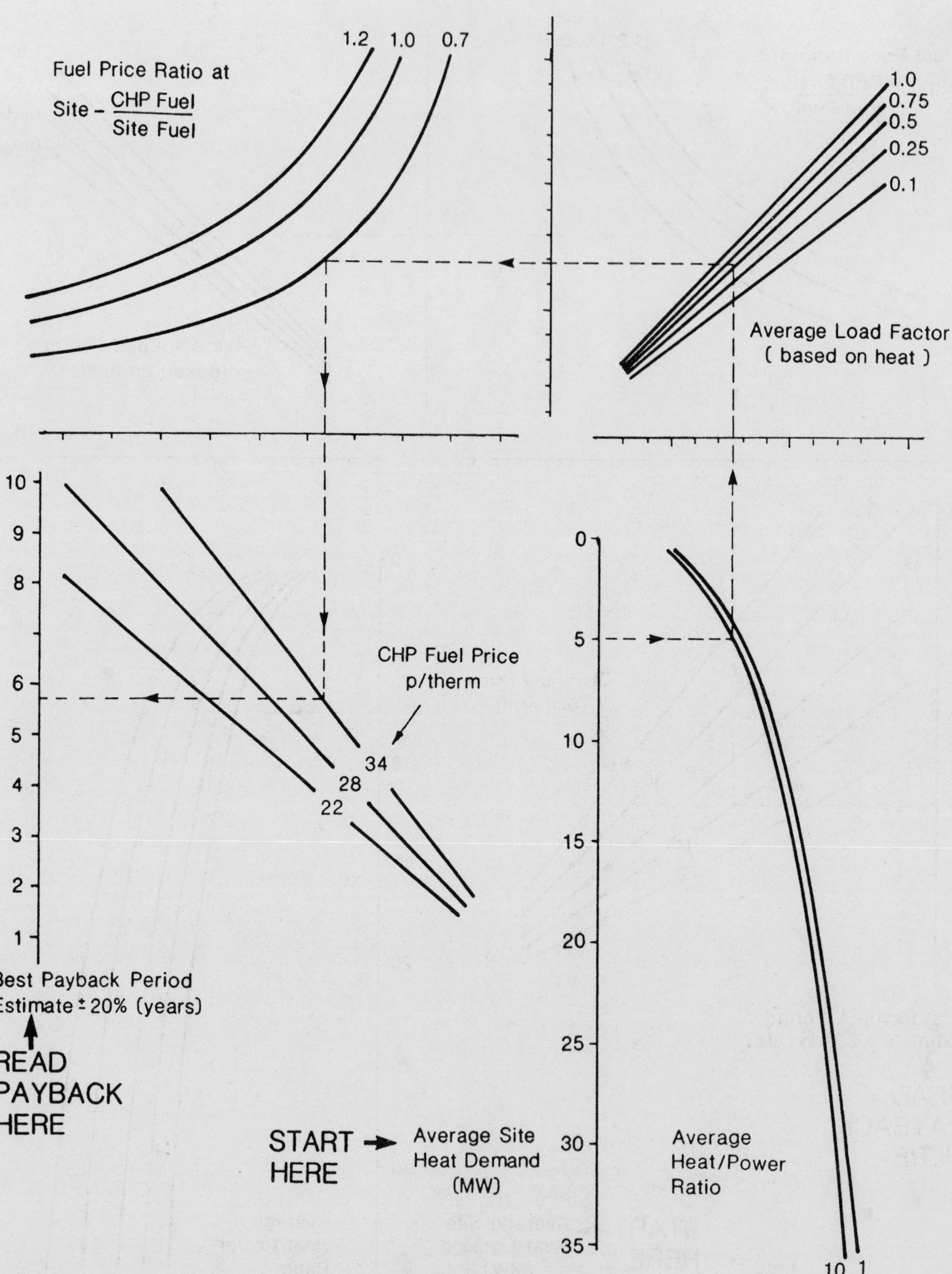

PRELIMINARY CHP ASSESSMENT CHART – SPARK IGNITION ENGINES (> 100kWe.)
Industrial Sites (Based on 1985 capital costs)

Fig. 4

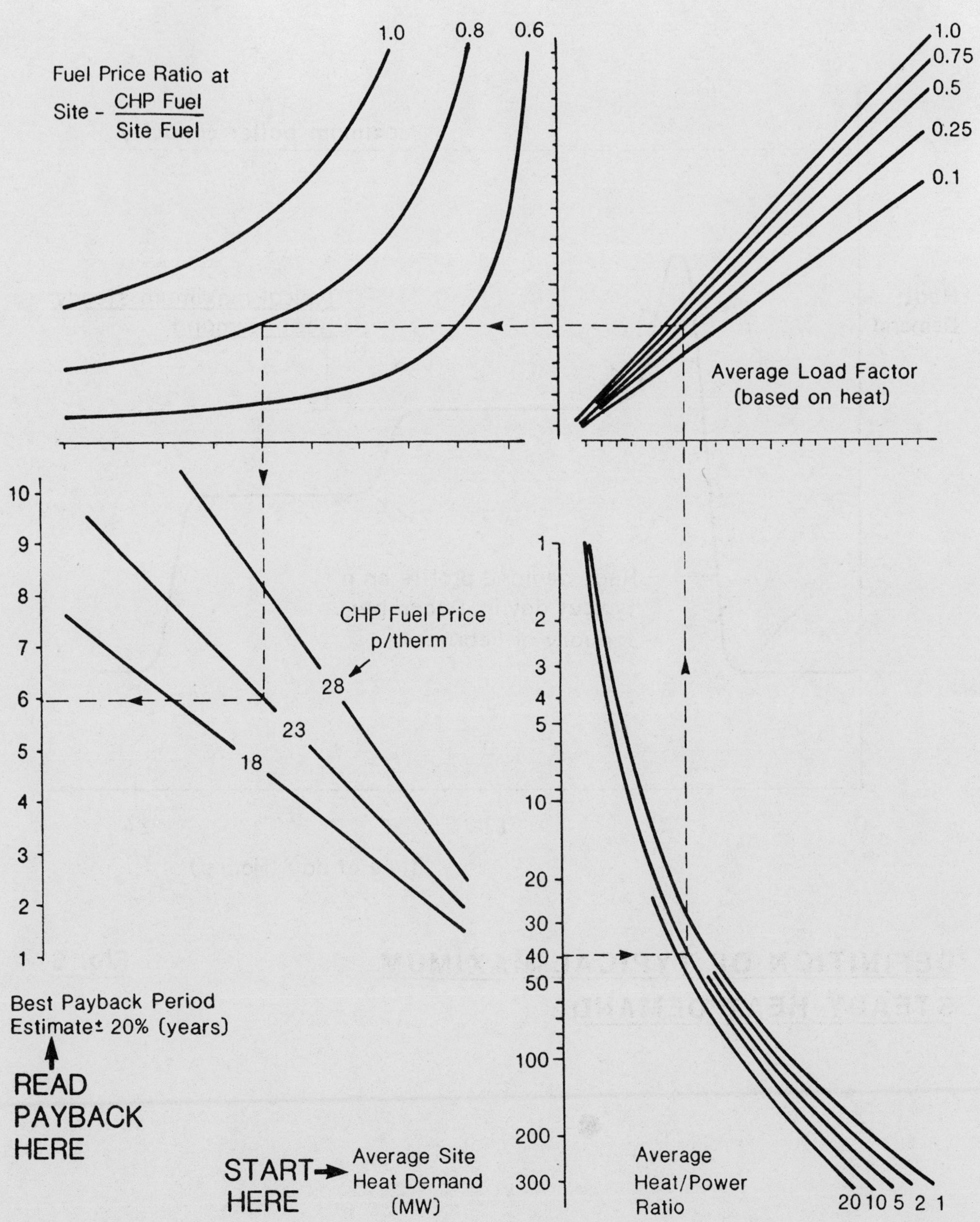

PRELIMINARY CHP ASSESSMENT CHART - COAL FIRED BOILERS + STEAM TURBINE PLANT **Fig. 5**
(Based on 1985 capital costs)

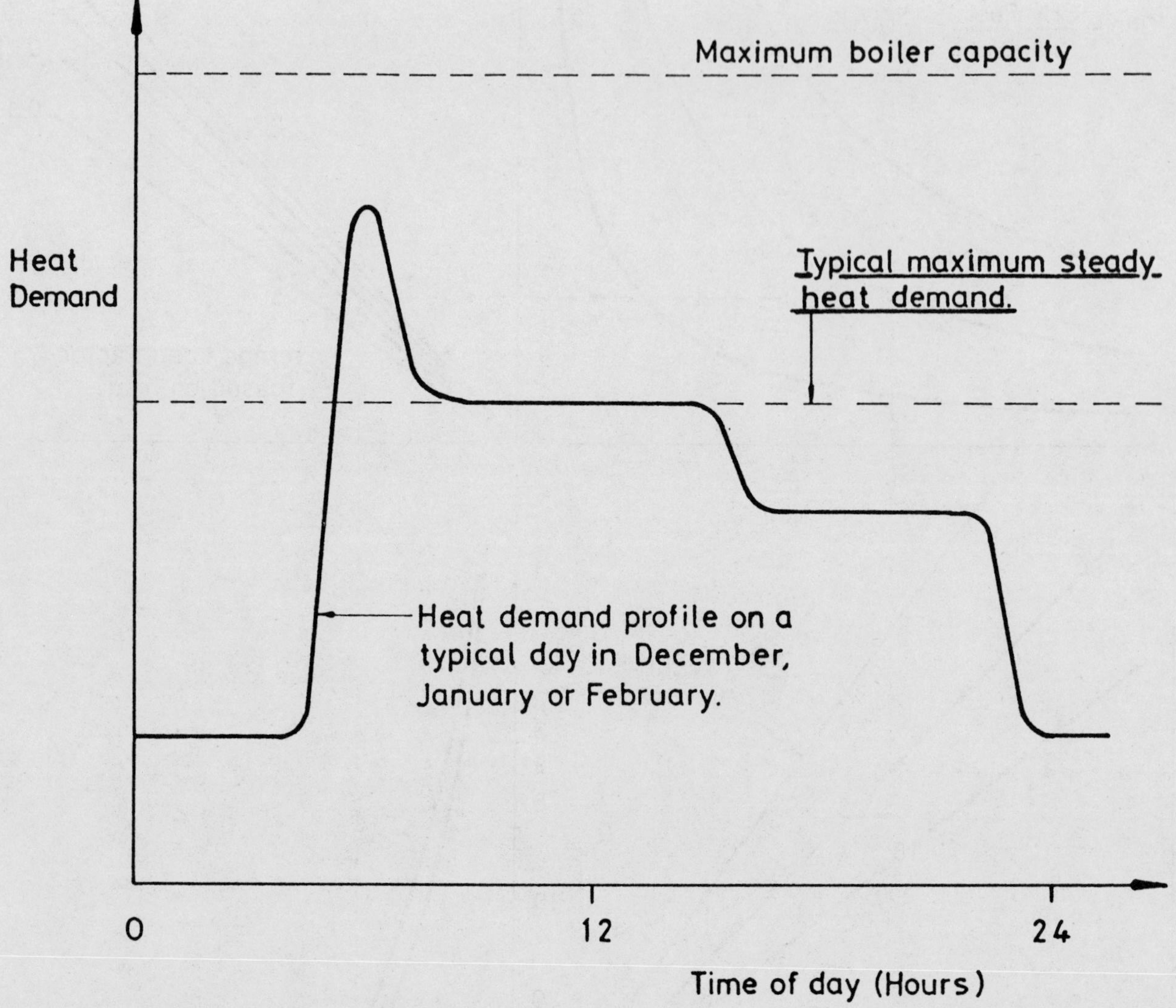

DEFINITION OF TYPICAL MAXIMUM STEADY HEAT DEMAND **Fig. 6**

3.4 Applicability of the Assessment Charts

15 The assessment charts were derived from correlations of the results of computer CHP assessments on over a hundred different industrial sites covering the full range of UK industry. It was found that the charts predicted the most cost effective payback period to within ±20% of that given by the computer assessment. It is therefore reasonable to suppose that they will give a payback to within ±20% on any industrial site, providing that the following points are observed:

- The ranges of each parameter must be within the ranges shown on each chart.

- They apply only to industrial sites, preferably those with steam or high temperature hot water demands. They do not apply to commercial or leisure buildings or any other premises where the heat is supplied as low temperature hot water.

- They apply only to CHP plant within the size range shown on each chart.

- Capital costs and electricity tariffs are those prevailing in 1985/86.

16 The detailed assessment of a CHP scheme is a complex affair and requires a large number of factors to be taken into account. Nevertheless from the five simple parameters described here a reasonable estimate of payback period can be made, as many other characteristics of industrial sites are contained inherently within the correlations.

3.5 Derivation of assessment charts - Basic correlations

17 The assessment charts were derived from the results of correlations of payback period with the parameters described earlier, for each type of CHP plant. Initially correlations of payback were obtained against four parameters: average site heat demand; site heat to power ratio; site load factor and the fuel price ratio - CHP plant fuel/fuel for existing boilers. The correlations were obtained by multiple linear regression of the most cost effective payback period, for a particular type of CHP plant at a particular site, against the four parameters for that site. For each type of plant over 50 sites were investigated. The resulting correlations were expressed in the form of the equation:

$$PB = K \times \frac{HP^{a} \cdot FP^{b}}{HT^{c} \cdot LF^{d}} \qquad (1)$$

where: PB = Best Payback Period
HP = Heat/Power Ratio
FP = Fuel Price Ratio CHP Fuel/Site Fuel
HT = Average Heat Demand
LF = Load Factor

K, a, b, c and d are constants for each correlation.

18 At this stage the correlations did not include (for simplicity) any allowance for changes in fossil fuel prices. The correlations were thus applicable only to the fuel prices prevailing in December 1984.

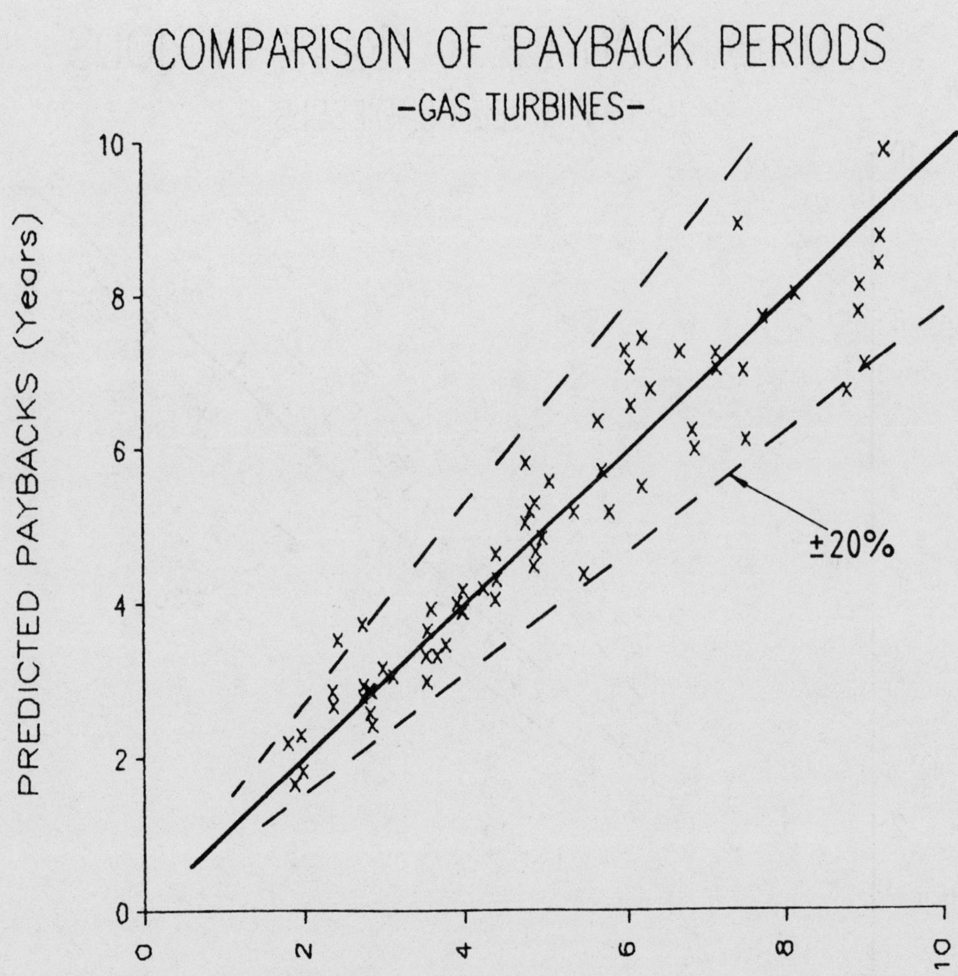

CORRELATION OF PAYBACK PERIODS FOR GAS TURBINE CHP PLANT **Fig. 7**

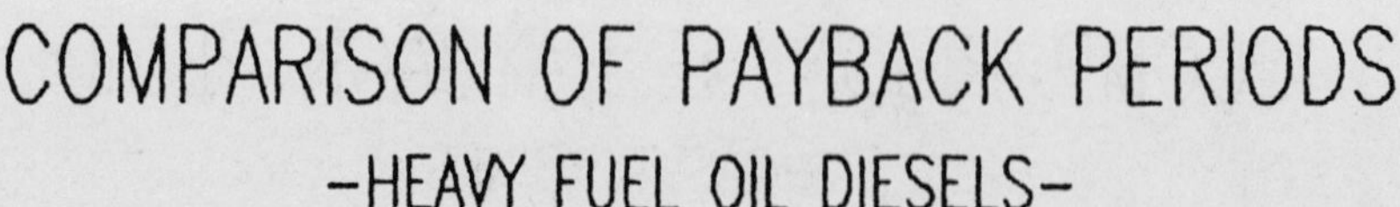

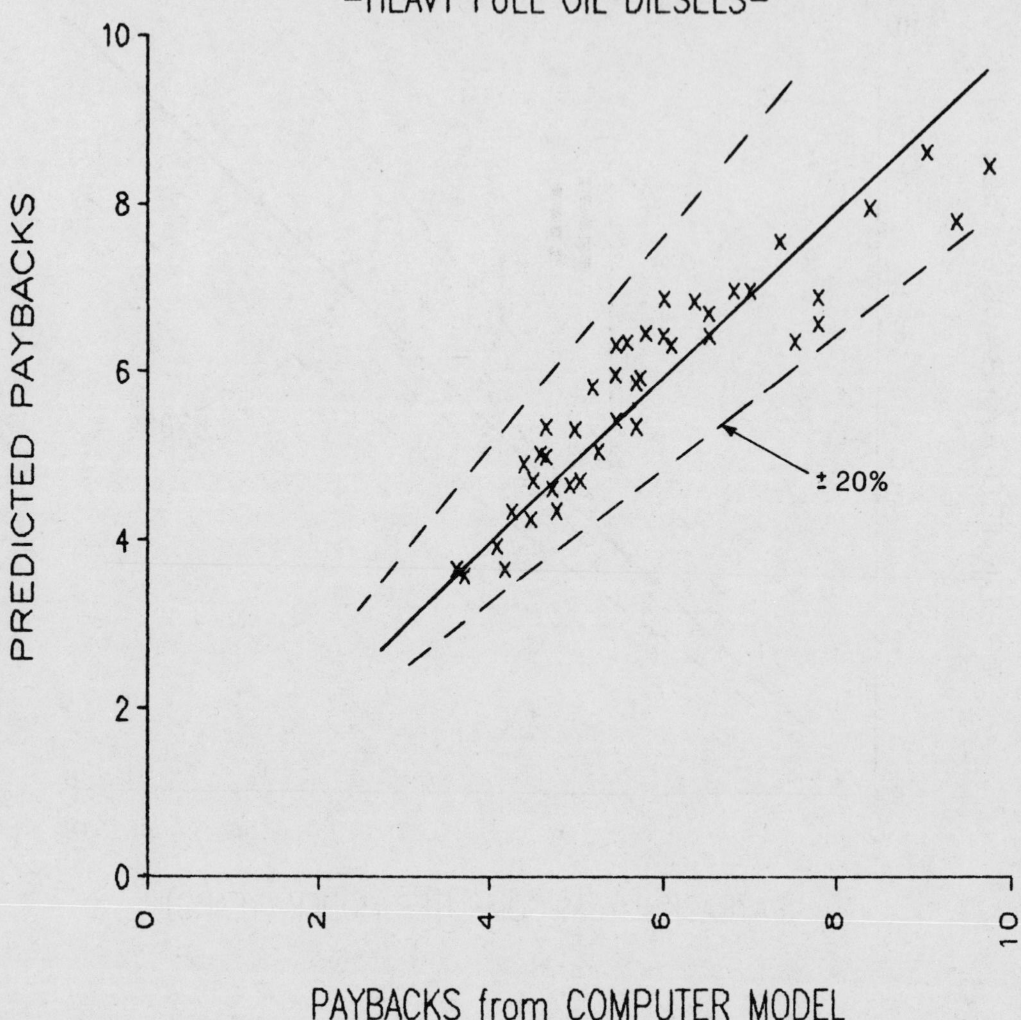

CORRELATION OF PAYBACK PERIODS FOR HEAVY FUEL OIL DIESEL PLANT **Fig. 8**

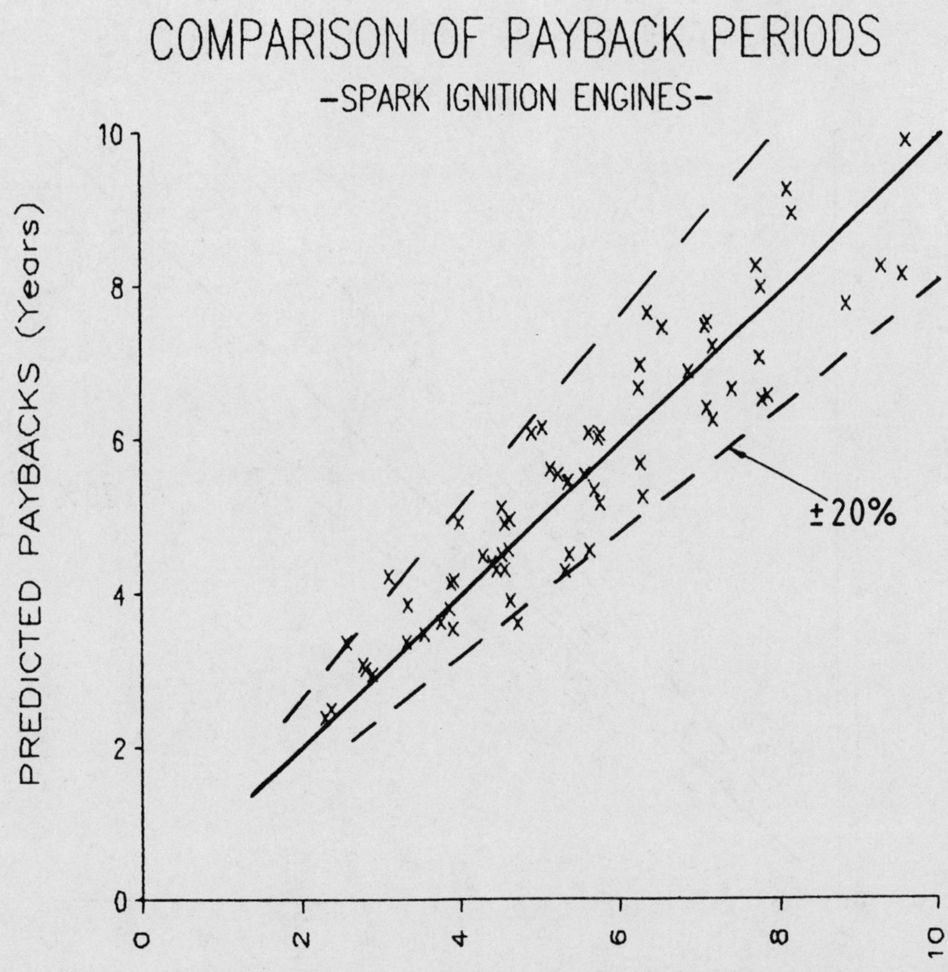

CORRELATION OF PAYBACK PERIODS FOR SPARK IGNITION ENGINE CHP PLANT **Fig. 9**

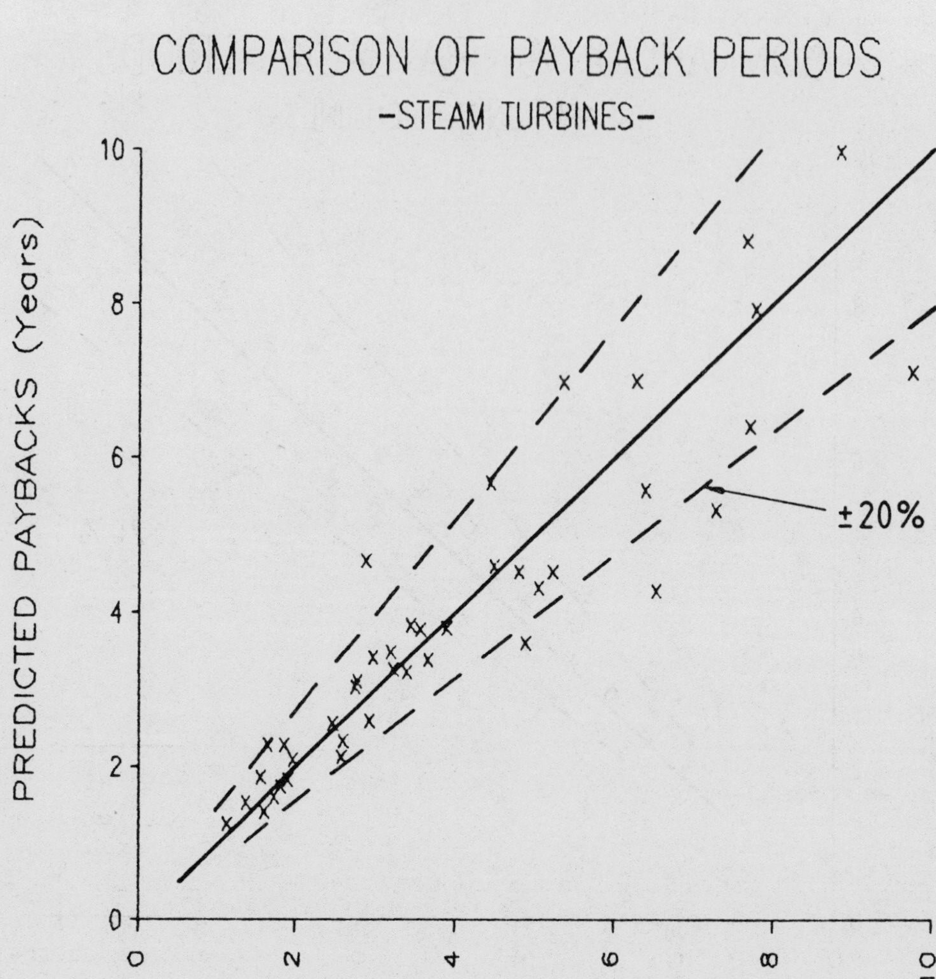

CORRELATION OF PAYBACK PERIODS FOR STEAM TURBINE CHP PLANT

Fig. 10

19 The constants in equation 1 are shown in Table 1 for each correlation. Also shown is the correlation coefficient which gives an indication of the strength of the correlation. In each case the coefficient is greater than 91% indicating a strong correlation. Figs. 7-10 show a comparison between the actual payback periods for industrial sites, calculated by the computer program, and those predicted by the correlations. It can be seen in each case that the paybacks predicted by the correlations are within ±20% of the values from the computer program.

20 The scatter in the payback period predictions arises from parameters that are not included in the correlations. These include such factors as: the proportion of electricity generated by the CHP plant which is exported; the temperature of the heat demand; the efficiency of the existing boiler plant, etc. These additional parameters were not included in the correlations because the extra complexity involved would not be justified by the marginal improvement in the accuracy of the correlations. The effect of some of these is discussed in Section 9.

Table 1
Results of Regressions

Types of CHP Plant	Values of Constants in Equation 1					Correlation
	k	a	b	c	d	
Gas Turbine	6.960	0.158	1.703	0.258	0.245	95%
Heavy Fuel-Oil Diesel	6.940	0.129	0.821	0.192	0.111	87%
Spark Ignition Engine	8.640	0.015	1.081	0.244	0.138	93%
Steam Turbine	40.60	0.083	3.210	0.440	0.186	91%

Adjustment for changing fuel prices

21 Following the sharp decrease in oil price it became apparent that an adjustment to allow for changing fuel prices had to be made. Consequently an extra parameter was added to the charts to allow for a change in the CHP plant fuel cost. This was achieved using the CHP computer program to investigate the effect on payback period of changes in fossil fuel price. The results were then incorporated in the charts, as shown in Figures 2 to 5.

4.0 DERIVATION OF GROSS POTENTIAL

4.1 Definition

22 The gross potential is defined as the total CHP capacity that it would be technically feasible to install within certain economic criteria.

23 Technical feasibility was assessed in terms of whether a given CHP plant was capable of supplying heat of a suitable quality to meet the heat demands of the site.

24 The economic criteria used were ranges of simple payback period for a CHP installation. The ranges chosen were payback periods of <1, <2, <3, <4, <5, <6, <8 and <10 years. These were selected to cover the following groups of potential installers of CHP plant:

- Manufacturing companies who would probably expect payback periods of less than 3 to 4 years. Equivalent to an internal rate of return of 20% or better.

- Utility companies who would be prepared to accept longer payback periods of up to 4 to 6 years. Equivalent to an internal rate of return of 10% or better.

- Organisations prepared to make long term investments with rates of return of 5% or more (equivalent to payback periods of up to about 8 years).

4.2 Method of Calculating the Gross Potential

25 The gross potential was calculated for each of the 50 selected industries and the results were summed to give an industry total. The method of calculation for each industry was as follows:

26 In each of the 50 industries considered several 'typical' sites were defined. These were considered to be representative of the industry as a whole (see Appendix 3). The CHP computer model was used to assess the economics of operating CHP plants of every type and size considered in this study (see Appendix 2) at each of the 'typical' sites. The results from the computer model were in the form of the printouts shown in Figure 11. For each 'typical' site the results from the computer model were then sorted into bands of payback period (eg. <1, <2, <3 years etc.). Within each band of payback period the CHP plants offering the greatest annual running cost savings were identified and a table of economic viability produced as shown in Table 2.

THE EXISTING FUEL IS 1

MAX ELEC DEMAND= 9426. MAX HEAT DEMAND (STEAM)= 14108. MAX DIRECT FIRING DEMAND= 0.

AVE ELEC DEMAND= 7739. AVE HEAT DEMAND (STEAM)= 11622. AVE DIRECT FIRING DEMAND= 0.

PLANT TYPE	SIZE BAND	UNIT RATING	NO. UNITS	ECON	FIRED BOIL	TCE SAVING	ELEC GENERATED [MWh]	ELEC EXPORTED [MWh]	CAPITAL COST [£]	NPV 5% [£]	EXISTING RUNNING COST [£/YR]	RUNNING COST SAVING [%]	PAY-BACK [YRS]
1	2	1520	1	1	0	1.235E+03	8.915E+03	0.000E+00	1.213E+06	-2.190E+05	3.644E+06	3.5	9.42
1	2	1520	1	1	1	1.763E+03	8.915E+03	0.000E+00	1.346E+06	-6.838E+04	3.644E+06	4.5	8.13
1	2	1520	2	1	0	2.449E+03	1.774E+04	0.000E+00	2.255E+06	-2.801E+05	3.644E+06	7.0	8.81
1	2	1520	2	1	1	2.575E+03	1.783E+04	0.000E+00	2.503E+06	-4.587E+05	3.644E+06	7.3	9.45
1	3	2970	1	1	0	2.922E+03	1.742E+04	0.000E+00	1.407E+06	8.090E+05	3.644E+06	7.9	4.90
1	3	2970	1	1	1	3.381E+03	1.742E+04	0.000E+00	1.583E+06	8.798E+05	3.644E+06	8.8	4.96
1	4	5145	1	1	0	5.752E+03	3.013E+04	0.000E+00	2.056E+06	2.154E+06	3.644E+06	15.0	3.77
1	4	5145	1	1	1	5.933E+03	3.018E+04	0.000E+00	2.286E+06	2.022E+06	3.644E+06	15.3	4.10
1	5	2840	1	1	0	2.567E+03	1.666E+04	0.000E+00	1.481E+06	4.508E+05	3.644E+06	6.9	5.92
1	5	2840	1	1	1	3.108E+03	1.666E+04	0.000E+00	1.626E+06	5.972E+05	3.644E+06	7.9	5.65
1	5	2840	2	1	0	4.187E+03	2.987E+04	0.000E+00	2.755E+06	5.656E+05	3.644E+06	11.8	6.41
1	5	2840	2	1	1	4.156E+03	3.041E+04	0.000E+00	3.025E+06	2.856E+05	3.644E+06	11.8	7.05
1	6	3140	1	1	0	3.557E+03	1.842E+04	0.000E+00	1.549E+06	9.742E+05	3.644E+06	9.0	4.74
1	6	3140	1	1	1	5.157E+03	2.331E+04	0.000E+00	1.677E+06	1.106E+06	3.644E+06	9.9	4.65
1	6	3140	2	1	0	5.547E+03	3.171E+04	0.000E+00	2.882E+06	1.271E+06	3.644E+06	14.8	5.36
1	6	3140	2	1	1	5.498E+03	3.220E+04	0.000E+00	3.119E+06	1.012E+06	3.644E+06	14.7	5.83
1	7	3790	1	1	0	4.189E+03	2.223E+04	0.000E+00	2.101E+06	8.892E+05	3.644E+06	10.6	5.42
1	7	3790	1	1	1	4.580E+03	2.223E+04	0.000E+00	2.256E+06	9.445E+05	3.644E+06	11.4	5.44
1	8	4825	1	1	0	8.826E+03	3.933E+04	0.000E+00	2.192E+06	2.122E+06	3.644E+06	15.3	3.92
1	8	4825	1	1	1	9.322E+03	3.933E+04	0.000E+00	2.356E+06	2.225E+06	3.644E+06	16.3	3.97

EXISTING FUEL COST (HCV) £ 2.68

CHP FUEL COST (HCV) £ 2.68

PLANT TYPE	SIZE BAND	UNIT RATING	NO. UNITS	ECON	FIRED BOIL	TCE SAVING	ELEC GENERATED [MWh]	ELEC EXPORTED [MWh]	CAPITAL COST [£]	NPV 5% [£]	EXISTING RUNNING COST [£/YR]	RUNNING COST SAVING [%]	PAY-BACK [YRS]
3	2	2000	1	1	1	2.588E+03	1.140E+04	0.000E+00	1.720E+06	-4.147E+05	3.644E+06	4.6	10.17
3	2	2300	1	1	1	2.899E+03	1.311E+04	0.000E+00	1.825E+06	-3.653E+05	3.644E+06	5.2	9.65
3	2	2600	1	1	1	3.210E+03	1.482E+04	0.000E+00	1.930E+06	-3.159E+05	3.644E+06	5.7	9.23
3	2	2600	2	1	1	5.619E+03	2.965E+04	0.000E+00	3.592E+06	-7.939E+05	3.644E+06	9.9	9.91
3	2	2900	1	1	1	3.521E+03	1.653E+04	0.000E+00	2.035E+06	-2.665E+05	3.644E+06	6.3	8.88
3	2	2900	2	1	1	6.046E+03	3.307E+04	0.000E+00	3.788E+06	-7.853E+05	3.644E+06	10.7	9.74
3	2	3200	1	1	1	3.832E+03	1.824E+04	0.000E+00	2.140E+06	-2.171E+05	3.644E+06	6.8	8.59
3	2	3200	2	1	1	6.462E+03	3.640E+04	0.000E+00	3.984E+06	-7.789E+05	3.644E+06	11.4	9.60
3	2	4200	1	1	0	3.775E+03	2.394E+04	0.000E+00	2.297E+06	-4.480E+05	3.644E+06	6.6	9.59
3	2	4200	1	1	1	4.795E+03	2.394E+04	0.000E+00	2.490E+06	-9.220E+04	3.644E+06	8.5	8.02
3	2	4200	2	1	0	6.670E+03	4.505E+04	2.331E+02	4.272E+06	-8.970E+05	3.644E+06	12.0	9.77
3	2	4200	2	1	1	7.533E+03	4.505E+04	2.331E+02	4.636E+06	-7.962E+05	3.644E+06	13.6	9.32
3	2	5200	1	1	0	4.569E+03	2.975E+04	0.000E+00	2.609E+06	-3.783E+05	3.644E+06	7.9	9.03
3	2	5200	1	1	1	5.637E+03	2.975E+04	0.000E+00	2.840E+06	-3.429E+04	3.644E+06	10.0	7.81
3	2	5200	2	1	0	7.461E+03	5.067E+04	2.598E+03	4.852E+06	-7.555E+05	3.644E+06	14.6	9.14
3	2	5200	2	1	1	8.216E+03	5.067E+04	2.598E+03	5.288E+06	-7.853E+05	3.644E+06	16.0	9.07
3	2	6200	1	1	0	5.347E+03	3.547E+04	0.000E+00	2.921E+06	-3.147E+05	3.644E+06	9.3	8.65
3	2	6200	1	1	1	6.351E+03	3.547E+04	0.000E+00	3.190E+06	-4.329E+04	3.644E+06	11.2	7.83
3	2	6200	2	1	0	7.970E+03	5.435E+04	6.267E+03	5.432E+06	-7.222E+05	3.644E+06	16.7	8.90
3	2	6200	2	1	1	8.636E+03	5.435E+04	6.267E+03	5.940E+06	-8.718E+05	3.644E+06	18.0	9.05

EXISTING FUEL COST (HCV) £ 2.68

CHP FUEL COST (HCV) £ 2.68

TYPICAL PRINTOUT FROM COMPUTER PROGRAM **Fig. 11**

Example of Table of Economic Viability

Table 2

S I Z E   B A N D   2

PRESENT BOILER FUEL	TYPE OF CHP PLANT	INSTALLED CAPACITY (no. of units * net kWe) of the CHP plant that achieves the specified payback period with the lowest running cost — PAYBACK PERIOD (years): <1	<2	<3	<4	<5	<6	<8	<10
natural gas	gas turbine						1*4825	1*4825	1*4825
	heavy fuel oil diesel engine								
	dual fuel reciprocating engine								
	spark ignition reciprocating engine							1*540	2*540
	steam turbine with coal fired boiler								
	combined cycle (gas turbine based)								
heavy fuel oil	gas turbine					1*4825	1*4825	1*4825	1*4825
	heavy fuel oil diesel engine								
	dual fuel reciprocating engine							1*6200	1*6200
	spark ignition reciprocating engine						2*540	2*540	2*1850
	steam turbine with coal fired boiler							1*600	1*600
	combined cycle (gas turbine based)								
coal	gas turbine							1*4825	1*4825
	heavy fuel oil diesel engine								
	dual fuel reciprocating engine								
	spark ignition reciprocating engine							1*540	2*540
	steam turbine with coal fired boiler								
	combined cycle (gas turbine based)								

Table 3

Example of Table of Economic Viability with Most Effective CHP Plant Selected.

SIZE BAND 2

PRESENT BOILER FUEL	TYPE OF CHP PLANT	INSTALLED CAPACITY (no. of units * net kWe) of the CHP plant that achieves the specified payback period with the lowest running cost — PAYBACK PERIOD (years) <1	<2	<3	<4	<5	<6	<8	<10	NO. OF SITES	CHP plant selected for the calculation of the total potential <1	<2	<3	<4	<5	<6	<8	<10
natural gas	gas turbine						1*4825	1*4825	1*4825	5						*	*	*
	heavy fuel oil diesel engine																	
	dual fuel reciprocating engine																	
	spark ignition reciprocating engine							1*540	2*540									
	steam turbine with coal fired boiler																	
	combined cycle (gas turbine based)																	
heavy fuel oil	gas turbine					1*4825	1*4825	1*4825	1*4825	3					*	*	*	*
	heavy fuel oil diesel engine																	
	dual fuel reciprocating engine							1*6200	1*6200									
	spark ignition reciprocating engine						2*540	2*540	2*1850									
	steam turbine with coal fired boiler							1*600	1*600									
	combined cycle (gas turbine based)																	
coal	gas turbine							1*4825	1*4825	0							*	*
	heavy fuel oil diesel engine																	
	dual fuel reciprocating engine																	
	spark ignition reciprocating engine							1*540	2*540									
	steam turbine with coal fired boiler																	
	combined cycle (gas turbine based)																	

27 Each 'typical' site represents a range of actual sites and these use a variety of boiler fuels. The price of the existing boiler fuel at a site has a major effect upon the economics of operating a CHP plant. So the model assessed each 'typical' site three times, using either gas, heavy fuel oil or coal as the existing boiler fuel. This is shown in Table 3 where the number of actual sites (represented by the 'typical' site) using each of the boiler fuels is shown.

28 From the tables of economic viability for each typical site (example in Table 3), the CHP plant, of any type, offering the greatest annual cost savings was selected for each band of payback period. These CHP plants are shown with an asterisk '*' in Table 3.

29 For each band of payback period, the gross potential for the sites represented by this typical site is obtained, by multiplying the CHP capacity of the most cost effective plant by the number of sites. And then summing for all the boiler fuels.

30 The gross potential for all of the 50 industries considered can then be calculated by summing the results for each of the 'typical' sites. The results are presented in the Technical Report.

4.3 <u>Calculations of Gross Potential</u> (with changing fuel prices)

31 The gross potential was initially calculated using the fuel prices prevailing in December 1984. However by the end of 1985 the price of heavy fuel oil had fallen substantially and this significantly affected the economics of CHP at some sites. Consequently the gross potential was reassessed using the fuel prices as of December 1985. These were:

Interruptible natural gas	28.3p/therm	£2.68/GJ
Heavy fuel oil	11p/litre	£2.68/GJ
Coal	£62.5/tonne	£2.20/GJ

The gross potential using these fuel prices is presented in the Technical Report, Section 7.

4.4 <u>Types of CHP Plant</u>

32 The gross potential identifies only the most cost effective CHP plant. Those plants which come a 'close second' are not included. The gross potential does not therefore provide a comparison of the relative cost effectiveness of the different types of CHP plant. This is discussed in a qualitative way in Section 9.0 of this Appendix.

4.5 Energy Savings

33 The CHP computer model also identified the primary energy savings that would result from a CHP installation. These were summed in the same way as the CHP plant capacity to identify the total primary energy savings that could be achieved.

4.6 Assumptions

34 The gross potential calculated by this method identi fies the most cost effective CHP plant installation for a site. No account is taken however of other competing investment opportunities, such as investment in energy conservation measures, which may be more cost effective and if implemented can affect CHP economics. This and other assumptions inherent in the calculation of gross potential are discussed in the Technical Report.

4.7 Retro-fit CHP Plant

35 The calculation of gross potential is based entirely upon the installation of new CHP plant of the six types considered. Small retro-fit CHP plant, some in direct drive applications, have also been investigated. These are not included in the gross potential but the potential for them is described more fully in Appendix 4.

5.0 CONTACTS WITH INDUSTRY

36 To obtain industrial views and comments regarding CHP, 40 site assessment forms were sent to selected individual within industrial organisations. Data from the 25 forms received back was used in the development of the industrial data base. CHP assessments were also carried out based on the site data from each of the forms. These assessments were returned to the individuals and 22 of them were contacted again to obtain the views of potential CHP installers.

37 The 22 respondents were interviewed by telephone and a copy of the questionnaire used in conducting the interviews included (Annex 1) at the end of this appendix. The interviews were carried out in mid 1985 and the CHP assessments discussed reflected the fuel prices at that time.

5.1 Nature of Sample

38 The 22 respondents were not selected at random; the majority of them were known to ETSU or the EEO previously, and may have been more interested in energy cost savings than the average industrialist.

39 Table 4 provides an overview of the sample of sites. The sites are listed in order of increasing annual energy expenditure. The sites are arbitrarily divided into three groups of roughly equal size:

Small energy users	8 sites
Medium energy users	7 sites
Large energy users	7 sites

The sample included sites involved in the production of chemicals, pharmaceuticals, food and drink, plastics conversion, paper, glass and electronic and mechanical engineering products.

40 The sites employed from 120 to over 6000 persons, with an average of around 1000.

41 Of the 22 sites, 13 used gas as the main fuel (60% by number), 6 used heavy fuel oil (27% by number), and 3 were using coal (13% by number). Gas was the main fuel at 7 of the 8 'small' sites in the sample.

42 Energy costs ranged from around £0.2 million per year to just under £5 million per year, with an average of £1.8 million.

43 The sample is not intended to reflect the structure of UK manufacturing industry, either in the balance of activities or in terms of its size distribution and pattern of energy consumption. The sites chosen for CHP assessment were biased in general towards those likely to prove attractive for CHP.

5.2 CHP Payback and Reactions

44 Tables 4 and 5 show data relevant to the CHP assess ment: payback, previous involvement with CHP or coal conversion studies, and reactions to and the likely consequence of the present assessment.

Previous Studies of CHP

45 The possibility of installing CHP plant had been examined previously at just over half the sites in the sample (55%). There is no apparent correlation with energy consumption or fuel.

46 In half the cases examined, the possibility of installing CHP plant was rejected on account of an unattractive payback. At a third of the sites, the assessment had not been completed at the time of the interview. At the remaining sites, action had been deferred (usually pending the conclusion of other studies such as detailed site energy audits).

Studies of Coal Conversion

47 Three of the sites in the sample were already burning coal as the predominant fuel, and the question was not relevant. The sample for this question thus comprised 19 sites.

48 At 15 sites burning gas or heavy fuel oil, the possibility of converting to coal had been examined. The exceptions were:

- the smallest site in the sample, burning gas:
- processed food site, currently burning heavy fuel oil, where aesthetic objections have been raised against coal;
- a material converter currently burning heavy fuel oil where coal is unattractive for technical reasons.

49 This means that coal conversion has been considered for virtually all the sites in the sample where it was a sensible technical option.

50 Of the 15 sites examined for coal, conversion had been rejected at 11 (73%) on account of unattractive payback. Decisions had been deferred pending the completion of other studies at four sites.

Sample Summary

Table 4

SAMPLE SUMMARY

Site Code Letter	INDUSTRY	No. of Employees @ site	Current Fuel	Energy Cost £ M	Best payback Identified by Assessment (yrs)
	Small energy users:				
A	PLASTICS PROCESSING	170	GAS	.2	8
B	MECHANICAL ENGINEERING	481	GAS	.3	12
C	FOOD	120	GAS	.6	5.7
D	OTHER	1000	GAS	.8	4.3
E	PLASTICS PROCESSING	400	GAS	.8	6
F	DRINKS	192	HFO	.8	2.9
G	CHEMICALS	400	GAS	.8	5.7
H	FOOD	200	GAS	.9	4.7
	Medium energy users:				
I	DRINKS	600	HFO	1	2.4
J	FOOD	1400	GAS	1.3	5.3
K	DRINKS	700	HFO/GAS	1.4	5
L	MECHANICAL ENGINEERING	1700	HFO	1.5	6.3
M	CHEMICALS	1180	COAL	1.8	5.3
N	CHEMICALS	425	COAL	2	3.3
O	OTHER	270	GAS	2	5.5
	Large energy users:				
P	FOOD	3000	GAS	2.4	5.1
Q	CHEMICALS	320	HFO	2.4	2
R	FOOD	1200	GAS	2.5	4.5
S	DRINKS	2000	GAS	3.5	3.3
T	FOOD	6370	COAL	3.7	4.6
U	OTHER	1000	HFO	4.5	12
V	CHEMICALS	670	GAS	4.8	3.7
	Average: small	370.38		.65	6.16
	: medium	896.43		1.57	4.73
	: large	2080		3.4	5.03
	: overall	1081.73		1.82	5.35

Response to CHP Assessment

Table 5

RESPONSE TO CHP ASSESSMENT - Question no. in brackets ()

Site Code Letter	INDUSTRY	(1) Previous CHP Studies?	Previous CHP study Result	(2) Coal Conversion Evaluated?	Coal Conversion Result?	(3) Study Data OK?	Response To Result	(4) Next Step?	(6) Chance For CHP?
	Small energy users:								
A	PLASTICS PROCESSING	YES	REJECT	NO		YES	WORSE	NOTHING	NIL
B	MECHANICAL ENGINEERING	YES	REJECT	YES	REJECT		EQUAL	NOTHING	NIL
C	FOOD	YES	REJECT	YES	REJECT	YES	EQUAL	NOTHING	NIL
D	OTHER	YES	INCOMPLETE	YES	REJECT	NO	BETTER	STUDY	FAIR
E	PLASTICS PROCESSING	NO		YES	REJECT	YES	WORSE	STUDY	POOR
F	DRINKS	NO		NO		NO	BETTER	STUDY	FAIR
G	CHEMICALS	YES	REJECT	YES	DEFER	YES	WORSE	NOTHING	NIL
H	FOOD	NO		YES	DEFER	YES	BETTER	STUDY	NIL
	Medium energy users:								
I	DRINKS	NO		YES	REJECT	NO	BETTER	OIL to GAS	POOR
J	FOOD	NO		YES	DEFER	NO	WORSE	STUDY	POOR
K	DRINKS	NO		YES	DEFER	YES	WORSE	NOTHING	NIL
L	MECHANICAL ENGINEERING	YES	INCOMPLETE	YES	REJECT	YES	EQUAL	STUDY	POOR
M	CHEMICALS	NO			ON COAL	YES	WORSE	STUDY	POOR
N	CHEMICALS	YES	INCOMPLETE		ON COAL	YES	BETTER	STUDY	GOOD
O	OTHER	YES	REJECT	YES	REJECT	GENERAL	WORSE	STUDY	FAIR
	Large energy users:								
P	FOOD	NO		NO		NK	BETTER	STUDY	POOR
Q	CHEMICALS	YES	INCOMPLETE	YES	REJECT	NO	BETTER	STUDY	GOOD
R	FOOD	NO		YES	REJECT	NK	EQUAL	NOTHING	NIL
S	DRINKS	YES	DEFER	YES	REJECT	YES	EQUAL	DEFER	NIL
T	FOOD	NO			ON COAL	YES	BETTER	STUDY	POOR
U	OTHER	YES	REJECT	NO		NO	WORSE	NOTHING	NIL
V	CHEMICALS	YES	DEFER	YES	REJECT	YES	EQUAL	STUDY	GOOD

51 The pattern of rejection/deferral does not appear to correlate with energy consumption or type of fuel used.

Reaction to the Present CHP Assessment

52 Before exploring detailed reactions, respondents were asked whether the data used as a basis for the assessment reflected adequately the pattern and cost of energy consumption at the site, and the choice facing the decision maker.

53 The response was affirmative in the case of 14 sites. At three sites the respondent was unable to comment, and in one case the site data was 'generalised' rather than specific. Thus the data and model adequately reflected conditions at the site in the view of over 75% of the respondents able to comment.

54 The three objections to the data set are summarised below:

- Two respondents expressed reservations regarding the assumed capital costs, expecting rather higher values in reality (though without access to the details their ability to comment was very limited).

- One respondent noted that site conditions were liable to change, and therefore the site data reflected the past rather than the future, although the results were still very useful.

- One respondent noted that the schemes considered by the model omitted one possibility relevant to the particular site - however, inclusion of this option would not have altered the results.

55 It may be assumed that in other respects the data base used in the assessment adequately reflects the site conditions.

First Reaction to the Assessment

56 The initial reaction to the computer assessment was probed. As shown in the table below, roughly a third of those interviewed found the results worse than expected, a third found them about as expected, and a third found them better than expected.

57 Narrowing the sample to those firms which had previously studied CHP for the site in question, suggests that about half found the results more or less in line with their own findings, about a third found them worse, and a quarter found that the computerised assessment of this project gave better results than expected on the basis of their own work.

First Reaction to Assessment

Table 6

FIRST REACTION TO THE CHP ASSESSMENT

	Small energy users:	
A	PLASTICS PROCESSING	Very disappointing result
B	MECHANICAL ENGINEERING	Knew that CHP was not appropriate to the site
C	FOOD	Payback not good enough to justify further interest
D	OTHER	Do not believe figures, should add £75K to costs, payback 5-6 years
E	PLASTICS PROCESSING	Very keen on GT/WHB when next reboilering, but payback very disappointing
F	DRINKS	Do not accept results; should use hfo/gas conversion as baseline
G	CHEMICALS	Results do not justify interest
H	FOOD	Expect payback of 4-5 years
	Medium energy users:	
I	DRINKS	Must consider oil/gas conversion as competing alternative; must include gas supply costs
J	FOOD	Worthwhile participating in project, but results disappointing; would need more info on heat loads
K	DRINKS	Very disappointed with results, was hoping for 2 year payback
L	MECHANICAL ENGINEERING	Surprised that viable options are either "very small" and "very large"; why?
M	CHEMICALS	Requires more careful study, but such paybacks hardly justify it
N	CHEMICALS	Capital costs too low, likely to double in reality
O	OTHER	Gas turbine looks good, worth further investigation
	Large energy users:	
P	FOOD	Company keen to save costs; payback limit of 4 years;stretching it under discussion, few good projects available
Q	CHEMICALS	Plant being rebuilt - data will change
R	FOOD	Confirms that CHP has no role to play at the site
S	DRINKS	Unlikely to take action in the next few years
T	FOOD	Very interesting
U	OTHER	If exhausting GT into furnaces, payback improves, but not good enough compared to oil/gas conversion
V	CHEMICALS	Plant expansion in progress, will not know heat/power needs for a while; in meantime, preoccupied

CHP assessment was:	All sites in sample	Those which had studied CHP before
Worse than expected	36%	33%
Equal to expectation	27%	42%
Better than expected	36%	25%

58 It is concluded that the model used in the current assessment gives a reasonable, if a shade conservative, view of the economic viability of CHP investment.

Anecdotal reactions to the assessment have been paraphrased and are summarised in Table 6.

<u>Likely Next Step</u>

59 Respondents were asked what the most likely next step would be regarding the implementation of CHP. The results are shown in Table 5 and are summarised below:

Action (if any)	Response: Number	Percent
No action	7	32%
Recommend further study	13	59%
Other (defer, oil to gas conversion)	2	9%

<u>Chances of Implementation</u>

60 Respondents were also asked to rate their assessment of the chances that a CHP scheme might be implemented at their site within the next few years. Only four choices were allowed:

- virtually zero chance (NIL)
- worse than even chance (POOR)
- better than even chance (FAIR)
- virtually certainty (GOOD)

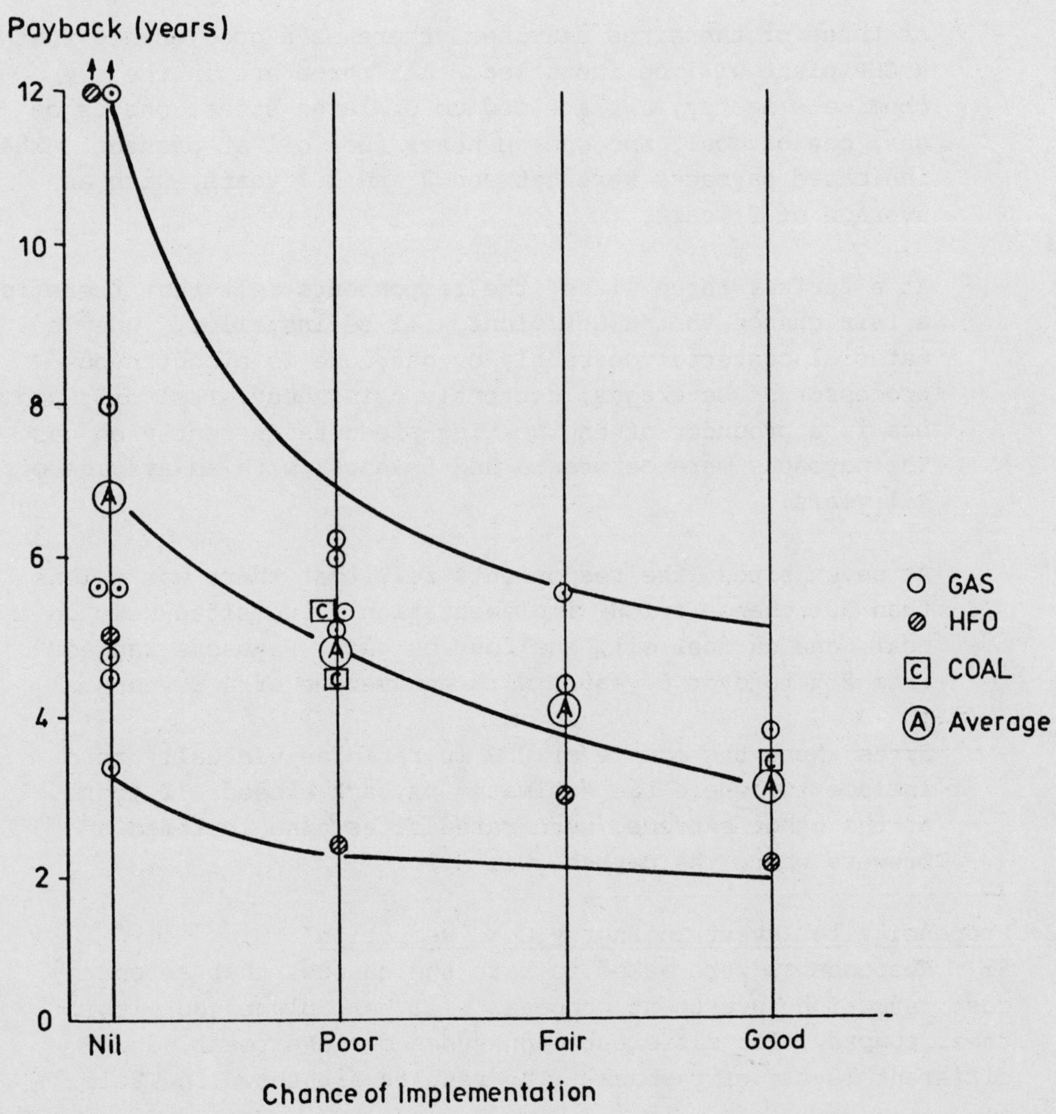

SUBJECTIVE PROBABILITY OF IMPLEMENTATION AS A FUNCTION OF PAYBACK PERIOD AND FUEL TYPE

Fig. 12

61 The results are shown in Table 5, and are also plotted schematically in Figure 12. The results are summarised below:

- At three of the sites examined, there is a good chance that a CHP plant will be installed. All three are in the chemicals sector, all are medium or large sites; one is on gas, one on coal, and one on heavy fuel oil at present. The indicated paybacks were between 2 and 3.7 years, with an average of 3 years.

- At a further three sites, the respondents felt that there is a fair chance that a CHP plant will be installed. One is a material converter currently on gas; one is a continuous processor of beverages, currently using heavy fuel oil; and one is a producer of engineering products currently on gas. The paybacks were between 3 and 6 years, with an average of 4.1 years.

- At seven sites, the respondents felt that there was a less than 50% chance of CHP implementation. Two sites were on coal, one on fuel oil, and four on gas. Paybacks ranged from 2.5 to over 6 years, with an average of 4.9 years.

- Sites where the chance of CHP is rated as virtually zero include two where the estimated payback exceeds 12 years. At the other extreme, zero rated sites also included a brewery where the payback was 3.3.

5.3 Propensity to Invest in Energy Cost Reduction

62 Respondents were asked to rate the chances that an energy cost reduction investment proposal will be implemented within their company, for different magnitudes of investment, and different levels of payback. The results are shown in Table 7.

63 The mean subjective probability of investment in cost saving is presented Table 4 for the three size groups, and for the sample overall.

64 The response does not correlate with the size of the site, suggesting that some other influence, such as management attitude and style, dominates. For this reason overall averages were used as a basis for subsequent analysis.

65 Figure 13 plots the overall average subjective probability to invest against payback for the three levels of cost reduction investment. These show a clear and consistent separation.

Subjective Assessment of Investment Probability — Table 7

SUBJECTIVE ASSESSMENT OF THE PROBABILITY OF INVESTMENT IN ENERGY COST REDUCTION

Site Code Letter	INDUSTRY	:CAPITAL INVESTMENT: £5 MILLION :payback periods (years)					:CAPITAL INVESTMENT: £1 MILLION :payback periods (years)					:CAPITAL INVESTMENT: £0.1 MILLION :payback periods (years)				
		:4-5	3-4	2-3	1-2	<1	:4-5	3-4	2-3	1-2	<1	:4-5	3-4	2-3	1-2	<1
	Small energy users:															
A	PLASTICS PROCESSING	0	0	0	0	0	0	10	30	40	50	0	10	30	40	50
B	MECHANICAL ENGINEERING	0	0	0	0	0	0	0	25	50	80	0	0	70	80	100
C	FOOD	0	0	0	0	0	0	0	0	0	0	0	0	15	50	75
D	OTHER	20	40	60	80	100	20	40	60	90	100	20	50	75	100	100
E	PLASTICS PROCESSING	0	0	0	0	20	0	0	0	75	90	0	0	75	80	100
F	DRINKS	50	75	100	100	100	50	75	100	100	100	50	75	100	100	100
G	CHEMICALS	5	25	50	75	80	5	25	50	75	80	5	25	50	75	100
H	FOOD	0	0	20	50	75	0	5	25	75	80	0	20	30	80	100
	Medium energy users:															
I	DRINKS	0	0	10	40	60	0	20	40	60	80	0	30	50	80	90
J	FOOD	10	20	50	70	90	10	20	50	70	90	10	20	50	70	90
K	DRINKS	0	0	10	50	80	0	0	10	50	80	0	0	20	60	90
L	MECHANICAL ENGINEERING	0	10	70	80	90	0	10	70	80	90	0	10	70	80	90
M	CHEMICALS	0	0	0	0	0	0	0	5	25	50	0	0	10	60	80
N	CHEMICALS	0	20	80	80	100	0	20	80	80	100	0	20	80	80	100
O	OTHER	0	0	0	30	50	0	20	40	65	70	10	30	50	70	90
	Large energy users:															
P	FOOD	10	50	75	100	100	15	60	80	100	100	25	70	90	100	100
Q	CHEMICALS	0	0	25	75	100	0	0	30	80	100	0	0	50	100	100
R	FOOD	0	0	0	40	60	0	0	0	50	70	0	0	0	60	80
S	DRINKS	0	0	10	50	75	0	0	10	50	75	0	0	25	75	100
T	FOOD	0	10	25	50	75	0	10	25	50	75	0	10	25	50	75
U	OTHER	0	0	0	25	50	0	0	0	50	75	0	0	0	75	80
V	CHEMICALS	0	50	70	80	100	0	50	0	0	100	0	50	70	80	100
	Average: small	9.38	17.5	28.75	38.13	46.88	9.38	19.38	36.25	63.13	72.5	9.38	22.5	55.63	75.63	88.13
	: medium	1.43	7.14	31.43	50	67.14	1.43	12.86	42.14	61.43	80	2.86	15.71	47.14	71.43	90
	: large	1.43	15.71	29.29	60	80	2.14	17.14	20.71	54.29	85	3.57	18.57	37.14	77.14	90.71
	: overall	4.32	13.64	29.77	48.86	63.86	4.55	16.59	33.18	59.77	78.86	5.45	19.09	47.05	74.77	89.55

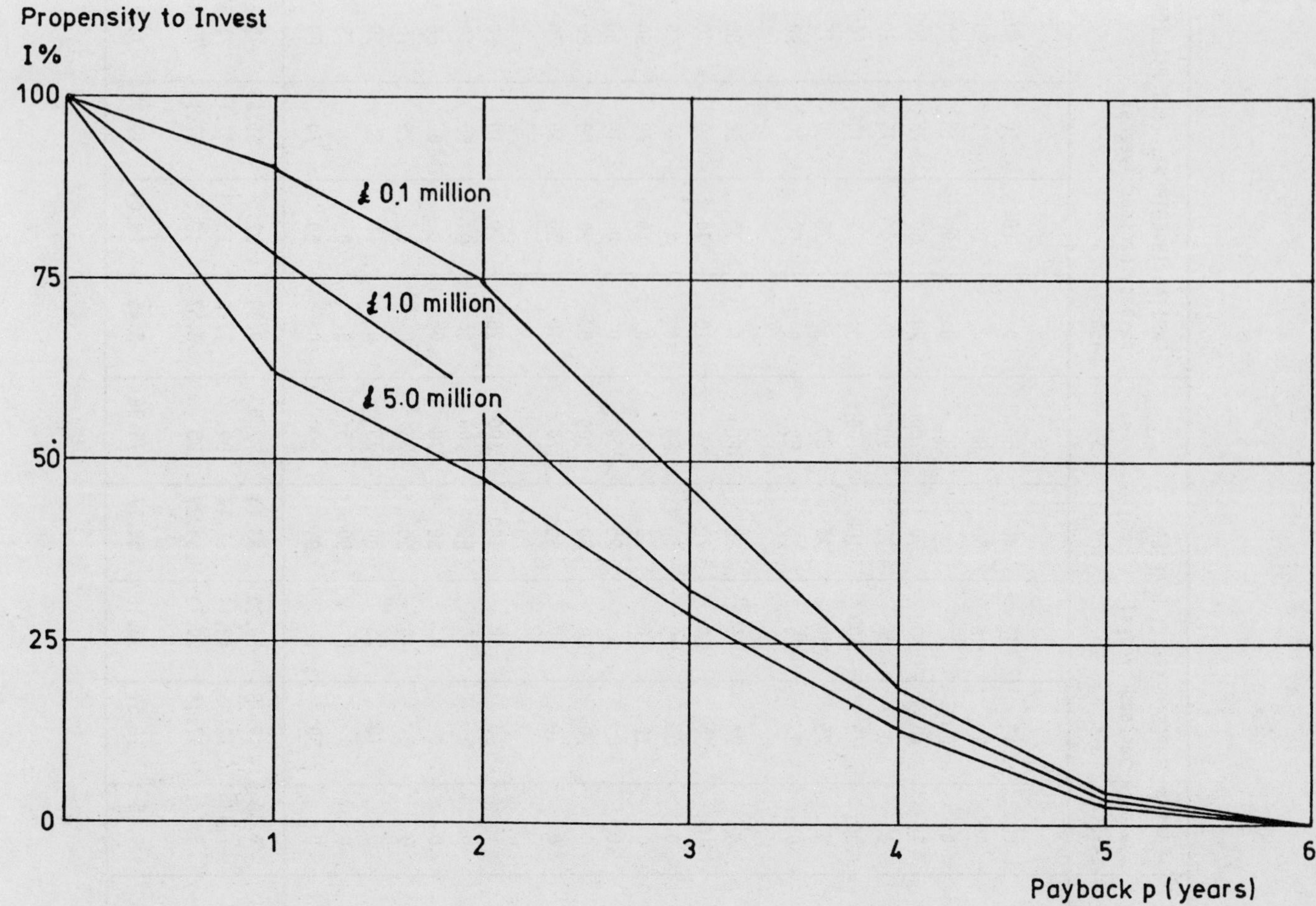

PROPENSITY TO INVEST VERSUS PAYBACK PERIOD AS A FUNCTION OF LEVEL OF CAPITAL INVESTMENT

Fig. 13

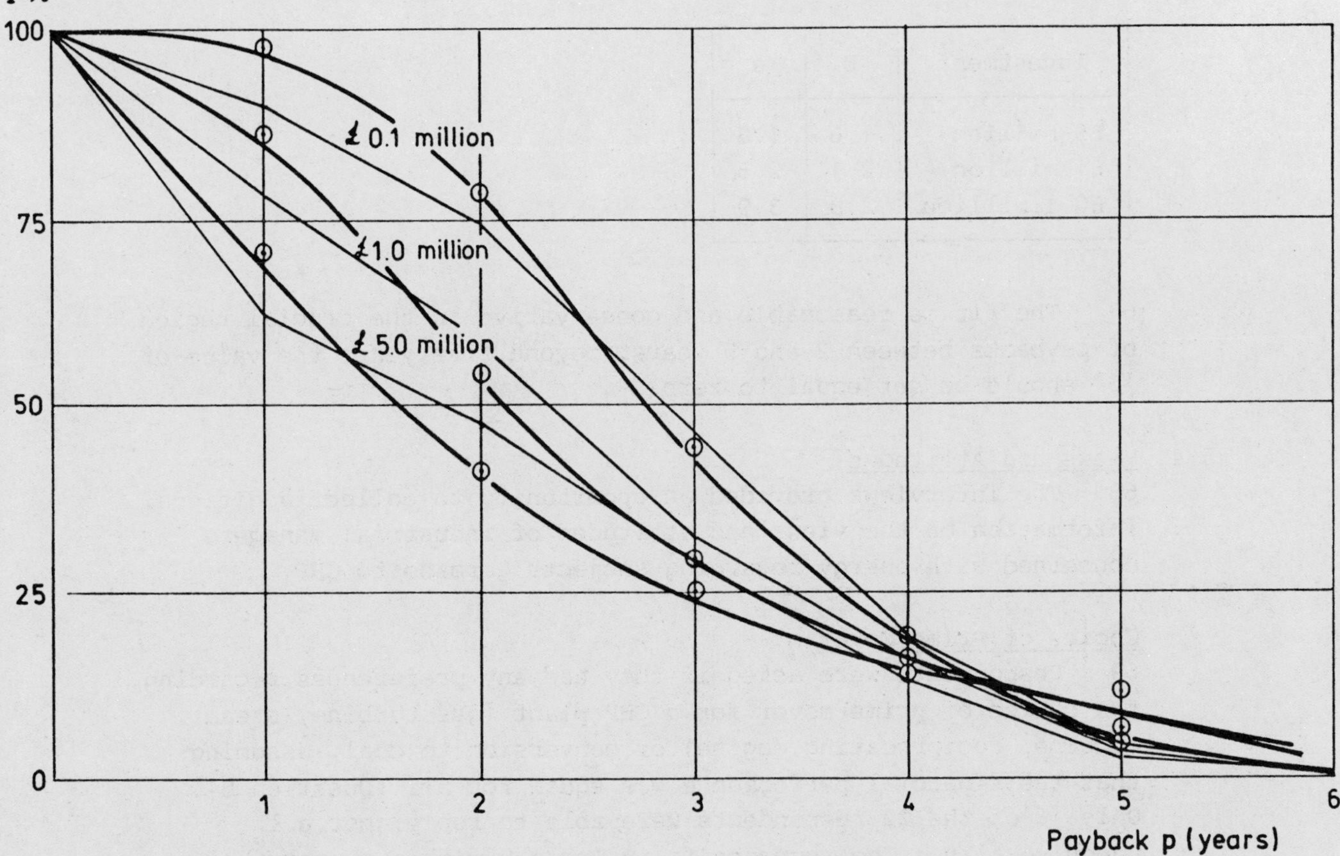

Function used:

$$I = \frac{100}{1+\left(\frac{p}{E}\right)^{a}}$$

where E and a are coefficients determined by the least squares method

MODELLING OF PROPENSITY TO INVEST VERSUS PAYBACK PERIOD **Fig. 14**

66 In Figure 14 the function $I = 100/(1 + (p/E)^a)$ is superimposed on the average obtained from the interviews using values of 'E' and 'a' giving the best fit determined by the least squares method. The calculated coefficients are shown below:

Investment	E	a
£5 million	1.6	1.8
£1 million	2.1	2.5
£0.1 million	2.8	3.9

67 The fit is reasonable and conservative in the crucial region of paybacks between 2 and 5 years; beyond five years the value of 'I' should be set equal to zero.

5.4 Views and Attitudes

68 The interviews provided an opportunity to collect additional information on the views and attitudes of industrial managers concerned with energy regarding subjects germane to CHP.

Choice of Prime Mover

69 Respondents were asked if they had any preferences regarding the choice of prime mover for a CHP plant (gas turbine, steam turbine, reciprocating engine) or conversion to coal, assuming that the financial performance was equal for all (Question 8). Only 16 of the 22 respondents were able to reply, not all unequivocally. The response is in Table 9 with a summary below:

Choice Ranked as:	1st		2nd		3rd	
	No.	%	No.	%	No.	%
Prefer gas turbines	4	25	5	31	2	12
Prefer steam turbines	3	18	3	18	2	12
Prefer engines	2	12	1	6	3	18
Prefer turbines*	1	6	1	6	1	6
Dislike gas turbines			2	12	1	6
Prefers coal conv'n	3	18	3	18	3	18
No strong preference	3	18	1	6	3	18

* steam or gas turbine

70 Comments regarding the choice of prime movers are summarised in paraphrased form in Table 10. There is evidence that the choice is influenced by the nature of the activities conducted at the site.

Comments on Implementation

Table 8

COMMENTS REGARDING IMPLEMENTATION

	Small energy users:	
A	PLASTICS PROCESSING	Payback completely unacceptable
B	MECHANICAL ENGINEERING	Inappropriate
C	FOOD	Unacceptable payback
D	OTHER	Firm very progressive, but limit investment to 2 years payback, though 4.4 might be acceptable
E	PLASTICS PROCESSING	Marginally acceptable payback, high cost
F	DRINKS	Very interested in CHP for cost reduction; implementing scheme at another site
G	CHEMICALS	Payback is unacceptable
H	FOOD	The company is not making a profit; could not justify investment with 4+ years payback
	Medium energy users:	
I	DRINKS	The industry has too many other problems to solve before it can consider diverting effort to CHP
J	FOOD	Company very flexible, might stretch payback if secondary benefits strong, but 5+ yr difficult
K	DRINKS	Payback puts CHP out of the question; coal conversion a better option; will adopt when reboilering
L	MECHANICAL ENGINEERING	Payback is likely to be too high
M	CHEMICALS	Trading conditions in the group are very difficult, competition for limited resources
N	CHEMICALS	Determined to reduce costs
O	OTHER	It is certainly the way to go in the long run
	Large energy users:	
P	FOOD	Difficult to implement on the site which is a private industrial estate
Q	CHEMICALS	CHP is the right course; it will take 5 years to implement
R	FOOD	Payback is unacceptable
S	DRINKS	Big investment underway at site, competes for cash, payback too low to secure higher ranking
T	FOOD	Must first undertake full PI study; then improve site; then decide if CHP is appropriate
U	OTHER	Unacceptable payback
V	CHEMICALS	Almost certain in the longer term, say by 1995

Preferences and Attitudes

Table 9

PREFERENCES AND ATTITUDES
Question numbers in brackets ()

Site Code Letter	INDUSTRY	(8) Choice for prime mover			(9) Inducement needed		(10) Electricity export		Utility companies		Cost Reduction Required %?
		1st	2nd	3rd	2nd to 1st		OK?	Inducement Required?	(13) Aware of Existence?	(14) Freedom from Worries?	
	Small energy users:										
A	PLASTICS PROCESSING						YES	N	NO	YES	10
B	MECHANICAL ENGINEERING						YES	N	YES	YES	10
C	FOOD	COAL	ST	GT			YES	N	NO		25
D	OTHER						YES	N	YES		25
E	PLASTICS PROCESSING						NO	MM	YES		25
F	DRINKS	ST	GT	GE			YES	N	YES		
G	CHEMICALS	GT	ST/GT	COAL	MB	HE	YES	N	YES		
H	FOOD	GT/ST/GE	GT/ST/GE	GT/ST/GE			YES	1/4	YES	YES	25
	Medium energy users:										
I	DRINKS	GE	GT/ST	COAL	1/2	MB	YES	N	YES	YES	10
J	FOOD	GT/ST/GE	COAL	GT/ST/GE			YES	N	NO		10
K	DRINKS	COAL	GT	ST			YES	N	NO		
L	MECHANICAL ENGINEERING	GE	GT	ST			NO		YES		
M	CHEMICALS						YES	N	YES		25
N	CHEMICALS	GT/ST	GE	GT/ST			NO		YES		
O	OTHER	COAL	ST	GT	1/2	1/2	YES	N	YES		10
	Large energy users:										
P	FOOD						YES	N	YES		
Q	CHEMICALS	ST	GT	COAL	MB	MB	YES	N	NO		
R	FOOD	GT/ST/GE	COAL	GT/ST/GE	MB		NO	MM	YES		25
S	DRINKS	GT	ST	GE	1/2	1/2	YES	N	YES		25
T	FOOD	ST	GT	GE	1/2	1/2	NO	N	YES		
U	OTHER	Site	COAL		HE		YES	N	NO		
V	CHEMICALS	GT	ST/GE	ST/GE			YES	N	YES		25

ST: steam turbine
GT: gas turbine
GE: gas engine
COAL: coal conversion

1/2: 0.5 year payback improvement
MB: much better
HE: hardly ever

1/4: 0.25 year payback improvement
MM: much more
N: none

Comments on Prime Movers — Table 10

COMMENTS REGARDING THE CHOICE OF PRIME MOVERS FOR CHP PLANT

	Small energy users:	
A	PLASTICS PROCESSING	
B	MECHANICAL ENGINEERING	Dislike all options, particularly coal
C	FOOD	Keen to convert to coal but payback not good enough
D	OTHER	Coal disliked for computer site
E	PLASTICS PROCESSING	No experience, but equally happy with all except coal
F	DRINKS	No experience with gas turbine as yet, but will come; dislike coal
G	CHEMICALS	Coal conversion lacks flexibility
H	FOOD	
	Medium energy users:	
I	DRINKS	Turbines are alien technology; know reciprocating compressors well; coal conv. senstitive to steam demand
J	FOOD	No experience; do not like coal at food site
K	DRINKS	With coal conversion there are no risks or unknowns
L	MECHANICAL ENGINEERING	Engines are easier to maintain at this site; coal conversion too expensive for a small site
M	CHEMICALS	
N	CHEMICALS	In USA gone for steam turbine, but run 7 days/week
O	OTHER	Used to running CHP plant; with coal you cannot be wrong; with GT it may be right today, wrong in 1990
	Large energy users:	
P	FOOD	No experience with any
Q	CHEMICALS	No experience with gas turbine or gas engine; would introduce alien technologies
R	FOOD	Strong dislike for coal at food sites
S	DRINKS	Coal conversion a good option, but only when reboilering anyway
T	FOOD	No experience with any, but steam turbines are more robust
U	OTHER	Like gas turbine technology, fits in with site; coal gasification disliked owing to tech'risk
V	CHEMICALS	Dislike coal, woory about high traffic levels, wrong image

71 Respondents were also asked to indicate what improvements in payback would be needed to alter the rankings (Question 9). This proved to be a difficult question, with frequent misunderstandings, and the results are ignored.

Attitude to the Export of Electricity

72 Respondents were asked if there were likely to be any objections from within their companies to the possible export of electricity from the site to the area electricity boards, assuming that this was financially advantageous to the company (Question 10). The results are shown in Table 9 with a summary below:

Attitude	No.	Percent
No objection	17	77
Likely to object	5	23

73 Paraphrased comments are summarised in Table 11. Objections tended to centre on the possibility of outside interference with running the company's business.

74 Again, an attempt was made to gauge what improvement would be needed in the payback to justify exporting electricity. Only three of the five 'objectors' commented, and the results are not meaningful.

Attitude to Utility Companies

75 Respondents were asked whether they were aware of the activities of the private utility companies (Question 13). Details are in Table 9 with a summary below:

Aware of Utility Companies	No.	Percent
Small energy users	6	75
Medium energy users	5	71
Large energy users	5	71
Overall	16	72

76 Although the majority of respondents were aware of the existence of CHP utility companies, only a minority had entered into negotiations with them. The majority appeared to have failed to appreciate utility companies' selling proposition, and

felt that, to a company operating well managed utility plant they had nothing to offer.

77 Comments regarding utility companies are summarised in paraphrased form in Table 12. Several respondents with direct responsibility for energy management or the operations of utilities expressed anxiety about job security in general, and security of their own position in particular.

78 Few were able to answer the question: what would they expect from an arrangement with a utility company? (Question 14). After prompting, four said that freedom from worries regarding the operation of a utility plant could be one of their aims. The majority, however, saw little point in it.

79 Respondents were asked what sort of financial inducement they would need to enter into an agreement with a utility company. Only 13 replies were received, as indicated in Table 9, with a summary below:

Expect a cost reduction of at least	No.	Percentage responding
10%	5	40
25%	8	60

Energy Price Movements

80 Respondents were asked to indicate their perception of likely energy price movements between 1985 and 1990, disregarding the effects of inflation. For each fuel a choice of three statements was offered: relative to the other fuels, the price will decrease by up to 10%, remain in line with others, or increase by up to 10%. Some of the respondents felt the choice restricting and stretched the limit to 20%.

81 The response (from 21 out of the 22 approached) is shown in Table 13, with a summary below:

Expect	Oil		Gas		Coal		Electricity	
	No.	%	No.	%	No.	%	No.	%
Increase	11	52	15	71	7	33	8	38
Stable	7	33	5	24	11	52	13	62
Decrease	3	14	1	5	2	10		

82 Comments regarding future energy price movements are summarised in paraphrased form in Table 14. The interviews were conducted at a time when the sale of British Gas Corporation to private ownership had been in the news and many respondents were preoccupied with the matter.

<u>Impact of Energy Costs on Operation</u>

83 About half the respondents contacted regarded energy costs as a 'major' concern of management at the site, the rest preferred the description 'minor concern' (Table 13, Question 16). The division is uneven in the three energy user categories.

User Category	Major Worry No.	%
Small energy users	6	75
Medium energy users	4	57
Large energy users	2	28
Overall	12	54

84 The apparent inverse correlation between energy expenditure and 'concern' over energy is surprising, but the sample is too small to give confidence in the results.

Comments on Electricity Exports — Table 11

COMMENTS REGARDING ELECTRICITY EXPORTS

	Small energy users:	
A	PLASTICS PROCESSING	OK if it does not cause additional problems
B	MECHANICAL ENGINEERING	OK if it helps with payback
C	FOOD	Would be happy to consider if price is right, but in past the offer was laughable
D	OTHER	
E	PLASTICS PROCESSING	Concerned with control problems of interfacing with Electricity Boards
F	DRINKS	All for it to improve economics; would prefer a better price, vitally important at other sites
G	CHEMICALS	With the new Energy Act, keen to exploit
H	FOOD	Acceptable, if capital investment is not greatly increased; otherwise "stick to knitting"
	Medium energy users:	
I	DRINKS	Been looking at it for anaerobic digester project
J	FOOD	Could create extra work for staff, but will consider on merit
K	DRINKS	Not familiar, but doubts if Electricity Boards pay enough to justify it
L	MECHANICAL ENGINEERING	No policy on matter, but unlikely to be favoured
M	CHEMICALS	Acceptable as long as the Electricity Boards offer reasonable terms
N	CHEMICALS	Would not want to be involved, Government might force company to act contrary to own interests
O	OTHER	Had trouble running CHP in parallel with grid; now resolved with better control
	Large energy users:	
P	FOOD	Interested in concept, visited ICI, impressed
Q	CHEMICALS	No problem, would be keen to do it
R	FOOD	Do not like to be involved in business outside food
S	DRINKS	Keen to export if it improves economics
T	FOOD	It would be costly at this site, needs massive civil works
U	OTHER	Would be happy with the idea
V	CHEMICALS	Would not want to be dependent on EB's but no objection if aids longer term economic benefits

Attitude to Utilities

Table 12

ATTITUDE TO UTILITY COMPANIES

	Small energy users:	
A	PLASTICS PROCESSING	Could be interesting
B	MECHANICAL ENGINEERING	Currently collecting proposal, do not expect they have much to offer but prepared to consider
C	FOOD	Do not see how they could run the plant better than the company itself
D	OTHER	No objection, but run very efficient plant themselves
E	PLASTICS PROCESSING	Do not know
F	DRINKS	Does not see what they can do for the company, but open minded and very interested
G	CHEMICALS	Prepared to consider, but for good effect power plant must be integrated, could be difficult
H	FOOD	Had proposal, very suspicious of hidden catches
	Medium energy users:	
I	DRINKS	Placed contract for a London site with AHS; prepared to consider for other sites
J	FOOD	Found the idea interesting but unable to comment
K	DRINKS	Would be worried about possible redundancies
L	MECHANICAL ENGINEERING	No contact with utilities, no policy
M	CHEMICALS	
N	CHEMICALS	No role, can do just as good a job themselves
O	OTHER	Interesting offer, but hard choice: if accept, miss out on gain, otherwise cannot afford to implement
	Large energy users:	
P	FOOD	Could offer a solution
Q	CHEMICALS	
R	FOOD	Do not like to be involved with other businesses at the company's sites
S	DRINKS	Not keen, worries about unions, job security, external dependence
T	FOOD	Do not know what they have to offer, but prepared to consider on merit
U	OTHER	Not acceptable to glass making
V	CHEMICALS	Would have to be offered an exceptional deal, otherwise unlikely to be interested

Expectations of Electricity Prices — Table 13

EXPECTATIONS OF ENERGY PRICE CHANGES AND IMPACT OF ENERGY COSTS
Question numbers in brackets ()

Site Code Letter	INDUSTRY	(15) OIL price change % (-10,0,+10)	GAS price change % (-10,0,+10)	COAL price change % (-10,0,+10)	ELECT'Y price change % (-10,0,+10)	(16) Energy Cost Impact?
	Small energy users:					
A	PLASTICS PROCESSING	0	20	0	0	MINOR
B	MECHANICAL ENGINEERING	10	10	0	0	MAJOR
C	FOOD	10	20	0	0	MAJOR
D	OTHER	10	0	10	10	MAJOR
E	PLASTICS PROCESSING	10	10	0	0	MAJOR
F	DRINKS	0	10	10	10	MAJOR
G	CHEMICALS	-10	10	0	0	MINOR
H	FOOD	-10	0	10	0	MAJOR
	Medium energy users:					
I	DRINKS	10	10	10	0	MAJOR
J	FOOD	10	10	10	10	MINOR
K	DRINKS	10	-10	-10	10	MAJOR
L	MECHANICAL ENGINEERING	0	10	10	10	MINOR
M	CHEMICALS	0	10	0	0	MAJOR
N	CHEMICALS	0	10	0	10	MINOR
O	OTHER	-10	0	0	0	MAJOR
	Large energy users:					
P	FOOD	0	0	10	10	MINOR
Q	CHEMICALS					MAJOR
R	FOOD	0	20	0	0	MINOR
S	DRINKS	20	10	0	0	MINOR
T	FOOD	10	10	0	10	MINOR
U	OTHER	10	0	0	0	MAJOR
V	CHEMICALS	10	10	-10	0	MINOR

Comments on Energy Prices — Table 14

COMMENTS ON LIKELY ENERGY PRICE MOVEMENTS TO 1990

	Small energy users:	
A	PLASTICS PROCESSING	Gas prices likely to go through roof
B	MECHANICAL ENGINEERING	Cheap imports will depress coal price; oil prices will recover
C	FOOD	Expect very rapid escalation in gas prices; oil prices are bound to recover by 1990
D	OTHER	Gas will still be very competitive in 1990
E	PLASTICS PROCESSING	Almost impossible to anticipate
F	DRINKS	Eventually gas prices will catch up; will become a premium fuel
G	CHEMICALS	Worried by effects of privatisation on gas price
H	FOOD	Market is set by oil price which depends on US$
	Medium energy users:	
I	DRINKS	Oil/gas parity likely to remain the same
J	FOOD	Energy sources compete with each other, must move in step provided OPEC or BG do not break ranks
K	DRINKS	Imports will depress coal prices in the south
L	MECHANICAL ENGINEERING	Oil likely to be stable; gas becomes premium; coal under labour cost pressure; elect'y follows coal
M	CHEMICALS	The anticipated rise in gas prices would not exclude consideration of CHP
N	CHEMICALS	Gas will become a premium fuel
O	OTHER	Very concerned about high gas prices in the midst of a Europe-wide surplus
	Large energy users:	
P	FOOD	Privatisation likely to restrain gas prices
Q	CHEMICALS	Do not know
R	FOOD	Privatisation of British Gas will result in a gas price explosion
S	DRINKS	Oil prices are likely to rebound by 1990
T	FOOD	Energy costs are unpredictable, must plan short term
U	OTHER	Oil prices will rebound by 1990, gas/electricity will remain competitive
V	CHEMICALS	Very worried about British Gas privatisation, will put up prices in the long run

Barriers to CHP

85 In Question 17, respondents were asked to consider what barriers impeded the implementation of CHP. The results are detailed in Table 15, with a summary below:

Barrier to CHP	Noted by No.	%
Poor payback	13	60
Capital costs are too high	8	36
Competing demands for capital	4	18
Uncertain fuel costs	3	14
Technical risks	4	18
Uncertain site conditions	7	32
Other	6	27

86 Among 'other' problems, respondents mentioned the following:

- The high cost of technical backup needed to introduce CHP to the site;
- The unsuitability of steam (in the context of a steam producing CHP plant) as a heating medium in some parts of the works, and concern over steam quality in process use;
- The electricity prices were too low owing to the relatively low rate of return expected from investment in electricity supply;
- The existence of a heat surplus at the site;
- Purely site specific problems (such as availability of space)

Awareness of Government Assistance with Studies

87 Respondents were asked whether they were aware of the availability of help from the Government with energy surveys and CHP studies, and whether they were likely to make use of the facility. the details in Table 16 are summarised below:

Government assistance for:	Aware No.	%	Likely to use No.	%
Energy surveys	22	100	22	100
CHP studies	17	77	12	55

Obstacles to Adoption of CHP — Table 15

MAIN OBSTACLES TO THE ADOPTION OF CHP
Question numbers in brackets ()

Site Code Letter	INDUSTRY	(17) POOR PAYBACK?	CAPITAL TOO HIGH?	COMPETING DEMAND FOR CAPITAL?	UNCERTAIN FUEL COSTS?	TECHNICAL RISK?	UNCERTAIN SITE CONDITIONS?	OTHER OBSTACLES?
	Small energy users:							
A	PLASTICS PROCESSING	Y	Y				Y	
B	MECHANICAL ENGINEERING	Y	Y			Y	Y	
C	FOOD	Y	Y					None
D	OTHER					Y		Fear high cost of technical backup
E	PLASTICS PROCESSING	Y	Y					
F	DRINKS							None
G	CHEMICALS	Y			Y			
H	FOOD	Y	Y	Y			Y	
	Medium energy users:							
I	DRINKS		Y	Y		Y	Y	
J	FOOD	Y					Y	Steam quality, seasonality
K	DRINKS	Y	Y			Y		Fear of upheavel
L	MECHANICAL ENGINEERING	Y						Cheap public electricity
M	CHEMICALS	Y	Y				Y	
N	CHEMICALS							None
O	OTHER			Y	Y			
	Large energy users:							
P	FOOD						Y	
Q	CHEMICALS			Y				
R	FOOD	Y			Y			Heat surplus at site
S	DRINKS							None
T	FOOD							Site specific problems
U	OTHER	Y						
V	CHEMICALS							None

(Y=Yes)

Awareness of Government Assisstance **Table 16**

AWARENESS OF GOVERNMENT ASSISTANCE WITH STUDIES
Question numbers in brackets ()

Site Code Letter	INDUSTRY	(5) ENERGY SURVEYS AWARE	ENERGY SURVEYS LIKELY TO USE	CHP FEASIBILITY STUDIES AWARE	CHP FEASIBILITY STUDIES LIKELY TO USE
	Small energy users:				
A	PLASTICS PROCESSING	Y	Y	N	N
B	MECHANICAL ENGINEERING	Y	Y	Y	N
C	FOOD	Y	Y	Y	N
D	OTHER	Y	Y	Y	Y
E	PLASTICS PROCESSING	Y	Y	Y	Y
F	DRINKS	Y	Y	Y	Y
G	CHEMICALS	Y	Y	Y	N
H	FOOD	Y	Y	Y	
	Medium energy users:				
I	DRINKS	Y	Y	N	Y
J	FOOD	Y	Y	N	Y
K	DRINKS	Y	Y	N	N
L	MECHANICAL ENGINEERING	Y	Y	Y	
M	CHEMICALS	Y	Y	Y	
N	CHEMICALS	Y	Y	Y	Y
O	OTHER	Y	Y	Y	Y
	Large energy users:				
P	FOOD	Y	Y	Y	Y
Q	CHEMICALS	Y	Y	Y	Y
R	FOOD	Y	Y	Y	N
S	DRINKS	Y	Y	Y	Y
T	FOOD	Y	Y	Y	Y
U	OTHER	Y	Y	N	
V	CHEMICALS	Y	Y	Y	Y

6.0 DERIVATION OF INVESTMENT POTENTIAL

Definition

88 The gross potential is an estimate of the CHP capacity that it is feasible to install within particular ranges of payback period. It gives an indication of what is technically feasible but not what might actually be installed. The contacts with industrialists provided a means of estimating what proportion of the gross potential might actually be installed. The term 'Investment Potential' was therefore defined. This is the CHP capacity that might be installed, given current investment trends in industry, within the next 10 years or so assuming that industry is aware of this opportunity.

89 The investment potential was estimated by combining the gross potential with the curves of propensity to invest, Figure 14.

6.1 Method of Estimating Investment Potential

90 To obtain a measure of the investment potential for CHP, the problem was regarded from the viewpoint of an individual decision maker presented with a choice of options. These include taking no action, and one or several investment possibilities giving different levels of payback.

91 It was assumed that a certain proportion of decision makers, as suggested by the propensity to invest as a function of the size of the investment, might decide to install the CHP plant giving the shortest payback.

92 Thus, the contribution to the installed CHP of the proportion investing in plant giving the shortest payback is:

$$N \times W1 \times R1$$

where N is the number of decision makers (sites) represented in each 'typical' site and boiler fuel case; W1 is the installed power of the plant giving the best payback, and R1 is the propensity to invest at a cost of W1.

93 In cases where there is an option to install a larger and more expensive CHP plant giving an inferior payback, a proportion of decision makers may decide to pursue this strategy. To determine their influence on the outcome, the propensity to invest appropriate to the higher level of investment was applied to the incremental installed MW.

94 Thus, the contribution to the installed CHP power of those investing in plant giving the next best payback is

N x (W2-W1) x R2

where W2 is the installed power of the plant giving the next best payback, and R2 is the propensity to invest at the cost of W2. Similar procedures were used to derive the contribution of decision makers opting for yet longer payback.

95 For the purposes of the calculation, it was assumed that each band of payback periods (say between 2 years and 3 years) may be represented by values appropriate to p(min) + 0.6, i.e. 1.6, 2.6, 3.6 etc.

96 The calculated values of the propensity to invest are shown below for a range of investments at the paybacks selected as representative of the discrete bands:

Propensity to Invest (Percent) vs. Payback and Investment:

Payback (years) Investment	1.6	2.6	3.6	4.6	5.6
£5.0 million	50.0	29.4	18.8	13.0	9.5
£2.5 million	56.1	31.6	18.9	12.2	8.4
£1.0 million	66.3	37.0	20.6	12.3	9.2
£0.5 million	73.4	41.5	22.2	12.5	7.6
£0.1 million	89.9	57.0	27.0	12.6	6.3

97 Investment in the smallest CHP plant in the Gross Potential is just under £0.5 million. Therefore the propensity to invest at a level of £0.1 million may be ignored for the present purposes.

98 Investment in the largest CHP plant studied is in the region of £10 million. For the present purposes, it was assumed that beyond £5 million, the propensity to invest is independent of the size of the investment.

99 In the range of investments relevant to this study, from £0.5 to £5+ million, and at payback periods of 3.6 years and over the change in the propensity to invest with the size of investment is small and can be ignored within the attainable accuracy. It was therefore decided to use the following values, regardless of investment size:

Payback Period	3.6	4.6	5.6
Propensity to Invest	0.18	0.12	0.06

100 At shorter payback periods, the variation of the propensity to invest with the scale of the investment is significant. Values appropriate to each case were therefore selected by hand.

101 The estimation of the associated capital investment and energy saving figures could follow the same procedure. In practice, it was more convenient to scale the gross potential totals for investment and energy saving at a payback of <8 years in the proportion of the investment potential.

6.2 Initial Calculation of Investment Potential

102 Section 7 of the Technical Report describes the gross potential assessed at the fuel prices prevailing in December 1985. These were:

Interruptible natural gas	28.3p/therm
Heavy fuel oil	11.0p/litre
Coal	£63.5/tonne

103 The gross potential calculated at these fuel prices was used as the basis for the initial calculation of the investment potential. The method used was as described above and the results are shown in Table 17. These results are discussed in more detail in Section 9 of the Technical Report.

Investment Potential for CHP — Table 17

Industry	Investment Potential MWe	No. of Sites	Capital Cost £ million	Primary Energy Savings ktce
Chemicals	64	6	32	102
Food/Drink	18	5	9	29
Engineering	16	7	8	26
Paper	17	2	8	27
Textiles	-	-	-	-
Rubber, Plastics etc.	10	1	5	16
TOTAL	125	21	£63 million	200

6.3 Changes in Investment Potential with Energy Price

104 Changes in the level and differentials of fuel and electricity prices will have a significant effect on the level of CHP investment. In general as fossil fuel prices fall relative to electricity the economics of CHP operations will improve. These changes, reflected in the "Gross Potential" for CHP, could be modelled in a quantitative manner.

105 Changes in energy prices will affect company profitability and liquidity in the short term. They will also create increased uncertainty regarding the longer term level of energy prices. Both these aspects will influence the financial criteria required from a CHP investment and hence affect the overall "Propensity to Invest" in CHP. These changes in corporate attitudes will also take place over a period of time and it is thus not possible to quantify how the propensity to invest in CHP will change with different energy prices.

106 The fall in fossil fuel prices which occurred during 1986 provides a major new element of uncertainty in estimating the CHP capacity likely to be installed.

107 In view of these problems an attempt was made to estimate the Investment Potential for CHP over a wide range of energy prices on the assumption that the financial criteria would remain unchanged. The method used involved assessing the effect of energy price changes on a number of individual sites, where CHP was a cost effective investment, and extrapolating the results to all industries under investigation.

108 First, the change in CHP payback was calculated on a number of individual sites for a range of fossil fuel and electricity prices. The effects of these changes were used to derive revised propensities to invest which were then applied to the original gross potential to estimate investment potential at the particular energy price scenario.

109 This method was used first to establish an investment potential based on a fossil fuel price of 28p/therm, a 5% increase on 1985 capital costs and 1986 electricity tariffs. Using this as a base for 1986, a range of fossil fuel prices from 12p/therm to 32p/therm and steadily decreasing electricity unit charges were used to produce the matrix of investment potential shown in Table 18.

Investment Potential for CHP Table 18

(for a range of energy prices)

Electricity Price	Prices for Gas and Heavy Fuel Oil [Pence/therm]					
	12	16	20	24	28	32
May 1986 Prices	708	497	367	247	161	125
May 1986 Prices Less 5%	603	432	285	195	122	86
May 1986 Prices Less 10%	497	340	209	139	82	42
May 1986 Prices Less 15%	412	254	153	104	53	23
May 1986 Prices Less 20%	261	165	107	57	27	8

110 In this analysis a single price for fossil fuel was assumed for all sites. This represented the price of natural gas and heavy fuel oil and sites currently using coal were not included in the analysis. Only 13% of the sites in the 50 industries under investigation burn coal. None of these sites appeared in the original assessment of investment potential (as described in Section 6.4) as coal was cheaper than natural gas or heavy fuel oil, the fuels for the most cost effective CHP plant. Omitting sites burning coal will not generally have a significant effect on the investment potential. However at low prices for oil and gas some potential may be missed. The matrix of investment potential is thus likely to be somewhat conservative at the low fossil fuel prices.

111 In addition to the matrix of investment potential shown in Table 18, a more detailed breakdown of the investment potential was made at a fossil fuel price of 20p/therm indicating the potential in particular industries, this is shown in Table 19. The effect of changes in capital cost has also been investigated by recalculating the matrix of investment potential with a 25% reduction in capital cost. These effects are described further in the Technical Report.

Investment Potential for CHP

Table 19

Prices for Gas and Heavy Fuel Oil - 20p/therm
Electricity Price - May 1986 Less 5%

INDUSTRY	MWe	No. of Sites
Chemical		
- Fertilisers	41	
- Pharmaceuticals	28	
- Synthetic Resins	21	
- Synthetic Rubber	12	
- Soap & Detergents	5	
- Photographic Materials	3	
- Organic Chemicals	1	
- Adhesives	1	
- Others	2	
TOTAL	114	13
Food Drink and Tobacco		
- Distilling	14	
- Brewing	16	
- Edible Oils	5	
- Miscellaneous Foods	2	
- Confectionery	4	
- Milk Products	4	
- Fruit and Veg. Processing	3	
- Tobacco	3	
- Fish Processing	2	
- Petfood	1	
- Others	1	
TOTAL	55	13
Paper		
- Paper and Board Manufacture	31	
- Paper Conversion	5	
TOTAL	36	5
Textiles		
- General Textiles	5	
TOTAL	5	4
Plastic & Rubber Processing		
- Rubber	19	
- Plastics	4	
TOTAL	23	2
General Engineering		
- Vehicles, ships and aircraft	50	
- Mechanical Engineering	1	
TOTAL	51	18
GRAND TOTAL	285	55

7.0 SENSITIVITY ANALYSES

112 The economics of operating CHP plant are influenced primarily by fossil fuel and electricity prices and capital costs. The effects of these influences are described elsewhere. However several other factors also affect the economics of a CHP operation; these include:

- Heat to power ratio at the site
- Site working hours
- Structure of the electricity export tariff.

113 Assessing the effect of these factors on the potential for the whole of industry would be complex and the effort could not be justified. Consequently sensitivity analyses were carried out on several typical sites and for particular CHP plant configurations. Those are listed in Table 20. These were selected to be broadly representative of those industrial sites where CHP may be particularly attractive. In the sensitivity analyses the base case was taken with a fossil fuel price of 28p/therm and 1985/86 electricity tariffs.

Table 20
CHP Plants used in the Sensitivity Analysis

Case	Site	CHP Plant	Existing fuel at the site
A	Medium sized	One gas turbine of 4.8 MWe with an after-fired heat recovery boiler	Natural gas
B	Large Distillery	One gas turbine of 5.15 MWe with an after-fired heat recovery boiler	Heavy Fuel-Oil
C	Large Distillery	One heavy fuel oil diesel engine of 3.3 MWe with an after-fired heat recovery boiler	Heavy Fuel-Oil

7.1 Sensitivity to Heat to Power Ratio

114 There is a general trend of falling heat to power ratios in UK industry. An analysis was carried out to investigate whether falling heat to power ratios would adversely affect the economics of CHP plants installed in 1985. In the main assessment of CHP potential it is assumed that heat and power demands do not change during the payback period of a CHP plant. This sensitivity analysis would indicate the validity of this assumption.

115 The method involved extrapolating the change in heat to power ratios already identified between 1979 and 1984 (see Appendix 3) to 1990. This indicated a possible fall of heat to power ratio of 15% between 1985 and 1990. The revised heat to power ratio was used to recalculate the annual savings for 1990 and the savings for the intervening years were calculated by interpolation. It was assumed that energy prices would not change in real terms between 1985 and 1990 for the purpose of identifying the changes attributable to the change in heat to power ratio. From these results revised payback periods were calculated allowing for these changes in heat to power ratio. The results are shown in Table 21.

Table 21

Effect of Future Changes in Heat to Power Ratio
(for CHP plant installed in 1985)

Case	Site	Plant	Fixed Heat/ Power Ratio: Payback Period (Years)	Fixed Heat/ Power Ratio: Internal Rate of Return (percentage)	Changing Heat/ Power Ratio: Payback Period (Years)	Changing Heat/ Power Ratio: Internal Rate of Return (percentage)
A	Papermill	Gas Turbine 4.80 MWe	5.0	15%	5.2	13%
B	Distillery	Gas Turbine 5.15 MWe	5.1	14%	5.1	14%
C	Distillery	HFO Diesel 3.30 MWe	6.3	9%	6.3	9%

116 It can be seen from Table 21 that a falling heat to power ratio will cause the payback periods to increase slightly. This is best shown in Case A where the heat output from the gas turbine is close to the maximum heat demand of the site. the falling heat demand reduces the maximum output of the CHP plant and hence reduces the annual savings in successive years. However the effect is nevertheless fairly small and does not significantly offset the payback. In cases B and C the heat demands are in excess of the heat output of the CHP plant and the 15% reduction in heat demands over the next 10 years does not affect the payback period.

7.2 Sensitivity to Working Hours

117 The working hours of an industrial site have an important effect on the economics of a CHP plant; the greater the annual utilisation of the plant the greater the annual savings.

118 The sensitivity to working hours was investigated by considering annual working of 8000, 6000, 4000 and 2000 hours for Cases A, B and C. The results are shown in Table 22. It was assumed that all fuel prices were fixed at the 1985 level and that the annual savings were constant during the payback period.

119 It can be seen from Table 22 that, in general, the payback period for CHP is approximately inversely proportional to the working hours; i.e. a halving of the working hours approximately doubles the payback period.

120 The potential for CHP is thus likely to be significantly reduced if, for instance, short time working such as a three day week was to be introduced.

7.3 Sensitivity to Electricity Export Tariffs

121 The payments made by the electricity board to a private generator for electricity exported from a CHP installation has an important influence upon the mode in which a CHP plant operates and the resultant cost savings.

122 Three examples of export tariff variations were considered:

- (a) a flat rate increase of 25%
- (b) a flat rate decrease of 25%
- (c) the removal of the capacity credit that is applied during the winter peak.

123 The capacity credit is a payment made by the electricity board to the private generator for the generating capacity made available to the board by the private generator. The capacity credit is paid only during the winter peak as this is the

Effect of Working Hours
(for CHP plant installed in 1985 and fixed fuel prices)

Case	Site	Plant	8000 hrs/year*		6000 hrs/year*		4000 hrs/year*		2000 hrs/year*	
			Payback Period (Yrs)	Internal Rate of Return	Payback Period (Yrs)	Internal Rate of Return	Payback Period (Yrs)	Internal Rate of Return	Payback Period (Yrs)	Internal Rate of Return
A	Papermill	Gas Turbine 4.8 MWe	5.0	15%	6.3	9%	8.9	2%	16.2	(negative)
B	Distillery	Gas Turbine 5.15 MWe	3.9	22%	5.0	15%	7.2	6%	13.1	(negative)
C	Distillery	HFO Diesel 3.3 MWe	5.0	15%	6.2	10%	8.2	4%	13.3	(negative)

The normal operating hours are 8000 hours/year for the Papermill and 6000 hours/year for the Distillery.

* The working hours refer to the process at the site and not necessarily the CHP plant.

occasion when the electricity industry needs to use its maximum capacity. The capacity credit payment is not made separately but included within the unit payments for electricity exported during the winter peak.

124 The electricity import tariffs used in the sensitivity analysis were the same as those used for the rest of the study. The export tariffs considered were as used in the rest of the study except that the unit payments were changed as described above and shown in Table 23. It was assumed that all other fuel prices were fixed at the 1985 level.

Table 23
Unit Payments for Exported Electricity

Period	Basic Unit Payments [p/kWh]	(a) Unit Payment +25% [p/kWh]	(b) Unit Payment -25% [p/kWh]	(c) Unit Payment Minus Capacity Credit [p/kWh]
00.03 to 07.30hrs	1.37	1.71	1.03	1.37
07.30 to 20.00hrs Mondays to Fridays in November and March	3.46	4.33	2.60	2.70
07.30 to 20.00hrs Mondays to Fridays in December to February	6.05	7.56	4.54	2.70
At all other times	2.43	3.04	1.82	2.43

125 The results of the sensitivity analysis are shown in Table 24.

Effect of Electricity Export Tariff
(for CHP plant installed in 1985 and fixed fuel prices)

Case	Site	Plant	Basic Unit Payment		Unit Payment +25%		Unit Payment -25%		Unit Payment Capacity Credit	
			Payback Period (Yrs)	Internal Rate of Return	Payback Period (Yrs)	Internal Rate of Return	Payback Period (Yrs)	Internal Rate of Return	Payback Period (Yrs)	Internal Rate of Return
A	Papermill	Gas Turbine 4.8 MWe	5.0	15%	4.3	19%	6.0	11%	5.7	12%
B	Distillery	Gas Turbine 5.15 MWe	5.0	15%	4.4	18%	5.9	11%	5.9	11%
C	Distillery	HFO Diesel 3.3 MWe	6.3	9%	5.6	12%	6.3	9%	6.3	9%

126 In Cases A and B the effect of the 25% increase or decrease in the electricity export tariff is to cause a decrease or increase in the payback period. The percentage change in the payback period is less than the percentage change in the export tariff as the savings attributable to the CHP plant are due only in part to the export of electricity. The effect of the removal of the capacity credit is similar, in Cases A and B, to a general 25% reduction in the export tariff. The effect though will depend upon the proportion of the exported electricity that is exported at peak times.

127 In Case C the effect of a 25% increase in the export tariff is to cause a decrease in the payback period of the same size as Case B. However the general 25% decrease in the export tariff and the removal of the capacity credit do not change the payback period. In the case of the removal of the capacity credit; electricity is only exported on non-working days and these are not 'peak' days. Consequently the removal of the capacity credit has no effect. In the case of the general reduction in the export tariff the effect is to make operation of the CHP plant non-cost effective on non-working days. As the savings were only small when the CHP plant operated on non-working days the effect of switching the plant off is also small (a 0.6% decrease in annual savings).

128 As demonstrated in Case C, the most cost effective mode of operation of a CHP plant may change with variations in the electricity tariff. This is best shown by the most cost-effective modes of operation of CHP plants at night for Cases A, B and C. This is shown in Table 25.

Table 25
Most Cost Effective Mode of Operation of CHP Plant at Night

Case	Base Case	Export Unit Payments +25%	Export Unit Payments -25%
A	Switched Off	Heat Matched	Switched Off
B	Heat Matched	Heat Matched	Electricity Matched
C	Switched Off	Switched Off	Switched Off

129 In Case A it was economic to operate the CHP plant at night when the export payment was increased by 25%. Similarly with Case B it was not economic to export electricity at night when the export payment was reduced by 25% yet it was still economic

to operate in an electricity matched mode. In Case C it was not economic to operate at night with any of the export tariffs.

130 The effect that changes in the export tariff have on the operation of any particular CHP installation will depend upon the proportion of the electricity that is exported and the cost effectiveness of operating the CHP plant in different modes. It is difficult to draw any general conclusions however the examples shown illustrate typical effects.

In addition a fuel cost adjustment of 0.26p/kWh was added on to all of these unit payments.

8.0 CONTACTS WITH UTILITY COMPANIES

8.1 Introduction

131 As has been indicated in other sections of the report, radical changes have occurred in relation to energy throughout UK industry since the first oil crisis of 1973. Both the technologies and the economics have undergone reappraisal, and the steep rises in energy costs have initiated a whole new industry of products and services all now aimed at reducing energy costs.

132 Twelve years ago, an energy management industry did not exist, but economic pressures have created a situation in which marginal costs are now also of vital interest to industry, whilst new ideas, new techniques and the complex amalgam of skills demand the attention of professional energy managers. It is against this background that energy management services are now offered by a number of companies ready to take the complete technical and financial responsibility for management of energy on industrial sites - as private utility companies.

133 Although aiming primarily at a heat service only, some of these companies are offering total energy management via CHP schemes. In almost every case, however, greater interest is shown in heat-service contracts.

8.2 Private (and Public) Utility Companies:

134 Initially, nine private companies known to be involved in the Heat-Service or Energy Management industry were contacted, of which five were thought to have a positive interest or involvement in CHP. Five companies expressed little or no interest in offering a CHP management service. Only one of the remaining companies is actually operating CHP plants, none in the UK. Two companies were (mid-1985) then negotiating contracts in the UK, one on two single sites, and one on a group scheme. Three companies expressed enthusiasm and were prepared to discuss the potential for them in CHP but, for all three, the activity appeared to be considered as a side-issue to their mainstream activities.

135 Discussions with a number of the gas and electricity area boards produced evidence of similar disinclination in relation to investment in CHP, although there are one or two known exceptions.

136 In summary, although the 1983 Energy Act is seen as opening a window on the CHP market for utility companies, there appears to have been little progress to date in this area. Investment in CHP by utility companies, either private or public, probably

represents the most significant prospect for increasing the spread of CHP within UK industry.

8.3 Attitudes towards investment in CHP

137 On the whole, utility companies are prepared to take a longer term view and hence to accept investment opportunities which yield a somewhat lower rate of return than manufacturing industry.

138 Broadly, the message received in 1985 was that CHP tends to be a fringe interest because other heat-service work is buoyant and because CHP contracts can be complex taking as long as 4-5 years to negotiate. A heat-service contract can be settled in 6-9 months.

139 A private utility company expects to obtain as good a rate of return as it would by investing at current rates of interest, typically (mid 1985) 12%, with a 10% minimum. Although margins may be discussed with individual clients, it is not a general or preferred policy.

140 Whereas their industrial clients would expect payback within 2-3 years, the private utility companies are more prepared to accept up to 5 years or even longer. Overhead costs appeared to be in the 5-7% range.

141 There appears to be a common approach to the sharing of savings with clients: typically a 50:50 split was mentioned. There are variations in the method of funding, with some more complicated then others, and reconciliation calculations in some instances result in very time-consuming negotiations, which may explain the long pre-contract run-in times.

8.4 Market Research into CHP by Utility Companies

142 Although there is evidence of some market research into CHP opportunities by utility companies, most appear to operate on 'feel' particularly with regard to 'where not be bother'. Comments received during this study indicate that UK industry generally is sceptical of the role for private utility companies

8.5 Perceived Potential and Scope for CHP

143 On the general aspect of potential and interest in the prospects for CHP, private utility companies do believe that there is a market in CHP, but not necessarily for them at the present time. The indications are that the achievement of 5 such contracts in five years would constitute a successful entry into the market.

8.6 Development of the Potential for CHP Via Utility Companies

144 The importance of developing the involvement of the utility companies may be judged by the USA experience: "The rapid growth of the US market can be gauged by the fact that in 1981 there were an estimated 20 companies in the field with total business of less than $1 million, whereas by 1984, 200 companies has completed projects worth well over $250 million". (Ref: Lessons from America No. 3, by the Association for the Conservation of Energy).

8.7 Group CHP Schemes

145 There are circumstances where the payback period for CHP at small industrial sites is too long to be considered attractive, but where the payback periods can be reduced if several sites are combined to form a group CHP scheme. Utility companies should be well suited to the installation and operation of group CHP schemes. They specialise in energy services and have the resources to initiate a group CHP scheme which may be too large an undertaking for any one site in a group.

146 An assessment of the national potential for group CHP schemes is beyond the scope of this study but a method of estimating the best payback period for a group CHP scheme has been established. This method is described in Appendix 5. The method is the same as the method for estimating the best payback period for CHP at individual sites described in section 3.2 except that the heat and power demands of the sites in the group are added together to simulate one large site.

147 Examples showing the improvement in payback period that can be achieved from group schemes are presented in Appendix 5 and several general conclusions can be drawn from the analysis:

a) The marginal benefits of group schemes are much greater when small sites rather than larger sites are combined. Larger individual sites can offer attractive payback periods for CHP plant on their own. Smaller sites with average heat demands below about 10 MW are rather unattractive individually and a combination of several sites can produce a much more attractive group scheme.

b) Where a large site is grouped together with several smaller sites the payback period for the resulting group CHP scheme will not be a significant improvement on the payback period for CHP at the large site. The payback period improvement on the small sites will however be much larger. It is situations such as this that may be attractive to utility companies; the larger sites do not stand to benefit significantly from a group CHP scheme involving several smaller sites and so they have no great incentive to become involved. Whereas the smaller sites may be too small to install a group CHP scheme involving the large site. A utility company however could install and operate a group CHP scheme and share the benefits out to each of the sites involved.

9.0 TECHNICAL COMMENTS ON CHP SYSTEMS

9.1 Factors Affecting CHP Economics

148 Site characteristics

Site Heat Demand - Figures 2 to 4 show how payback period decreased in a non-linear manner as average site heat demand increases. This reflects the fact that capital cost decreases with unit size for CHP plants; as shown in Figure 15. The same effects shown in Figures 2 to 4 also apply to steam turbines but they cannot be seen so readily from Figure 5 on account of the logarithmic scale for the heat demand.

Site Heat to Power Ratio - A site heat to power ratio close to that of the CHP system will mean that the CHP system will be able to provide most of the site heat and power demands, and hence achieve high savings as the CHP plant will have a high utilisation. If the CHP plant produces an excess of power which is exported, then the CHP system may be less cost effective. This is because exporting electricity is not as cost effective as displacing imported electricity.

Site Load Factor - The payback period for a CHP plant will depend on its utilisation. It follows that the greater the load factor the better the payback period and this can be seen clearly for all types of CHP plant in Figures 2 to 5. Load factors close to 1 are found when most of the heat demand is required for continuous process applications where the heat requirement does not vary with the weather conditions. At the other extreme where the heat demand is mainly for space heating, as in much of the mechanical engineering industry, then the load factors, as defined for the assessment charts, are in the range 0.2 to 0.3 depending on the shift operating pattern.

Gas Supply Pressure - For most gas fired CHP plant, and for gas turbines in particular, gas compression is often required to raise the pressure of the gas supply to the level required by the CHP plant. In the case of some gas turbines the power required to compress gas from a supply pressure of 1-2 psig to 250 psig (18.2 bar abs) can be up to 10% of the gas turbine power output. If the gas could be supplied at a higher pressure the compression power would be lower and the CHP plant would have a higher net electrical output. This would increase the savings and also reduce the capital cost since a smaller and cheaper compressor would be required.

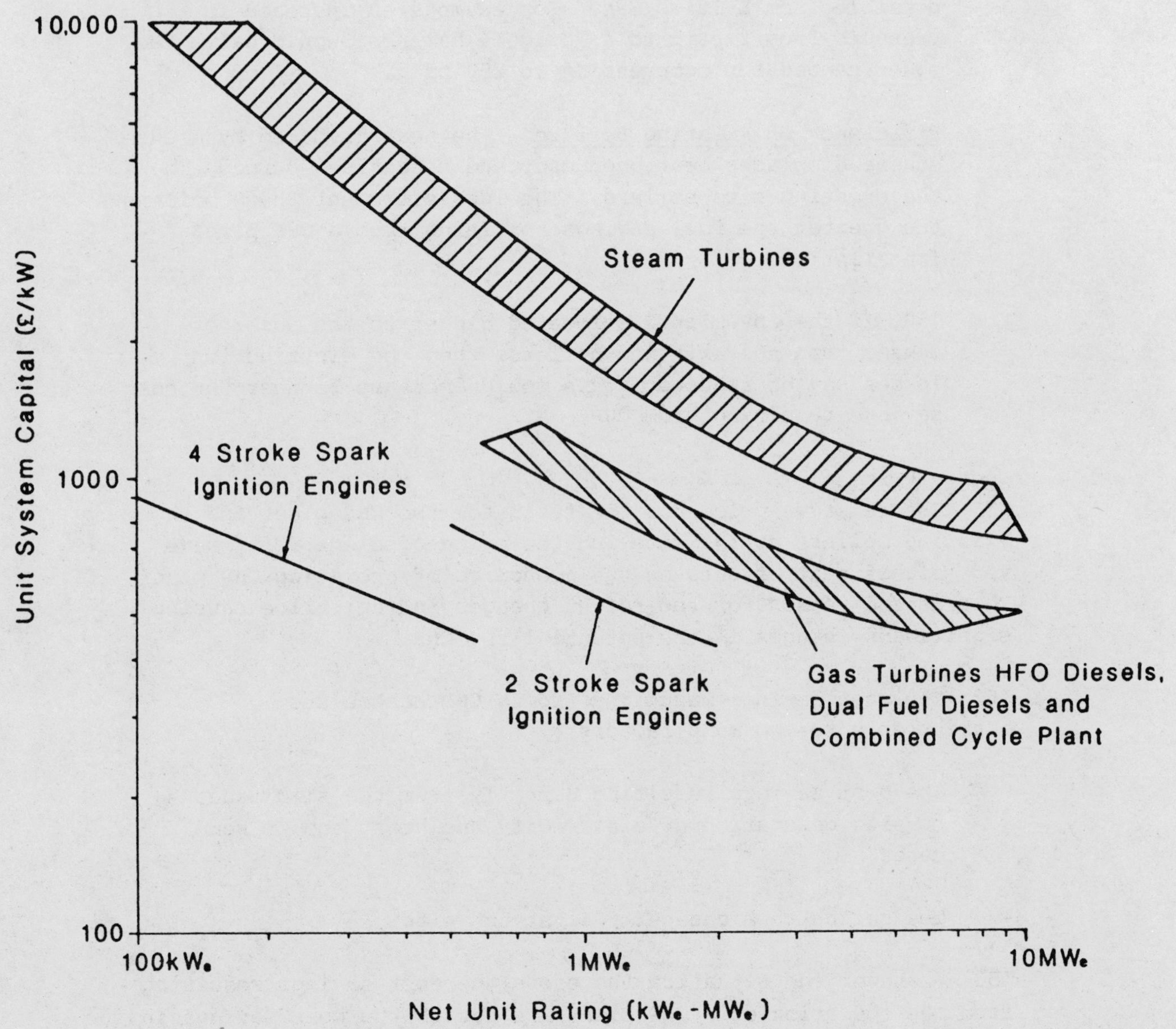

CHP PLANT INSTALLED CAPITAL COSTS **Fig. 15**

149 It is thus in the interest of the potential gas turbine operator to request as high a gas supply pressure as possible from British Gas. For example an increase in pressure from 2 psig to 45 psig (4 bar abs) would halve the power needed for compression to 250 psig.

Efficiency of Existing Boilers - The heat produced by a CHP scheme displaces heat produced, and hence fuel burned, in the existing site boilers. The less efficient these boilers the greater the fuel savings resulting from a CHP plant installation.

150 If the CHP plant is located closer to the point of demand than the existing boilers, then the distribution losses may be reduced. this would increase further the cost savings resulting from CHP.

Energy Prices

151 The relative prices of the fuels for the CHP plant and the existing boilers at the site and the price of electricity have very significant effects on the economics of operating CHP plant. This is the reason for the recent changes in fuel price causing significant changes in the potential for CHP.

152 The cost savings resulting from a CHP scheme are attributable to two main factors:

- The cost savings resulting directly from the simultaneous on-site generation of electricity and heat, and in some cases,

- The switch to a cheaper fuel at the site.

153 Whatever the situation therefore the cost savings resulting from the operation of a CHP plant are due to the cost savings in the purchase of energy at a site. Where the CHP plant fuel is the same as the fuel used in the existing boilers at the site, the savings arise solely from the benefits of the on-site generation of heat and electricity. If the CHP plant uses a cheaper fuel than that used for the existing boilers at the site then there will be an additional saving as heat will be produced from a cheaper fuel. However if the CHP plant fuel is more expensive then the fuel used for the existing boilers then the payback period will be increased.

Effect of the Electricity Price

154 Electricity produced on-site in a CHP plant is generated at

the expense of a fossil fuel. The price differential between the fossil fuel and the electricity displaced or exported is thus of crucial importance.

155 If the electricity tariff increases relative to the price of fossil fuels the effect is to increase the savings resulting from the on-site generation of electricity. A lowering of the electricity tariff has the reverse effect. It is therefore more profitable to operate the CHP plant at times of the day when the price of electricity is highest. At night it may be possible to purchase electricity more cheaply from the electricity board and this means that in some circumstances it may be cost effective to switch the CHP plant off and purchase electricity from the public supply.

156 In general the tariffs for exported electricity are lower than those for imported electricity and when a situation arises where the CHP plant can generate more electricity than the site requires, the cost benefit of exporting electricity is not as great as when displacing imported electricity. This may mean that it is better to run a CHP plant a part load and produce "top-up" heat from existing boiler plant rather than run the CHP plant at full load and export the surplus electricity. At certain times of peak national electricity demand it is in the interest of the electricity supply industry to encourage a private generator of electricity to operate CHP plant in order to reduce the load on the electricity grid. During these times the credit paid for exporting electricity can be much higher than the import tariff and it can be very profitable to export electricity.

<u>CHP System Efficiency</u>

157 The benefits of a CHP plant result from the fundamental feature that both the heat and power from the prime mover are utilised. It is therefore important that the heat from the prime mover is in a form that can be used efficiently.

158 In general the electricity produced by a CHP plant is of greater value than the heat and so there is the desire to improve the electrical efficiency of the prime mover. A high overall system efficiency is of crucial importance to the economic operation of a CHP plant. If greater electrical efficiency is achieved by degrading the heat output, with a consequential fall in the CHP system efficiency, then the CHP plant may be less cost-effective. This is a particularly important point as many prime movers used in CHP applications are originally designed for electricity generation applications, where the quality of the heat output is of little significance and a high electrical generation efficiency is crucial.

Benefit of the After-firing of Exhaust Gases

159 When the exhaust from the CHP plant prime mover has a sufficiently high oxygen content it is possible to burn extra fuel in the exhaust gases. The extra heat is recovered in the waste heat boiler. As the volume and temperature of the flue gases should not increase significantly the waste heat boiler dry stack loss does not increase and so the heat in the fuel used for after-firing is recovered at about 90-93% efficiency based on higher calorific value. It is therefore more attractive to supplement the heat generated by a CHP plant by after-firing the exhaust gases than to use a conventional boiler where efficiency may be only 70-80% based on higher calorific value.

160 The use of after-firing will always result in greater savings unless the fuel used for after-firing is more expensive than that used for the existing boilers. The capital cost of the plant will increase however as the waste heat boiler will be larger and an after-burner is required.

161 Factors Affecting the System Efficiency of CHP Plants

Gas Turbines - A high grade heat output and a high overall system efficiency are of prime importance in producing an economically attractive CHP installation. For this reason gas turbines are particularly attractive for CHP as, in addition to being compact and reliable, all of the heat is rejected at a high temperature, usually above 450°C.

After-firing of the exhaust gases, will improve the overall system efficiency.

Internal Combustion Engines - have the disadvantage that about half of the heat output is in the form of hot water, usually at a temperature of about 80°C. Unless this heat can be utilised the overall CHP system efficiency is low and, in spite of generating electricity more efficiently than any other prime mover, they may be economically less attractive then gas turbine based schemes. Some internal combustion engines are available where low pressure steam is generated in the jacket cooling system. These systems tend to be rather specialised at present but do produce a CHP system with more flexibility.

162 With the introduction of diesel engine CHP systems with after-firing of the exhaust gases the overall CHP system efficiencies can be quite high. Further development of industrial internal combustion engine CHP to produce a greater proportion of the heat output at higher temperatures should improve the attractiveness of these CHP systems by avoiding the requirement for a low temperature heat demand.

Steam Turbine CHP Systems - have a high heat to power output ratio and do not usually produce CHP schemes as economically attractive as those based on prime movers which generate a greater amount of electricity. The overall CHP system efficiencies can approach those of boilers themselves. In spite of this the high capital cost per unit of electrical output results in steam turbine based CHP systems which are often less cost-effective then gas turbine or internal combustion engine systems. The big advantage of steam turbine based systems is that the boiler can burn a wide range of fuels, including cheap waste materials and very cost effective schemes can be devised under these circumstances.

9.2 Comparison of CHP Systems

163 Internal combustion engine CHP schemes often have lower overall system efficiencies than gas turbine CHP systems and it is often found that the latter will be the more cost effective. But the ratio between fossil fuel and electricity prices also has an effect. Internal combustion engines, and diesel engines in particular, generate electricity more efficiently than gas turbines. Consequently if the fuel for a CHP plant decreases in price relative to electricity, the economics of an internal combustion engine CHP plant may improve more rapidly than a gas turbine plant. Even though the internal combustion engine system may have a lower overall efficiency.

164 The most cost effective CHP installation at any particular site can only be selected by a detailed study of the site. However bearing in mind the factors described above some generalisations can be made about the systems most suitable for many industrial sites.

165 Firstly, gas turbine CHP systems are often the most cost effective at large sites where all of the heat output can be used efficiently to produce process steam.

166 Secondly, on smaller sites, those with average heat demands less than about 6 MW, internal combustion engine based CHP systems will often be the most cost effective. This will be most noticeable at lower fossil fuel prices, below about 20-25p/therm (1.9-2.4 £/GJ). Internal combustion engines have a lower heat to power ratio than gas turbines, especially for unit sizes below 3 MWe. They are therefore more likely to be able to match the heat demands of small sites.

167 These conclusions only apply generally to large 'whole site' CHP schemes. There will often be particular situations where

small CHP plants can be cost effective in providing heat for particular processes, even when fuel prices are much higher.

9.3 Split Between Different CHP Plant Types in the Potential for CHP

168 For the reasons described in the previous section, the potential for CHP identified at fossil fuel prices of 28p/therm (£2.68/GJ) is likely to be based mainly on gas turbine systems. This can be seen in section 7 of the Technical Report. As fuel prices fall relative to the price of electricity internal combustion engine CHP plants and diesel engine plants in particular will become more cost effective at some sites. And at smaller sites internal combustion engine CHP plants will also be cost effective. The effect of falling fuel prices therefore will be to increase the proportion of internal combustion engine CHP plants in the Potential for CHP.

9.4 Effect of Capital Cost

169 The payback period for a CHP installation is directly proportional to the capital cost. It can be be seen from Figure 15 that the capital costs for gas turbine CHP plant reach a minimum at about £400-500/kWe for gas turbines > 3MWe in size. The smaller gas turbines are more expensive and generate electricity less efficiently. There are however many smaller industrial sites with steam demands of less than 5MWe. Smaller, more efficient gas turbines would be cost effective at these sites provided that the capital cost is low enough. The potential for CHP could therefore be increased if small gas turbine CHP plants were available with similar capital costs to the larger plants.

170 The situation, shown in Figure 15, where capital costs per kWe decrease with unit size can produce a curious effect. This occurs when it is found that an 'oversized' CHP plant can have a shorter payback period than one which gives the greatest annual savings per kWe installed. A large plant may operate with a reduced load factor and may not increase the annual savings in proportion to its size yet, the lower capital cost per kWe results in a shorter payback period.

9.5 Alternative Fuels

171 Since the start of the study the prices of fuels have changed considerably. The fuels chosen for each type of CHP plant were the cheapest at the start of the study. However now the situation is slightly different as heavy fuel oil is comparable in price with coal. This means that steam turbine CHP plants using oil fired boilers may be more cost effective than those using coal fired boilers. It is not felt that this will cause any significant change to the potential for CHP as steam turbine plants do not feature very prominently.

INDUSTRIAL COMBINED HEAT AND POWER
THE POTENTIAL FOR NEW USERS

APPENDIX 2
CHP PLANT DATA BASE

CONTENTS

1.0 INTRODUCTION

1 The data base of CHP plant comprised the following information:

Plant performance data;
Plant capital, operating and maintenance costs;
Fuel and Electricity prices.

2 This appendix contains details of the information in the data base and a description of the computer model used in the economic assessment of CHP plant. The selection of plant for the data base is described in the introduction and the detailed data is presented in the following sections:

2.0 CHP Plant Performance Data
3.0 CHP Plant Capital and Maintenance Costs
4.0 Fossil Fuel Prices
5.0 Electricity Tariffs
6.0 Computer Model for Assessment of CHP Plants.

1.1 Types of Industrial CHP Plant

3 There are three types of "prime mover" systems commonly used for industrial combined heat and power schemes; gas turbine engines, internal combustion reciprocating engines, and steam turbines:

- Gas Turbines
  A gas turbine CHP plant is shown schematically in Fig. 1. The high temperature products of combustion which would be rejected in an electricity-only scheme, are used to produce heat in the form of steam or hot water. Alternatively gas turbine exhaust can be used directly in a process requiring a hot gas stream, such as dryer. Gas turbines are available across the size range considered in this study (100 kW to 10 MW) however the turbines below 1 MW have low electrical generation efficiencies. Gas turbines can operate on gaseous fuels or light distillate fuel oils.

- Internal Combustion Reciprocating Engines
  Internal combustion engines may be compression ignition (diesel) or spark ignition and can be used with a variety of fuels both gaseous and liquid. they are available right across the size range used in this study: spark ignition gas engines are available up to 2 MW; and larger slow speed diesels are available in sizes up to 10 MW (the maximum size considered in this study) and above. An internal combustion engine CHP plant is shown schematically in Fig. 2. Heat can be recovered from the exhaust gases and also from the engine

cooling system. The hot exhaust gases can be used to generate steam or hot water whereas the cooling system can only produce hot water up to 80°C unless a specialised cooling system is utilised.

- <u>Steam Turbine Plant</u>
  Steam turbine plant is used in the majority of electricity generating stations in the world. When producing electricity only the exhaust steam is condensed at low pressure to maximise the electricity output. However if the steam is exhausted from the turbine at a higher temperature and pressure it can be used for industrial processes. A typical back pressure steam turbine CHP plant is shown schematically in Fig. 3. Steam turbines are available across the full range of output considered in this study.

- <u>Combined Cycle Plant</u>
  There are many possibilities for combining different power cycles to maximise electricity generation. The most common involves passing the exhaust gas from a gas turbine(s) through a boiler to raise high pressure steam which is passed through a back pressure steam turbine. A typical combined cycle plant is shown in Fig. 4.

- <u>After-firing of Exhaust Gas</u>
  When a CHP plant produces a hot exhaust gas stream with a high oxygen content it is possible to increase the temperature of the exhaust gas by after-firing with extra fossil fuel. This is an efficient method of producing heat as the stack losses from the waste heat boiler do not increase significantly when after-firing is employed. The heat added by after-firing is thus recovered in the boiler at about 90-93% efficiency (based on higher calorific value). Gas turbines and compression ignition internal combustion engines have a sufficiently high exhaust gas oxygen content to permit after-firing.

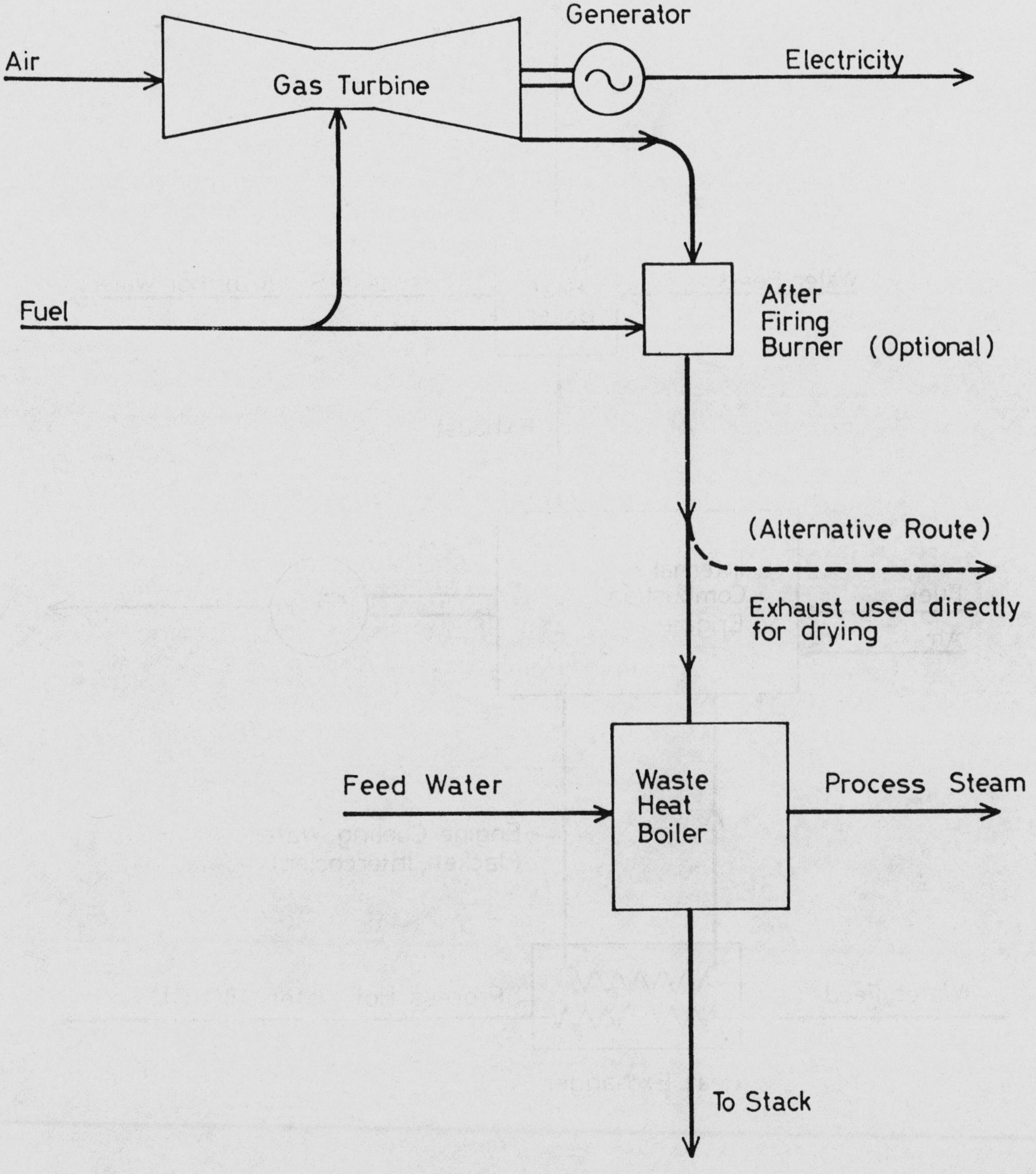

GAS TURBINE CHP PLANT

Fig. 1

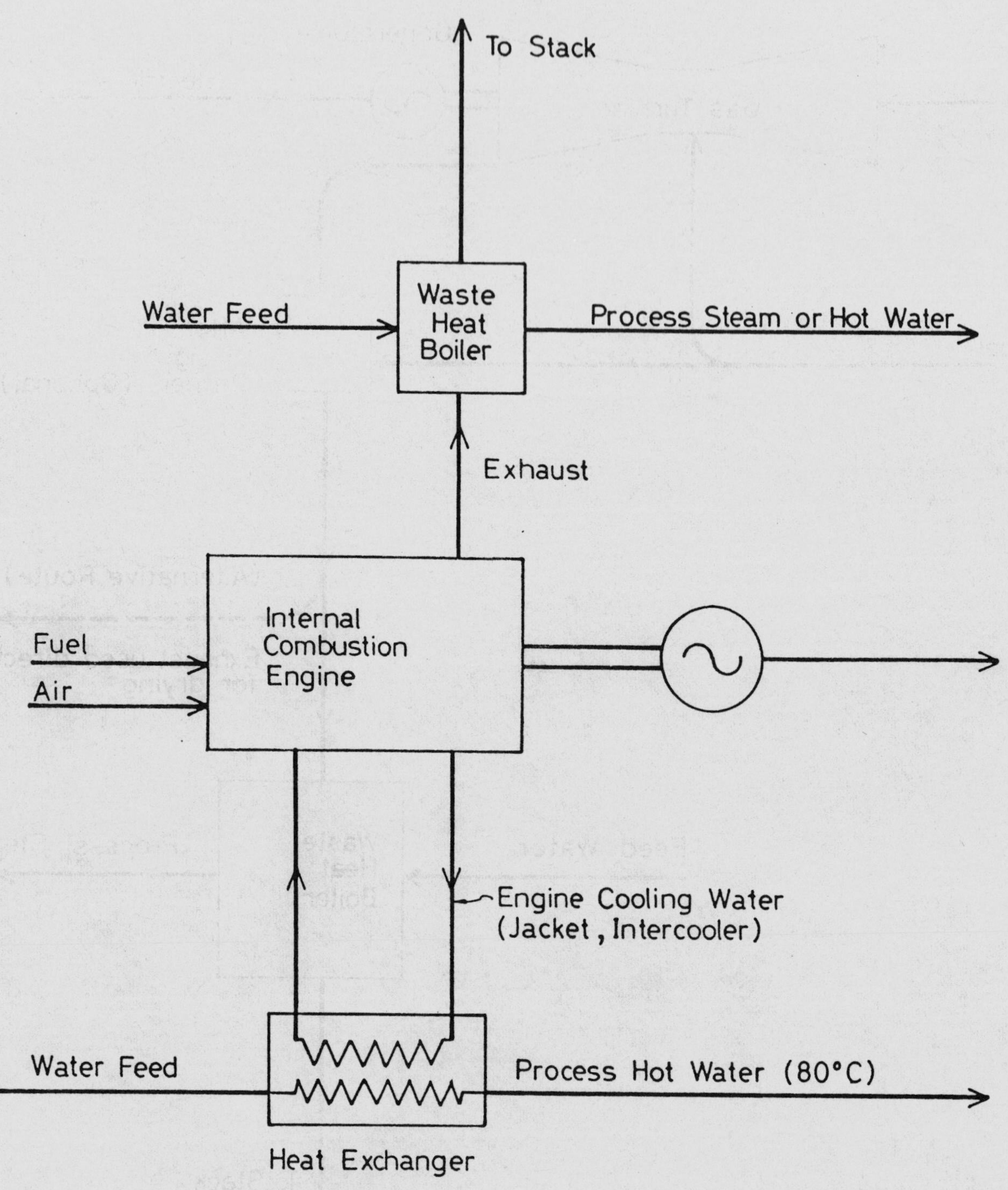

INTERNAL COMBUSTION ENGINE CHP PLANT

Fig. 2

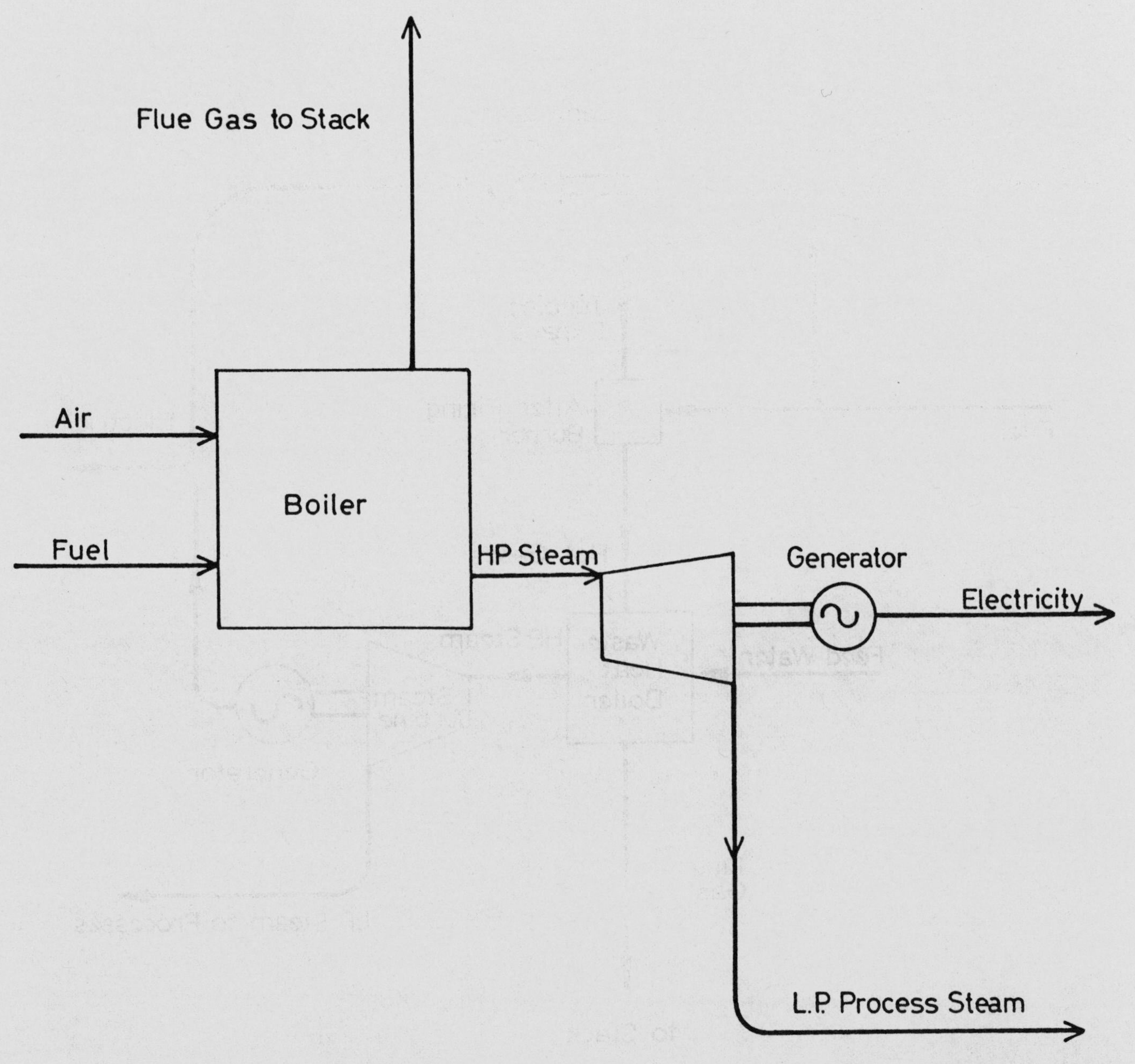

STEAM TURBINE CHP PLANT **Fig. 3**

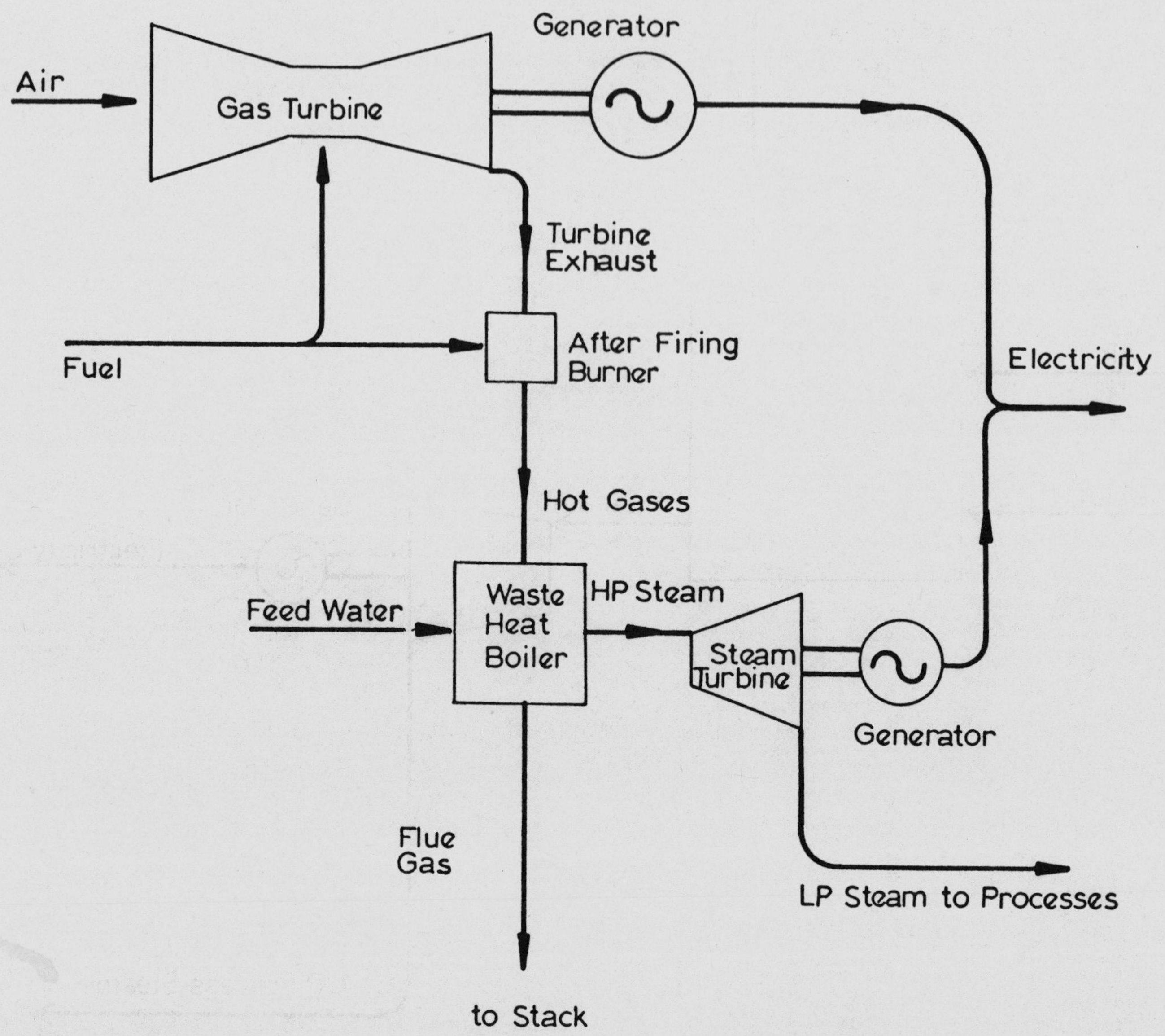

COMBINED CYCLE CHP PLANT **Fig. 4**

1.2 Selection of CHP Plant

4 From the three prime mover categories, six basic CHP plant configurations were selected for the assessment of CHP potential. Only prime movers within the size range of 100 kWe to 10 MWe were included in the data base as this size range was most appropriate to the industrial sites considered. These are shown in the table below and the reasons for their selection are described.

CHP Plant Selected for the Study

Plant Type	Fuel Type	After-firing of Exhaust Gas	Size Range (MW net electrical)
1. Gas Turbines	Natural Gas	Yes	0.5-10 MWe
2. Diesel Engines	Heavy Fuel Oil	Yes	0.65-10 MWe
3. Dual-Fuel Diesel Engines	Natural Gas & Light Oil	Yes	0.65-6.4 MWe
4. Industrial Spark Ignition Engines	Natural Gas	No	0.1-2.0 MWe
5. Back Pressure Steam Turbines	Coal	-	0.1-10 MWe
6. Combined Cycle	Natural Gas	Yes	4.0-8.0 MWe

i) Gas Turbines

Gas turbines can operate on liquid or gaseous fuels. At the time the data base was compiled (Dec' 84) natural gas was the cheapest and cleanest industrial fuel suitable for gas turbines and so it was selected for the study.

There are two main types of gas turbine; those derived from aircraft engines and purpose built industrial machines. Within the size range used in this study the aero-derived gas turbines have a higher shaft efficiency then the industrial machines. However, the industrial gas turbines tend to require less maintenance. Net electrical efficiencies (allowing for auxiliaries and based on higher calorific value) of 20-25% are achievable from gas turbines of 2 MW electrical output and above. Below this size, however, the electrical efficiency may fall below 15%. As gas turbines are manufactured in a relatively few discrete sizes, ten examples were selected for the study, representing industrial and aero-derived turbines across the size range of 0.5 to 10 MW.

A disadvantage in the use of gaseous fuel for a gas turbine is that it must be injected into the engine at a high pressure. In the case of the aero-derived gas turbines, the gas must be supplied at 18-20 bar absolute. This can require a substantial amount of power for compression, up to 10% of the turbine output in some cases, if the gas is supplied at a low pressure to the site. The power for compression can be greatly reduced if the gas is supplied at a higher pressure and the capital cost of the compression equipment will also be reduced. Frequently British Gas can provide the gas at a higher supply pressure to a site and it is in the interests of the industrial consumer to request a higher pressure whenever this would benefit the CHP installation. For the purposes of this study it is assumed that gas is supplied at 26" W.G. (0.065 bar gauge).

ii) <u>Diesel Engines</u> - Heavy fuel oil and dual-fuel
Heavy fuel oil and natural gas are the cheapest industrial fuels for diesel engines, in the size range considered for the study, and so only diesel engines using these fuels have been studied. In the size range of 100 kW to 10 MW medium speed 4-stroke diesel engines are the most common. These engines are produced by several manufacturers in a wide range of sizes and are commonly available from 500 kWe to 10 MWe. Unlike gas turbines the variation in engine performance between different manufacturers is not great and, rather than selecting particular engines, average data has been derived to cover engines of any size across the range.

Dual-fuel diesel engines, which utilise natural gas as the main fuel, also require a light oil pilot fuel (5-10% of total fuel input).

iii) <u>Spark Ignition Gas Engines</u>
Natural gas is the cheapest industrial fuel available for spark ignition engines. Four stroke and two stroke engines are available and for the purposes of this study four stroke engines from 100 kWe to 600 kWe and two stroke engines from 500 kWe to 2 MWe are considered.

iv) <u>Steam Turbines</u>
The most common steam turbine CHP plant is a steam raising boiler and a back pressure steam turbine exhausting at the site distribution pressure, as illustrated in Fig. 3. At

some sites where steam is distributed at several pressures a passout turbine may be appropriate, however this would be very site specific and so this option was not included in the study. High pressure steam boilers operating at pressures from 32 to 64 bar absolute were selected as a reasonable optimum as more electricity can be generated from a particular flow of steam when the pressure ratio across the turbine is larger. Back pressures ranging from 2 to 18 bar absolute were considered as this is the range applicable to most industrial sites. Only coal fired boilers were investigated as coal was substantially the cheapest industrial boiler fuel at the time the data base was compiled (Dec. 1984). Single stage and multi stage steam turbines were included. In general, single stage turbines are limited to power outputs below 2 MW and pressure ratios of less than 3:1. For other conditions multi stage turbines are more economical; sizes up to 10 MW were considered.

It was assumed that a new high pressure boiler would always be required when a steam turbine was installed, as the existing boiler would not be able to produce the higher pressure steam for the turbine in most cases. (See Section 1.5 however).

v) Combined Cycle Plant

Of the several possible configurations of combined cycle plant available the type included in the study comprised a gas turbine with high pressure waste heat boiler providing steam for a back pressure steam turbine, as illustrated in Figure 4. Twelve combined cycle plants were included in the study ranging in size from 4.0 to 8.0 MWe. They are based on four typical gas turbines within the size range of the study. A lower limit of 4 MWe was selected; below this size the gas turbines have lower efficiencies and the steam turbines become more expensive per unit of output which makes the combined cycle plant less cost effective. The fuel selected for the combined cycle plant was natural gas, the cheapest appropriate fuel available.

The combined cycle plants chosen all utilised after-firing of the gas turbine exhaust gases as these plants were found to be most cost effective. When combined cycle plants are used for generating electricity only after-firing reduces the system efficiency, in a CHP plant however after-firing increases the system efficiency.

1.3 Sources of Data

5 Plant manufacturers and suppliers were approached and

invited to supply performance details and budget costs of their CHP equipment. It was not possible to contact every supplier of CHP equipment but several were approached for each type and size range of plant. The companies who provided the information used in the study are listed in the acknowledgement at the end of this report.

6 Details of maintenance costs were obtained from the equipment suppliers but data from existing schemes was also used where possible. In the case of diesel engines it was recommended by the manufacturers that this information was obtained from the Working Cost Report of the Institute of Diesel and Gas Turbine Engineers.

7 In all cases, except for gas turbine plant where individual engines were studied, information was obtained from several suppliers and used as the basis for average performance and costs typical of the different types of CHP plant. The data presented therefore may not equate exactly with any particular equipment from a supplier.

8 Capital cost estimates were also available from recent studies for actual CHP projects. This was used in conjunction with the data provided by the equipment manufacturers.

1.4 <u>Plant Performance and Cost Details Included in the Data Base</u>

9 The CHP plant data base contained the following performance data for each plant configuration considered. These were compiled for both full load and part load operation of the plant.

i) Maximum continuous rating of plant
ii) Fuel consumption (including pilot fuel and after-firing fuel as appropriate)
iii) Fuel type
iv) Exhaust flow rate
v) Exhaust temperature
vi) Proportion of heat recoverable from the cooling water (Internal Combustion Engines only)
vii) Steam flow rate ┐ Steam
viii) Heat in steam turbine exhaust ├ Turbines
ix) Steam turbine inlet and outlet conditions ┘ Only
x) Net electrical efficiency of prime mover
xi) Total CHP system thermal efficiency
xii) System Heat/Power ratio.

10 This information is presented in detail in Section 2 of this appendix for each plant used in the study. It includes allowances for all auxiliaries and losses associated with the installation.

11 The CHP plant data base also contains the capital costs for complete CHP installations for each plant in the study. These include the following main items:

i) CHP plant prime mover.
ii) Waste heat recovery plant.
iii) Fuel supply system (including gas compressor for gas turbine plant).
iv) Electrical and Mechanical installation and commissioning.
v) Civil works (including new buildings).
vi) Design fees.
vii) Contingencies.

12 The costs are based on December 1984 prices and are suitable for use as budget costs for installations at an average site. Details of these costs are provided in Section 3 of this appendix.

1.5 Retrofit CHP Plant

13 The six main types of CHP plant described already were used in the main assessment of the potential for CHP. In addition three examples of retrofit plant were analysed to demonstrate how, in some circumstances, CHP plant can be fitted onto existing plant. Two of these plants are also examples of CHP plant where the prime mover drives an item of mechanical plant directly.

The retrofit plants considered were:

a) Spark ignition engines driving directly air compressors with heat recovery from the engine and compressor.

b) Steam turbines fitted on to existing low pressure steam boilers driving directly items of plant.

c) Steam turbine sets generating electricity fitted onto existing low pressure shell and tube packaged boilers.

14 The performance and cost data types of plant is presented in this appendix. The economic analysis of these types of plant is presented in full in Appendix 4. These retrofit plants were not considered in the main assessment of the potential for CHP.

2.0 CHP PLANT PERFORMANCE DATA

2.1 General

15 This section describes some general points concerning CHP plant performance data.

16 Assumptions - Conservative assumptions have been made in all calculations concerning CHP plant performance. In many cases the plant may achieve better performance and this would improve the economic attractiveness of CHP.

17 Calorific Value - For the purposes of consistency all efficiencies (of prime movers and boilers) are expressed as a proportion of the higher (or gross) calorific value (HCV or GCV) of the fuels concerned. It is customary for the performance of gas turbine and reciprocating engines to be described in terms of the lower (or net) calorific value. Boiler performance is usually described in terms of the fuel higher (or gross) calorific value. For the purposes of this study higher calorific value is used as fuel is always purchased on the basis of higher calorific value.

18 Heat Capacity of Exhaust Gases - In order to determine the mean specific heat capacity of exhaust gases when calculating the heat recovered in waste heat boilers, the charts shown in Fig. 5a and Fig. 5b were used. These take into account changes in specific heat capacity with gas composition and temperature.

19 Ambient Conditions - CHP plant performance was considered only at standard ambient conditions. These were 15°C, 1 standard atmosphere for gas turbines and 27°C, 1 standard atmosphere for internal combustion engines.

20 Standby Fuel - In all cases it was assumed that the system would revert to the existing boilers at times when the natural gas supply was interrupted. In certain instances it may be economically attractive to have a standby fuel installation. This would result in improved economics for the system as a whole.

21 Full Load Performance - Full load performance was considered to be the maximum continuous rating (MCR) of the plant.

22 Part Load Performance - The variation in plant efficiency, exhaust temperature and mass flow rate etc. at part load were considered. For the purposes of simplicity a linear variation was assumed over the turndown range considered for each type of plant. This linearising resulted, in most cases, in a

discrepancy of less than 5% in the performance at the minimum electrical output.

23 Electricity Generators - In all cases losses of 5% were assumed for generators unless a more precise figure had been provided by the supplier of the generating set. 415V generators were used up to unit ratings of 600 kW. Above 600 kW generators of 11 kV were used.

24 Steam Flows - For reference purposes a graph is shown in Figure 6 comparing heat flow with the equivalent flow rate of steam.

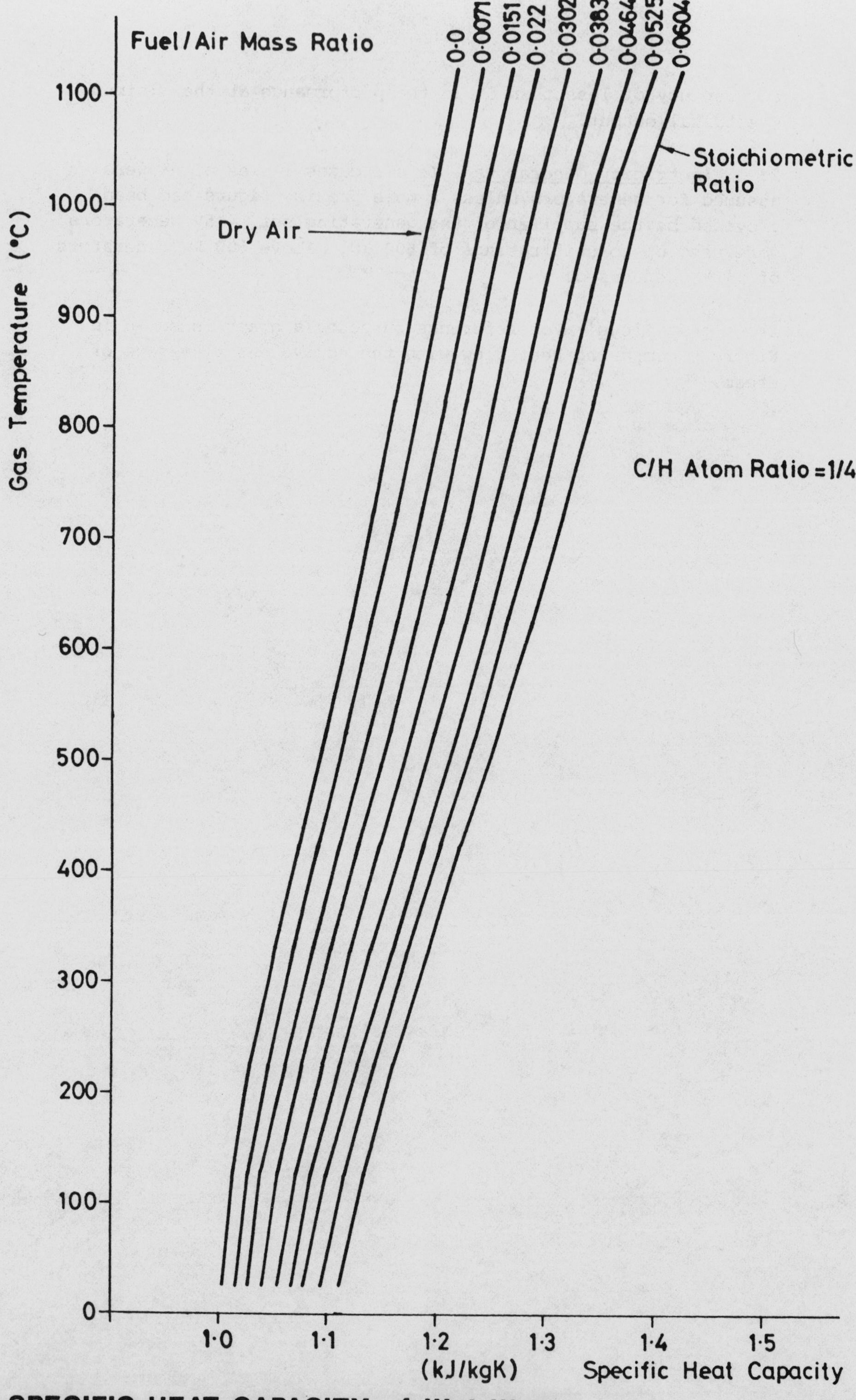

SPECIFIC HEAT CAPACITY of Wet Waste Gases from Combustion of Natural Gas **Fig. 5a**

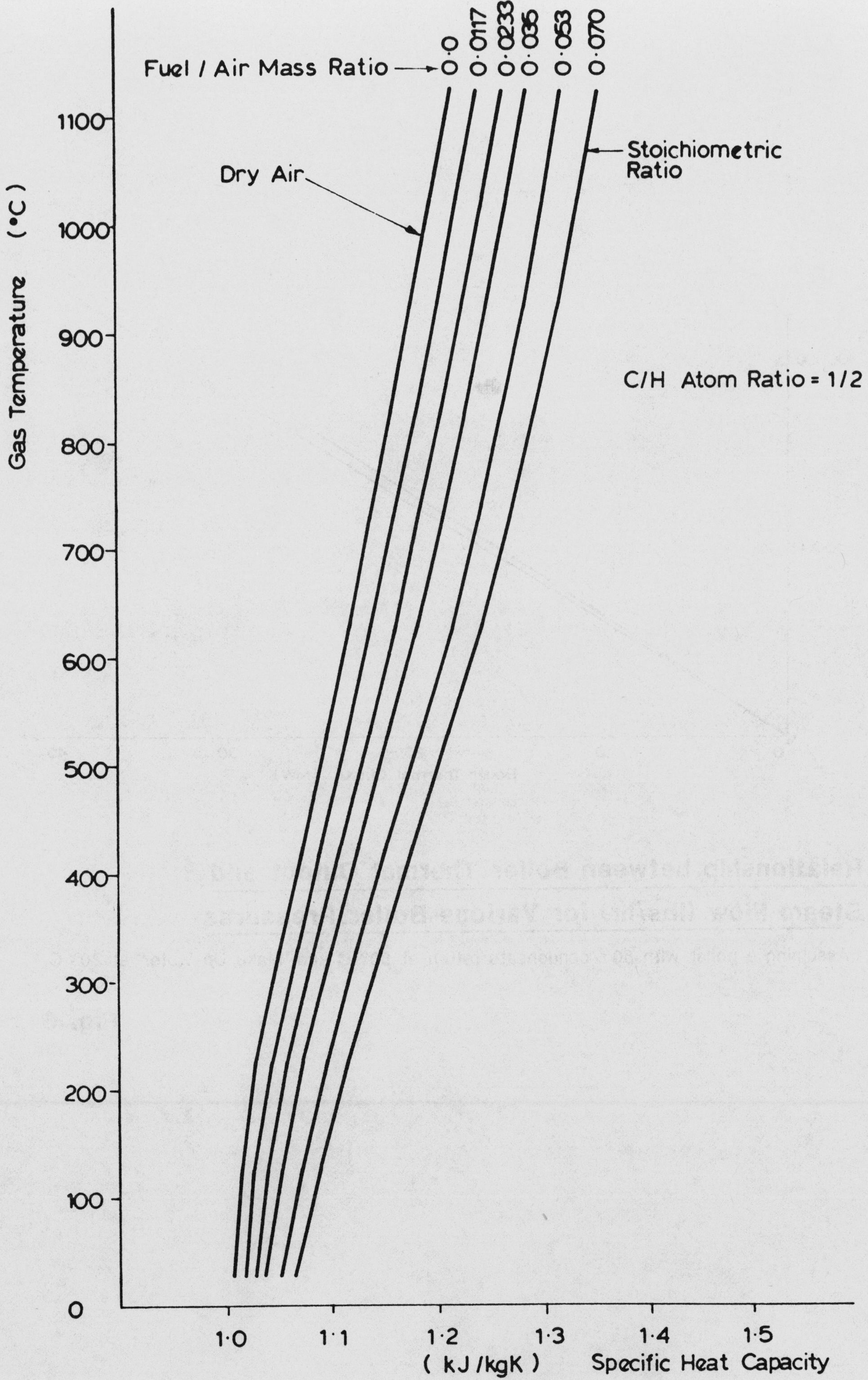

SPECIFIC HEAT CAPACITY of Wet Waste Gases from Combustion of a Liquid Hydrocarbon Fuel

Fig. 5b

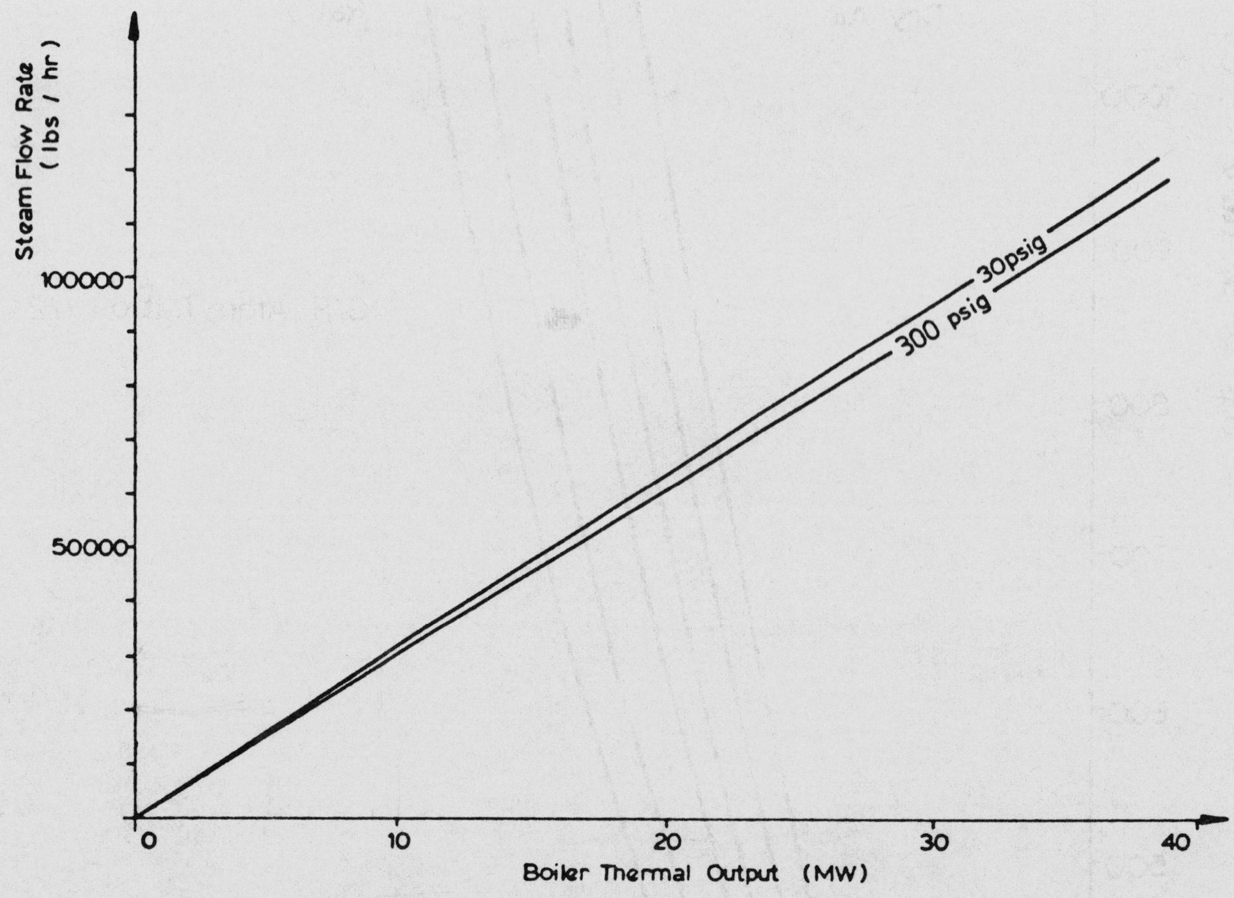

Relationship between Boiler Thermal Output and Steam Flow (lbs/hr) for Various Boiler Pressures

(Assuming a boiler with 50% condensate return at 80° C and 'Make Up Water' at 20° C)

Fig. 6

2.2 Gas Turbine Plant

25 The performance data for the gas turbines considered are shown in Tables 1-10. Key terms are explained below.

26 Size range - Ten specific gas turbines typical of those available in the size range 100kW-10MW were used. Six of these machines were purpose designed industrial gas turbines, and four were based on aero-derived gas generators.

27 After-Firing Temperature - In the case of gas turbines it was assumed that the exhaust gases could be after-fired up to 800 ^{o}C. Firing above this temperature could require a change in boiler design to a more expensive construction and was not considered. Water tube boilers were used for gas turbines as shell and tube boilers generally cannot accept the large exhaust gas mass flow rates without imposing an unacceptably large back pressure on the turbine.

28 Net Electrical Output - The electrical output of the CHP plant is specified at the maximum continuous rating after all the losses and auxiliaries have been taken into account.

29 Minimum Electrical Output - A turndown ratio of 2:1 was assumed for all gas turbine plant. If the electrical demand fell below 50% then the machine could be switched off or machine load maintained by exporting electricity.

30 Exhaust Gas Specific Heat Capacity - The mean specific heat capacities for fired and unfired exhaust gases were determined from Fig. 5.

31 Gas Compressor Power - It was assumed that natural gas was supplied to a site at 26" W.G.-pressure. (0.065 bar gauge) Gas compressors were therefore required to boost the pressure for the gas turbines. Gas inlet pressures of 6-13 bar absolute are required for the industrial gas turbines and 18-20 bar absolute for the aero-derived gas turbines. The power required to compress gas from 1 to 18 bar absolute is in the order of 10% of the gas turbine output. Significant reductions in the power for compression can be made if the gas is supplied at a higher pressure. For instance supplying the gas at 3 bar gauge (45 psig) will half the compression power requirement for compression to 18 bar absolute. This will increase the net electrical output by about 5%. British Gas can, in some cases, increase the gas supply pressure and it is in the interests of the industrial consumer to request a higher pressure when considering a gas turbine installation.

32 Single Shaft/Split Shaft - A single shaft gas turbine has the power turbine coupled to the compressor turbine. Consequently the compressor runs at constant speed for all loads. This means that the air mass flow rate through the gas turbine will not change significantly with load. In a split shaft machine the power turbine is independent and the compressor can run at varying speeds depending on the load. In general a split shaft machine has a higher part load efficiency than a single shaft machine.

33 Fuel consumption and Exhaust Mass Flow Rates - These quantities are expressed per kW of net electrical output.

34 Net Electrical Efficiency of the Gas Turbine - This is expressed as:

$$\frac{\text{Net Electrical Output from Generator}}{\text{Fuel into the Gas Turbine (HCV)}}$$

35 CHP System Efficiency - This representative system efficiency is defined as:

$$\frac{\text{Net Electrical Output + Heat Recovered}}{\text{Total Fuel into CHP System (HCV)}}$$

The heat recovered in a waste heat boiler varies with the exhaust gas mass flow rate and the flue gas temperature at the boiler exit.

36 A design flue gas temperature of 150°C was assumed for the purpose of calculating this representative system efficiency. More detailed calculations of boiler performance were undertaken when assessing a CHP plant at a particular site. See section 2.8.2. A flue gas temperature of 150°C is conservative and lower temperatures may be achieved in practice with a resulting improvement in economics.

37 Heat/Power Ratio - This is defined as:

$$\frac{\text{Heat Recovered}}{\text{Net Electrical Output}}$$

38 Auxiliaries - An allowance of 4% of gross electrical output was made for the turbine and boiler auxiliaries associated with the CHP plant. This included:

Pumps and Air blast fans for oil coolers
Control Equipment
Building ventilation
Nominal electrical services in the building
Waste heat boiler gas blower
Waste heat boiler feed and circulating water pumps

N.B. The gas compressor for the gas turbine was considered separately.

39 Intake and Exhaust Losses - A pressure loss of 100mm W.G. was assumed for the air intake of the gas turbine and a loss of 200mm W.G. for the exhaust duct and waste heater boiler.

Table 1

PLANT	GAS TURBINE	TYPE I

Net Electrical Output (MCR) 1080 [kW]

Maximum Afterfiring Temperature 800 [deg C]
Minimum Electrical Output 50 [% of MCR]

Exhaust Gas Mean Specific Heat Capacity - Fired 1.14 [kJ/kgK]
- Unfired 1.07 [kJ/kgK]

Gas Compressor Power Demand 75 [kW]

Comments:
Split Shaft Industrial Gas Turbine

Electrical Output		MCR	50% MCR
Turbine Fuel Consumption (HCV)	[kW/kWe]	7.80	9.75
Max After Firing Fuel Cons'n (HCV)	[kW/kWe]	4.58	9.23
Exhaust Gas Temperature	[deg C]	507.0	415.0
Exhaust Gas Mass Flow Rate	[kg/kWhe]	38.60	61.76
Net Electrical Efficiency (HCV) of Gas Turbine	[%]	12.82	10.26

Total CHP System Efficiency at MCR (Waste Heat Boiler Flue Gas Temp 150 deg C)		Heat/Power Ratio at MCR
Unfired =	64.3 % (HCV)	4.0
Fired =	71.0 % (HCV)	7.8

Steam Flow from Waste Heat Boiler at MCR (100 psig saturated)

Unfired = 13000 [lbs/hr]

Fired = 26000 [lbs/hr]

Table 2

PLANT	GAS TURBINE	TYPE II

Net Electrical Output (MCR)	1520 [kW]
Maximum Afterfiring Temperature	800 [deg C]
Minimum Electrical Output	50 [% of MCR]
Exhaust Gas Mean Specific Heat Capacity - Fired	1.14 [kJ/kgK]
Exhaust Gas Mean Specific Heat Capacity - Unfired	1.07 [kJ/kgK]
Gas Compressor Power Demand	95 [kW]

Comments:
Split Shaft Industrial Gas Turbine

Electrical Output		MCR	50% MCR
Turbine Fuel Consumption (HCV)	[kW/kWe]	6.33	7.77
Max After Firing Fuel Cons'n (HCV)	[kW/kWe]	3.60	7.08
Exhaust Gas Temperature	[deg C]	508.0	426.7
Exhaust Gas Mass Flow Rate	[kg/kWhe]	30.40	48.65
Net Electrical Efficiency (HCV) of Gas Turbine	[%]	15.80	12.87

Total CHP System Efficiency at MCR (Waste Heat Boiler Flue Gas Temp 150 deg C)		Heat/Power Ratio at MCR
Unfired =	65.9 % (HCV)	3.2
Fired =	71.8 % (HCV)	6.1

Steam Flow from Waste Heat Boiler at MCR (100 psig saturated)

Unfired =	14000 [lbs/hr]
Fired =	28000 [lbs/hr]

Table 3

PLANT	GAS TURBINE	TYPE III

Net Electrical Output (MCR) 2970 [kW]

Maximum Afterfiring Temperature 800 [deg C]
Minimum Electrical Output 50 [% of MCR]

Exhaust Gas Mean Specific Heat Capacity - Fired 1.14 [kJ/kgK]
- Unfired 1.06 [kJ/kgK]

Gas Compressor Power Demand 175 [kW]

Comments:
Split Shaft Industrial Gas Turbine

Electrical Output		MCR	50% MCR
Turbine Fuel Consumption (HCV)	[kW/kWe]	5.25	6.73
Max After Firing Fuel Cons'n (HCV)	[kW/kWe]	3.09	6.29
Exhaust Gas Temperature	[deg C]	496.0	406.7
Exhaust Gas Mass Flow Rate	[kg/kWhe]	25.06	41.10
Net Electrical Efficiency (HCV) of Gas Turbine	[%]	19.05	14.86

Total CHP System Efficiency at MCR (Waste Heat Boiler Flue Gas Temp 150 deg C)		Heat/Power Ratio at MCR
Unfired =	66.7 % (HCV)	2.5
Fired =	72.6 % (HCV)	5.1

Steam Flow from Waste Heat Boiler at MCR (100 psig saturated)

Unfired = 23000 [lbs/hr]

Fired = 46000 [lbs/hr]

Table 4

PLANT	GAS TURBINE	TYPE IV

Net Electrical Output (MCR) 5150 [kW]

Maximum Afterfiring Temperature 800 [deg C]
Minimum Electrical Output 50 [% of MCR]

Exhaust Gas Mean Specific Heat Capacity - Fired 1.14 [kJ/kgK]
- Unfired 1.07 [kJ/kgK]

Gas Compressor Power Demand 336 [kW]

Comments:
Split Shaft Industrial Gas Turbine

Electrical Output		MCR	50% MCR
Turbine Fuel Consumption (HCV)	[kW/kWe]	4.27	5.76
Max After Firing Fuel Cons'n (HCV)	[kW/kWe]	2.45	6.28
Exhaust Gas Temperature	[deg C]	483.0	377.0
Exhaust Gas Mass Flow Rate	[kg/kWhe]	19.36	38.71
Net Electrical Efficiency (HCV) of Gas Turbine	[%]	23.42	17.36

Total CHP System Efficiency at MCR (Waste Heat Boiler Flue Gas Temp 150 deg C)		Heat/Power Ratio at MCR
Unfired =	67.4 % (HCV)	1.9
Fired =	73.0 % (HCV)	3.9

Steam Flow from Waste Heat Boiler at MCR (100 psig saturated)

Unfired = 29000 [lbs/hr]

Fired = 62000 [lbs/hr]

Table 5

PLANT	GAS TURBINE	TYPE V

Net Electrical Output (MCR) 2840 [kW]

Maximum Afterfiring Temperature 800 [deg C]
Minimum Electrical Output 50 [% of MCR]

Exhaust Gas Mean Specific Heat Capacity - Fired 1.15 [kJ/kgK]
- Unfired 1.09 [kJ/kgK]

Gas Compressor Power Demand 280 [kW]

Comments:
Single Shaft Aero-devivative Gas Turbine

Electrical Output		MCR	50% MCR
Turbine Fuel Consumption (HCV)	[kW/kWe]	4.89	6.35
Max After Firing Fuel Cons'n (HCV)	[kW/kWe]	2.31	6.33
Exhaust Gas Temperature	[deg C]	513.0	384.8
Exhaust Gas Mass Flow Rate	[kg/kWhe]	19.78	39.56
Net Electrical Efficiency (HCV) of Gas Turbine	[%]	20.45	15.75

Total CHP System Efficiency at MCR (Waste Heat Boiler Flue Gas Temp 150 deg C)		Heat/Power Ratio at MCR
Unfired =	64.0 % (HCV)	2.1
Fired =	69.8 % (HCV)	4.0

Steam Flow from Waste Heat Boiler at MCR (100 psig saturated)

Unfired = 18000 [lbs/hr]

Fired = 35000 [lbs/hr]

Table 6

PLANT GAS TURBINE **TYPE** VI

Net Electrical Output (MCR) 3140 [kW]

Maximum Afterfiring Temperature 800 [deg C]
Minimum Electrical Output 50 [% of MCR]

Exhaust Gas Mean Specific Heat Capacity - Fired 1.15 [kJ/kgK]
- Unfired 1.10 [kJ/kgK]

Gas Compressor Power Demand 280 [kW]

Comments:
Single Shaft Aero-devivative Gas Turbine

Electrical Output		MCR	50% MCR
Turbine Fuel Consumption (HCV)	[kW/kWe]	4.50	5.86
Max After Firing Fuel Cons'n (HCV)	[kW/kWe]	1.94	5.58
Exhaust Gas Temperature	[deg C]	533.0	394.4
Exhaust Gas Mass Flow Rate	[kg/kWhe]	17.88	35.76
Net Electrical Efficiency (HCV) of Gas Turbine	[%]	22.22	17.06

Total CHP System Efficiency at MCR (Waste Heat Boiler Flue Gas Temp 150 deg C)		Heat/Power Ratio at MCR
Unfired =	67.8 % (HCV)	2.1
Fired =	72.0 % (HCV)	3.6

Steam Flow from Waste Heat Boiler at MCR (100 psig saturated)

Unfired = 19000 [lbs/hr]

Fired = 35000 [lbs/hr]

Table 7

PLANT	GAS TURBINE	TYPE VII

Net Electrical Output (MCR)	3790 [kW]
Maximum Afterfiring Temperature	800 [deg C]
Minimum Electrical Output	50 [% of MCR]
Exhaust Gas Mean Specific Heat Capacity - Fired	1.15 [kJ/kgK]
Exhaust Gas Mean Specific Heat Capacity - Unfired	1.10 [kJ/kgK]
Gas Compressor Power Demand	280 [kW]

Comments:
Split Shaft Aero-devivative Gas Turbine

Electrical Output		MCR	50% MCR
Turbine Fuel Consumption (HCV)	[kW/kWe]	4.62	5.27
Max After Firing Fuel Cons'n (HCV)	[kW/kWe]	1.81	4.02
Exhaust Gas Temperature	[deg C]	550.0	429.0
Exhaust Gas Mass Flow Rate	[kg/kWhe]	17.64	27.87
Net Electrical Efficiency (HCV) of Gas Turbine	[%]	21.65	18.99

Total CHP System Efficiency at MCR (Waste Heat Boiler Flue Gas Temp 150 deg C)		Heat/Power Ratio at MCR
Unfired =	67.4 % (HCV)	2.1
Fired =	71.4 % (HCV)	3.6

Steam Flow from Waste Heat Boiler at MCR (100 psig saturated)

Unfired =	24000 [lbs/hr]
Fired =	42000 [lbs/hr]

Table 8

PLANT	GAS TURBINE	TYPE VIII

Net Electrical Output (MCR) 4830 [kW]

Maximum Afterfiring Temperature 800 [deg C]
Minimum Electrical Output 50 [% of MCR]

Exhaust Gas Mean Specific Heat Capacity - Fired 1.15 [kJ/kgK]
- Unfired 1.10 [kJ/kgK]

Gas Compressor Power Demand 420 [kW]

Comments:
Split Shaft Aero-devivative Gas Turbine

Electrical Output		MCR	50% MCR
Turbine Fuel Consumption (HCV)	[kW/kWe]	3.83	4.83
Max After Firing Fuel Cons'n (HCV)	[kW/kWe]	1.52	3.25
Exhaust Gas Temperature	[deg C]	543.0	445.3
Exhaust Gas Mass Flow Rate	[kg/kWhe]	14.47	23.44
Net Electrical Efficiency (HCV) of Gas Turbine	[%]	26.11	20.70

Total CHP System Efficiency at MCR (Waste Heat Boiler Flue Gas Temp 150 deg C)		Heat/Power Ratio at MCR
Unfired =	70.6 % (HCV)	1.7
Fired =	73.7 % (HCV)	2.9

Steam Flow from Waste Heat Boiler at MCR (100 psig saturated)

Unfired = 25000 [lbs/hr]

Fired = 44000 [lbs/hr]

Table 9

PLANT	GAS TURBINE	TYPE IX

Net Electrical Output (MCR) 8610 [kW]

Maximum Afterfiring Temperature 800 [deg C]
Minimum Electrical Output 50 [% of MCR]

Exhaust Gas Mean Specific Heat Capacity - Fired 1.14 [kJ/kgK]
- Unfired 1.08 [kJ/kgK]

Gas Compressor Power Demand 870 [kW]

Comments:
Single Shaft Industrial Gas Turbine

Electrical Output		MCR	50% MCR
Turbine Fuel Consumption (HCV)	[kW/kWe]	5.62	7.30
Max After Firing Fuel Cons'n (HCV)	[kW/kWe]	3.40	8.83
Exhaust Gas Temperature	[deg C]	485.0	373.5
Exhaust Gas Mass Flow Rate	[kg/kWhe]	27.17	54.33
Net Electrical Efficiency (HCV) of Gas Turbine	[%]	17.79	13.70

Total CHP System Efficiency at MCR (Waste Heat Boiler Flue Gas Temp 150 deg C)		Heat/Power Ratio at MCR
Unfired =	65.4 % (HCV)	2.7
Fired =	71.8 % (HCV)	5.5

Steam Flow from Waste Heat Boiler at MCR (100 psig saturated)

Unfired = 71000 [lbs/hr]

Fired = 146000 [lbs/hr]

Table 10

PLANT	GAS TURBINE	TYPE X

Net Electrical Output (MCR)	400 [kW]
Maximum Afterfiring Temperature	800 [deg C]
Minimum Electrical Output	50 [% of MCR]
Exhaust Gas Mean Specific Heat Capacity - Fired	1.14 [kJ/kgK]
Exhaust Gas Mean Specific Heat Capacity - Unfired	1.08 [kJ/kgK]
Gas Compressor Power Demand	25 [kW]

Comments:
Single Shaft Industrial Gas Turbine

Electrical Output		MCR	50% MCR
Turbine Fuel Consumption (HCV)	[kW/kWe]	8.45	11.49
Max After Firing Fuel Cons'n (HCV)	[kW/kWe]	5.30	14.63
Exhaust Gas Temperature	[deg C]	500.0	370.0
Exhaust Gas Mass Flow Rate	[kg/kWhe]	44.19	89.35
Net Electrical Efficiency (HCV) of Gas Turbine	[%]	11.83	8.70

Total CHP System Efficiency at MCR (Waste Heat Boiler Flue Gas Temp 150 deg C)		Heat/Power Ratio at MCR
Unfired =	65.6 % (HCV)	4.5
Fired =	72.1 % (HCV)	8.9

Steam Flow from Waste Heat Boiler at MCR (100 psig saturated)

Unfired = 5000 [lbs/hr]

Fired = 11000 [lbs/hr]

2.3 Heavy Fuel Oil Diesels

40 The performance data for the heavy fuel oil diesel engines are shown in Tables 11 and 12. Key terms are explained below:-

41 Size Range - In the size range 100 kWe to 10MWe there are two main types of heavy fuel oil diesel engine, these are:

(a) 650kWe to 4MWe - 750 rpm engines - 4 stroke turbocharged

(b) 4MWe to 10MWe - 500 rpm engines - 4 stroke turbocharged

42 Within these divisions the performance of each type of engine was found to vary very little and almost any rating could be achieved by selecting an appropriate number of cylinders for the machine. Tables 11 and 12 indicate the typical performance of engines in each type. As the performance in each size range was independent of engine rating and was fairly consistent over several manufacturers it was not necessary to consider specific machines.

43 After-Firing Temperature - The exhaust gases from diesel engines can be after-fired up to 1200°C without the need for supplementary combustion air. Shell and tube boilers can be used for diesel engines as the exhaust gas flow rates are lower and the engines are not as sensitive to exhaust gas backpressure as gas turbines.

44 Net Electrical Output - The electrical output of the CHP plant is specified at maximum continuous rating after all the losses and auxiliaries have been taken into account.

45 Minimum Electrical Output - A turndown ratio of 2:1 was assumed for all diesel engine plant. If the electrical demand fell below 50% then the machine could be switched off or machine load maintained by exporting electricity.

46 Exhaust Gas Specific Heat Capacity - The mean specific heat capacities for fired and unfired exhaust gases were determined from Fig. 5a.

47 Fuel Consumption and Exhaust Mass Flow Rate - These quantities are expressed per kW of net electrical output

48 Hot Water Recovery - This is expressed as a proportion of the net electrical output from the engine.

50 Net Electrical Efficiency of Diesel Engine This is expressed as:-

$$\frac{\text{Net Electrical Output from Generator}}{\text{Fuel into the Diesel Engine (HCV)}}$$

51 CHP System Efficiency - This representative system efficiency is defined as:-

$$\frac{\text{Net Electrical Output + Heat Recovered}}{\text{Total Fuel into CHP system (HCV) (engine and boiler)}}$$

52 The heat recovered in a waste heat boiler varies with the exhaust gas mass flow rate and flue gas temperature. A flue gas temperature of 220°C was assumed for the purposes of calculating the representative system efficiency. More detailed calculations of boiler performance were undertaken when assessing CHP at a particular site. The flue gas temperature of 220°C is somewhat conservative and it may be possible to achieve lower flue gas temperatures without causing significant problems of acid corrosion in the boiler and flues. This would improve the cost effectiveness of the plant.

53 Heat to Power Ratio - This is defined as:-

$$\frac{\text{Heat Recovered (Steam and Hot Water)}}{\text{Net Electrical Output}}$$

54 Auxiliaries - An allowance of 4% of gross electrical output was made for the engine and boiler auxiliaries associated with the CHP plant. This included:-

- All water and oil cooling pumps
- Air blast coolers for the oil and low temperature water cooling
- Control equipment
- Building Ventilation
- Nominal electrical services in the building
- Waste heat boiler feed pumps
- Fuel supply system

55 Fuel System - The performance was based upon the use of a heavy fuel oil with a viscosity not exceeding 4200 seconds Redwood No. 1 at 100°F. A fuel cleaning system was included to treat the fuel before it was used in the engine.

56 Heat Recovery - Heat was recovered from an exhaust gas waste heat boiler and also as hot water from the jacket cooling water and the high temperature stage of the charge air cooler. It was assumed that the hot water was recovered at 80°C. At part loads

the temperature of the charge air after the turbocharger is very much reduced and in some cases no hot water at 80°C can be

recovered from the first stage of the intercooler.

57 It was assumed that the temperature of the cooling water from the second stage of the intercooler, oil cooler and other systems (i.e. valve cooling) was too low to allow economic heat recovery.

Table 11

PLANT — **HEAVY FUEL OIL DIESEL TYPE I**

Net Electrical Output (MCR) 650-4000 [kW]

Maximum Afterfiring Temperature 1200 [deg C]
Minimum Electrical Output 50 [% of MCR]

Exhaust Gas Mean Specific Heat Capacity - Fired 1.23 [kJ/kgK]
- Unfired 1.07 [kJ/kgK]

Comments:
Speed 750 rpm

Electrical Output (at mid size range)		MCR	50% MCR
Engine Fuel Consumption (HCV)	[kW/kWe]	2.85	3.08
Max After Firing Fuel Cons'n (HCV)	[kW/kWe]	2.73	3.49
Exhaust Gas Temperature	[deg C]	390.0	343.2
Exhaust Gas Mass Flow Rate	[kg/kWhe]	8.48	10.35
Hot Water Recovery (80 deg C)	[kW/kWe]	.46	.51
Net Electrical Efficiency (HCV) of Prime Mover	[%]	35.05	32.45

Total CHP System Efficiency at MCR — **Heat/Power Ratio at MCR**

(Waste Heat Boiler Flue Gas Temp 220 deg C)
(Hot Water Recovery included)

	Efficiency	Heat/Power Ratio at MCR
Unfired =	65.9 % (HCV)	.88
Fired =	76.0 % (HCV)	3.24

Steam Flow from Waste Heat Boiler at MCR (100 psig saturated)

(1MWe Engine)

Unfired = 2000 [lbs/hr]

Fired = 10000 [lbs/hr]

Table 12

PLANT	HEAVY FUEL OIL DIESEL	TYPE II

Net Electrical Output (MCR) 4-10 [MW]

Maximum Afterfiring Temperature 1200 [deg C]
Minimum Electrical Output 50 [% of MCR]

Exhaust Gas Mean Specific Heat Capacity - Fired 1.23 [kJ/kgK]
- Unfired 1.08 [kJ/kgK]

Comments:
Speed 500 rpm

Electrical Output (at mid size range)		MCR	50% MCR
Engine Fuel Consumption (HCV)	[kW/kWe]	2.67	2.72
Max After Firing Fuel Cons'n (HCV)	[kW/kWe]	2.62	2.36
Exhaust Gas Temperature	[deg C]	340.0	384.2
Exhaust Gas Mass Flow Rate	[kg/kWhe]	7.76	7.29
Hot Water Recovery (80 deg C)	[kW/kWe]	.41	.34
Net Electrical Efficiency (HCV) of Prime Mover	[%]	37.47	36.73

Total CHP System Efficiency at MCR		Heat/Power Ratio
(Waste Heat Boiler Flue Gas Temp 220 deg C) (Hot Water Recovery included)		at MCR
Unfired =	63.1 % (HCV)	.68
Fired =	74.8 % (HCV)	2.96

Steam Flow from Waste Heat Boiler at MCR (100 psig saturated)

(5MWe Engine)

Unfired = 10000 [lbs/hr]

Fired = 45000 [lbs/hr]

2.4 Dual - Fuel Diesel Engine

58 The performance data for dual-fuel engines are shown in Tables 13 and 14. The terms used in the tables are similar to those used for heavy fuel oil diesels in Section 2.3. Those terms which are different are explained below:

59 Size Range: In the size range 100 kW to 10 MW there are two types of dual-fuel diesel engine commonly used. These are:

a) 800 kW - 2100 kW - 600 rpm engine, 4 stroke - turbocharged
b) 2000 kW - 6400 kW - 500 rpm engine, 4 stroke - turbocharged

60 Within these divisions the differences in performance of each type of engine were found to be small and almost any rating could be achieved by selecting an appropriate number of cylinders for the machine. Tables 13 and 14 indicate the performance typical of engines in each type. As the performance in each size range was largely independent of engine rating it was not necessary to consider specific machines.

61 Fuel System - The fuel was natural gas with a light gas oil pilot fuel. The pilot fuel consumption is in the order of 5-10% of the total fuel input. The natural gas is compressed and introduced into the charge air after the turbocharger. At times of interruption of the gas supply the engine runs on 100% oil.

62 After-firing Temperature - The exhaust gases from a dual-fuel engine can be after fired up to 1200°C without the need for supplementary combustion air. Shell and tube boilers are suitable for the lower exhaust gas flow rates from dual-fuel engines. These engines are not as sensitive to exhaust back pressure as gas turbines.

63 Exhaust Gas Specific Heat Capacity - The mean specific heat capacities for fired and unfired exhaust gases were determined from Fig. 5.

64 CHP System Efficiency - In the use of dual-fuel engines a flue gas temperature of 150°C from the waste heat boiler was assumed for the purposes of calculating the representative CHP system efficiency. This is a rather conservative temperature and lower temperatures may be achieved in practice.

65 Auxiliaries - An allowance of 4% of gross electrical output

was made for the engine and boiler auxiliaries associated with the CHP plant. This included:-

- All water and oil cooling pumps
- Air blast coolers for oil and low temperature water cooling control equipment
- Building ventilation
- Nominal electrical services
- Waste heat boiler gas blower and water feed pumps
- A further allowance of 2% of gross electrical output was made for natural gas compression for the engine.

66 Heat Recovery - A similar heat recovery system to heavy fuel oil diesel engines was assumed.

Table 13

PLANT	DUAL-FUEL DIESEL	TYPE I

Net Electrical Output (MCR) 650-2000 [kW]

Maximum Afterfiring Temperature 1200 [deg C]
Minimum Electrical Output 50 [% of MCR]

Exhaust Gas Mean Specific Heat Capacity - Fired 1.26 [kJ/kgK]
- Unfired 1.11 [kJ/kgK]

Comments:
Speed 600 rpm

Electrical Output (at mid size range)		MCR	50% MCR
Engine Fuel Consumption (HCV)	[kW/kWe]	3.14	4.14
Pilot Fuel Consumption (HCV)	[kW/kWe]	.31	.62
Max After Firing Fuel Cons'n (HCV)	[kW/kWe]	2.69	4.28
Exhaust Gas Temperature	[deg C]	425.0	340.0
Exhaust Gas Mass Flow Rate	[kg/kWhe]	8.17	11.93
Hot Water Recovery (80 deg C)	[kW/kWe]	.42	.45
Net Electrical Efficiency (HCV) of Prime Mover	[%]	28.99	21.01

Total CHP System Efficiency at MCR			Heat/Power Ratio
(Waste Heat Boiler Flue Gas Temp 150 deg C) (Hot Water Recovery included)			at MCR
Unfired	=	60.8 % (HCV)	1.10
Fired	=	71.0 % (HCV)	3.36

Steam Flow from Waste Heat Boiler at MCR (100 psig saturated)

(1MWe Engine)

Unfired = 3000 [lbs/hr]

Fired = 10000 [lbs/hr]

Table 14

PLANT **DUAL-FUEL DIESEL** **TYPE II**

Net Electrical Output (MCR) 2000-6400 [kW]

Maximum Afterfiring Temperature 1200 [deg C]
Minimum Electrical Output 50 [% of MCR]

Exhaust Gas Mean Specific Heat Capacity - Fired 1.26 [kJ/kgK]
- Unfired 1.11 [kJ/kgK]

Comments:
Speed 500 rpm

Electrical Output (at mid size range)		MCR	50% MCR
Engine Fuel Consumption (HCV)	[kW/kWe]	2.83	3.00
Pilot Fuel Consumption (HCV)	[kW/kWe]	.18	.36
Max After Firing Fuel Cons'n (HCV)	[kW/kWe]	2.31	5.17
Exhaust Gas Temperature	[deg C]	450.0	490.0
Exhaust Gas Mass Flow Rate	[kg/kWhe]	7.20	16.85
Hot Water Recovery (80 deg C)	[kW/kWe]	.43	.39
Net Electrical Efficiency (HCV) of Prime Mover	[%]	33.17	29.71

Total CHP System Efficiency at MCR (Waste Heat Boiler Flue Gas Temp 150 deg C) (Hot Water Recovery included)		Heat/Power Ratio at MCR
Unfired =	69.1 % (HCV)	1.08
Fired =	75.6 % (HCV)	3.02

Steam Flow from Waste Heat Boiler at MCR (100 psig saturated)

(5MWe Engine)

Unfired = 15000 [lbs/hr]

Fired = 45000 [lbs/hr]

2.5 Industrial Spark Ignition Gas Engines

67 The performance data for spark ignition gas engines are shown in Tables 15 and 16. The terms used in the tables are similar to those used for heavy fuel oil diesels and dual-fuel engines. Those terms which differ are explained below.

68 Size Range - Two ranges of spark ignition gas engine were considered:

a) 100 kW - 550 kW - 4 stroke engines 1500 rpm
b) 500 kW - 2000 kW - 2 stroke engines 1000 rpm

Within these divisions the performance of each type of engine was found not to vary significantly with engine size. Almost any rating could be achieved by selecting a machine with a suitable number of cylinders from a particular manufacturer.

69 Only industrial spark ignition engines in the size range 100 kWe to 2 MWe were considered. The small packaged CHP plants, often based on vehicle-derived engines, were not included.

70 After-firing - As spark ignition engines operate with an air fuel ratio close to stoichiometric it is not possible to consider after-firing the exhaust gases.

71 Exhaust Gas Specific Heat Capacity - The mean specific heat capacities for unfired exhaust gases were determined from Fig. 5.

72 CHP System Efficiency - for the purpose of calculating a representative system efficiency a flue gas temperature from the waste heat boiler of 150°C was assumed. This is a rather conservative assumption and lower temperatures can be achieved in some applications.

73 Auxiliaries - An allowance of 2% of gross electrical output was made for the engine and boiler auxiliaries associated with the CHP Plant. These included:

- All water and oil cooling pumps
- Air blast coolers for oil cooling
- Control equipment
- Building ventilation
- Heat recovery auxiliaries

74 In the case of the turbocharged 2-stroke engines the gas has to be supplied to the engine atevated pressure and a further 2% of electrical output was allowed for a gas compressor.

75 Fuel System Natural gas fuel was used for the spark ignition engines.

76 Heat Recovery - Heat was recovered from the exhaust gas waste heat boiler as steam or hot water. In addition hot water at 80°C was recovered from the jacket and oil cooling system. Heat from the charge cooler was not recovered.

Table 15

PLANT — **SPARK IGNITION GAS ENGINE** — **TYPE I**

Net Electrical Output Range (MCR) 100-560 [kW]

Minimum Electrical Output 50 [% of MCR]

Exhaust Gas Mean Specific Heat Capacity - 1.18 [kJ/kgK]

Comments:
4-Stroke Spark Ignition Engine

Electrical Output (at mid size range)		MCR	50% MCR
Fuel Consumption (HCV)	[kW/kWe]	3.49	4.12
Exhaust Gas Temperature	[deg C]	600.0	540.0
Exhaust Gas Mass Flow Rate	[kg/kWhe]	5.62	5.62
Hot Water Recovery (80 deg C)	[kW/kWe]	.90	1.44
Net Electrical Efficiency (HCV) of Engine	[%]	28.67	24.30

Total CHP System Efficiency at MCR	Heat/Power Ratio
(Waste Heat Boiler Flue Gas Temp 150 deg C) (Hot Water Recovery included)	at MCR
77.8 % (HCV)	1.71

Steam Flow from Waste Heat Boiler at MCR (100 psig saturated)

(500 kWe Engine)

1000 [lbs/hr]

Table 16

PLANT	SPARK IGNITION GAS ENGINE	TYPE II

Net Electrical Output Range (MCR) 500-2000 [kW]

Minimum Electrical Output 50 [% of MCR]

Exhaust Gas Mean Specific Heat Capacity - 1.10 [kJ/kgK]

Comments:
2-Stroke Spark Ignition Engine

Electrical Output (at mid size range)		MCR	50% MCR
Fuel Consumption (HCV)	[kW/kWe]	3.52	4.57
Exhaust Gas Temperature	[deg C]	400.0	392.0
Exhaust Gas Mass Flow Rate	[kg/kWhe]	9.73	11.67
Hot Water Recovery (80 deg C)	[kW/kWe]	.79	1.11
Net Electrical Efficiency (HCV) of Engine	[%]	28.42	21.87

Total CHP System Efficiency at MCR	Heat/Power Ratio
(Waste Heat Boiler Flue Gas Temp 150 deg C) (Hot Water Recovery included)	at MCR
71.6 % (HCV)	1.52

Steam Flow from Waste Heat Boiler at MCR (100 psig saturated)

(1000 kWe Engine)

2000 [lbs/hr]

2.6 High Pressure Coal Fired Boilers and Steam Turbines

77 The performance data for steam turbines are shown in Tables 17 to 35. The terms used in the tables are explained more fully below:-

78 Type of Steam Turbine - Single stage and multi-stage steam turbines operating in the back-pressure mode only were considered. Three types of boiler were considered with outlet conditions of:

i) 450 psig/750°F (32 bar abs/400°C)
ii) 650 psig/850°F (46 bar abs/454°C)
iii) 900 psig/900°F (63 bar abs/482°C)

79 Back-pressures of 250, 150 100, 80, 30 psig were considered. When analysing any particular site a back-pressure corresponding to the site steam distribution pressure was chosen.

80 Size ranges - single stage back-pressure steam turbines are only suitable for outputs up to 2MW. They are also more suitable for steam pressure ratios of less than 3:1. For pressure ratios greater than 3:1 and for outputs greater than 2 MW multistage steam turbines are the more cost effective. Multistage turbines with back-pressures above 11 bar absolute were not considered as they are not generally available for such pressures. In all, 19 different combinations of back-pressure and turbine type were considered.

81 Minimum Turbine Output - Steam turbines can be operated down to zero shaft output. However the efficiency is low at outputs below about 10% of MCR. A net electrical output turndown ratio of 10:1 was chosen for all steam turbine plant. If the electrical demand fell below 10% of MCR then the steam turbine could be switched off or turbine load maintained by exporting electricity.

82 Turbine Efficiency - This is defined as:-

$$\frac{\text{Net Electrical Output}}{\text{Gross Shaft Output}}$$

The net electrical output includes allowances for the efficiency of the electrical generator, the turbine and gearbox mechanical losses, the turbine auxiliaries and the boiler auxiliaries. A steam turbine with a relatively high specific steam mass flow rate has a reduced efficiency since the boiler auxiliaries become a relatively larger proportion of the gross electrical output.

83 Useful Heat in the Turbine Exhaust - This is defined as the enthalpy of the turbine exhaust steam relative to saturated water at 50 C. In some cases the steam is superheated at the turbine exhaust and de-superheating may be required. This would not change the heat output from the turbine exhaust. At part loads the specific enthalpy of the exhaust steam increased as the overall isentropic efficiency of the turbine decreased.

84 Fuel Consumption - The fuel consumption represents the coal burned in the boiler used to raise steam in order to produce the electrical output from the turbines. A boiler feedwater temperature of 50°C was assumed (This corresponds to 50% condensate return at 80°C with make up water at 20°C.) This is a somewhat conservative assumption as over 95% condensate return can be achieved with a good system.

85 A boiler system efficiency of 78.5% was assumed for modern high pressure boilers.

86 Steam Mass Flow Rate - this is the steam mass required to generate 1 kilowatt hour of electricity. The decrease of efficiency at low loads is indicated by the increase in the specific steam mass flow rate.

87 Net Electrical Efficiency - This is defined as:-

$$\frac{\text{Net Electrical Output}}{\text{Fuel Consumption (HCV)}}$$

88 CHP System Efficiency - This is defined as:-

$$\frac{\text{Net Electrical Output and Useful Heat in Turbine Exhaust}}{\text{Fuel Consumption (HCV)}}$$

The CHP system efficiency is always lower than the boiler efficiency, the difference represents the losses in the turbine, the generator and the auxiliaries.

89 Heat/Power Ratio - This is defined as:

$$\frac{\text{Useful heat in Turbine Exhaust}}{\text{Net Electrical Output}}$$

90 Turbine Performance - The isentropic turbine efficiency at full load and the no load steam consumption were determined from manufacturers data.

91 Auxiliaries and Losses - The auxiliaries and losses at MCR were assumed to be:-

- Generator Loss - 5% of Turbine shaft power
- Turbine Auxiliaries and Losses
  - 5% of Turbine shaft power
- Gearbox Loss - 2% of Turbine Shaft Power
- Boiler Auxiliaries - 20kW per kg/s of steam flow. These are additional auxiliaries needed to provide the high pressure feedwater and water treatment.

92 Part Load - It was assumed that the total auxiliaries and losses at zero electrical output were 75% of those at MCR. The minimum electrical load of the turbine was assumed to be 10% of MCR.

93 System Pressure Loss - A pressure loss of 10% was assumed between the boiler and the turbine governor valve.

Table 17

PLANT	STEAM TURBINE	TYPE I

Net Electrical Output Range (MCR) 100-1000 [kW]

Boiler Outlet Conditions	32 [bar abs]	450 [psig]
	400 [deg C]	750 [deg F]
Steam Turbine Back Pressure	18.2 [bar abs]	250 [psig]
Minimum Turbine Output [% of MCR]	10 [%]	
Turbine Efficiency (mech + elec + allowance for boiler auxillaries)	67.6 [%]	
Useful Heat in Turbine Exhaust (at MCR)	2929 [kJ/kg]	

Comments:
Single Stage Steam Turbine

Electrical Output (at 1 MWe)		MCR	50% MCR
Fuel Consumption (HCV)	[kW/kWe]	59.3	87.2
Steam Mass Flow Rate	[kg/kWhe]	55.4	81.4
Net Electrical Efficiency (HCV) of Steam Turbine + Boiler	[%]	1.7	1.1

Total CHP System Efficiency at MCR	Heat/Power Ratio
77.7 % (HCV)	45.1 (at MCR)

Steam Flow For 1 MWe Turbine (at MCR) 122000 [lbs/hr]

Table 18

PLANT	STEAM TURBINE	TYPE II

Net Electrical Output Range (MCR) 100-1700 [kW]

Boiler Outlet Conditions	32 [bar abs] 400 [deg C]	450 [psig] 750 [deg F]
Steam Turbine Back Pressure	11.3 [bar abs]	150 [psig]
Minimum Turbine Output [% of MCR]	10 [%]	
Turbine Efficiency (mech + elec + allowance for boiler auxillaries)	75.0 [%]	
Useful Heat in Turbine Exhaust (at MCR)	2869 [kJ/kg]	

Comments:
Single Stage Steam Turbine

Electrical Output (at 1 MWe)		MCR	50% MCR
Fuel Consumption (HCV)	[kW/kWe]	32.9	44.7
Steam Mass Flow Rate	[kg/kWhe]	30.7	41.8
Net Electrical Efficiency (HCV) of Steam Turbine + Boiler	[%]	3.0	2.2

Total CHP System Efficiency at MCR	Heat/Power Ratio
77.5 % (HCV)	24.5 (at MCR)

Steam Flow For 1 MWe Turbine (at MCR) 67000 [lbs/hr]

Table 19

PLANT	STEAM TURBINE	TYPE III	
Net Electrical Output Range (MCR)	100-1600 [kW]		
Boiler Outlet Conditions	46 [bar abs] 454 [deg C]	650 [psig] 850 [deg F]	
Steam Turbine Back Pressure	18.2 [bar abs]	250 [psig]	
Minimum Turbine Output [% of MCR]	10 [%]		
Turbine Efficiency (mech + elec + allowance for boiler auxillaries)	74.0 [%]		
Useful Heat in Turbine Exhaust (at MCR)	2984 [kJ/kg]		

Comments:
Single Stage Steam Turbine

Electrical Output (at 1 MWe)		MCR	50% MCR
Fuel Consumption (HCV)	[kW/kWe]	36.8	50.4
Steam Mass Flow Rate	[kg/kWhe]	33.2	45.5
Net Electrical Efficiency (HCV) of Steam Turbine + Boiler	[%]	2.7	2.0

Total CHP System Efficiency at MCR	Heat/Power Ratio
77.6 % (HCV)	27.5 (at MCR)

Steam Flow For 1 MWe Turbine (at MCR) 73000 [lbs/hr]

Table 20

PLANT	STEAM TURBINE	TYPE IV
Net Electrical Output Range (MCR)	100-2200 [kW]	
Boiler Outlet Conditions	64 [bar abs] 482 [deg C]	900 [psig] 900 [deg F]
Steam Turbine Back Pressure	18.2 [bar abs]	250 [psig]
Minimum Turbine Output [% of MCR]	10 [%]	
Turbine Efficiency (mech + elec + allowance for boiler auxillaries)	77.8 [%]	
Useful Heat in Turbine Exhaust (at MCR)	2976 [kJ/kg]	

Comments:
Single Stage Steam Turbine

Electrical Output (at 1 MWe)		MCR	50% MCR
Fuel Consumption (HCV)	[kW/kWe]	26.7	35.8
Steam Mass Flow Rate	[kg/kWhe]	23.8	31.9
Net Electrical Efficiency (HCV) of Steam Turbine + Boiler	[%]	3.7	2.8

Total CHP System Efficiency at MCR	Heat/Power Ratio
77.5 % (HCV)	19.7 (at MCR)

Steam Flow For 1 MWe Turbine (at MCR) 58000 [lbs/hr]

Table 21

PLANT	STEAM TURBINE	TYPE V

Net Electrical Output Range (MCR) 600-4000 [kW]

Boiler Outlet Conditions	32 [bar abs]	450 [psig]
	400 [deg C]	750 [deg F]
Steam Turbine Back Pressure	11.3 [bar abs]	150 [psig]
Minimum Turbine Output [% of MCR]	10 [%]	
Turbine Efficiency (mech + elec + allowance for boiler auxillaries)	78.0 [%]	
Useful Heat in Turbine Exhaust (at MCR)	2869 [kJ/kg]	

Comments:
Multi Stage Steam Turbine

Electrical Output (at 1 MWe)		MCR	50% MCR
Fuel Consumption (HCV)	[kW/kWe]	24.1	32.3
Steam Mass Flow Rate	[kg/kWhe]	22.5	30.2
Net Electrical Efficiency (HCV) of Steam Turbine + Boiler	[%]	4.2	3.1

Total CHP System Efficiency at MCR	Heat/Power Ratio
78.6 % (HCV)	17.9 (at MCR)

Steam Flow For 1 MWe Turbine (at MCR) 49000 [lbs/hr]

Table 22

PLANT	STEAM TURBINE	TYPE VI

Net Electrical Output Range (MCR) 600-5500 [kW]

Boiler Outlet Conditions	32 [bar abs]	450 [psig]
	400 [deg C]	750 [deg F]
Steam Turbine Back Pressure	7.9 [bar abs]	100 [psig]
Minimum Turbine Output [% of MCR]	10 [%]	
Turbine Efficiency (mech + elec + allowance for boiler auxillaries)	80.9 [%]	
Useful Heat in Turbine Exhaust (at MCR)	2753 [kJ/kg]	

Comments:
Multi Stage Steam Turbine

Electrical Output (at 1 MWe)		MCR	50% MCR
Fuel Consumption (HCV)	[kW/kWe]	17.5	22.8
Steam Mass Flow Rate	[kg/kWhe]	16.4	21.3
Net Electrical Efficiency (HCV) of Steam Turbine + Boiler	[%]	5.7	4.4

Total CHP System Efficiency at MCR	Heat/Power Ratio
77.2 % (HCV)	12.5 (at MCR)

Steam Flow For 1 MWe Turbine (at MCR) 36000 [lbs/hr]

Table 23

PLANT	STEAM TURBINE	TYPE VII

Net Electrical Output Range (MCR)	600-6100 [kW]	
Boiler Outlet Conditions	32 [bar abs]	450 [psig]
	400 [deg C]	750 [deg F]
Steam Turbine Back Pressure	6.5 [bar abs]	80 [psig]
Minimum Turbine Output [% of MCR]	10 [%]	
Turbine Efficiency (mech + elec + allowance for boiler auxillaries)	81.3 [%]	
Useful Heat in Turbine Exhaust (at MCR)	2725 [kJ/kg]	

Comments:
Multi Stage Steam Turbine

Electrical Output (at 1 MWe)		MCR	50% MCR
Fuel Consumption (HCV)	[kW/kWe]	15.8	20.1
Steam Mass Flow Rate	[kg/kWhe]	14.8	18.7
Net Electrical Efficiency (HCV) of Steam Turbine + Boiler	[%]	6.3	5.0

Total CHP System Efficiency at MCR	Heat/Power Ratio
77.1 % (HCV)	11.2 (at MCR)

Steam Flow For 1 MWe Turbine (at MCR) 32000 [lbs/hr]

Table 24

PLANT	STEAM TURBINE	TYPE VIII

Net Electrical Output Range (MCR) 600-7100 [kW]

Boiler Outlet Conditions	32 [bar abs]	450 [psig]
	400 [deg C]	750 [deg F]
Steam Turbine Back Pressure	5.1 [bar abs]	60 [psig]
Minimum Turbine Output [% of MCR]	10 [%]	
Turbine Efficiency (mech + elec + allowance for boiler auxillaries)	82.3 [%]	
Useful Heat in Turbine Exhaust (at MCR)	2681 [kJ/kg]	

Comments:
Multi Stage Steam Turbine

Electrical Output (at 1 MWe)		MCR	50% MCR
Fuel Consumption (HCV)	[kW/kWe]	13.58	17.11
Steam Mass Flow Rate	[kg/kWhe]	12.68	15.98
Net Electrical Efficiency (HCV) of Steam Turbine + Boiler	[%]	7.37	5.84

Total CHP System Efficiency at MCR	Heat/Power Ratio
76.9 % (HCV)	9.44 (at MCR)

Steam Flow For 1 MWe Turbine (at MCR) 27000 [lbs/hr]

Table 25

PLANT	STEAM TURBINE	TYPE IX

Net Electrical Output Range (MCR) 600-8980 [kW]

Boiler Outlet Conditions	32 [bar abs]	450 [psig]
	400 [deg C]	750 [deg F]
Steam Turbine Back Pressure	3.0 [bar abs]	30 [psig]
Minimum Turbine Output [% of MCR]	10 [%]	
Turbine Efficiency (mech + elec + allowance for boiler auxillaries)	83.3 [%]	
Useful Heat in Turbine Exhaust (at MCR)	2594 [kJ/kg]	

Comments:
Multi Stage Steam Turbine

Electrical Output (at 1 MWe)		MCR	50% MCR
Fuel Consumption (HCV)	[kW/kWe]	10.73	13.20
Steam Mass Flow Rate	[kg/kWhe]	10.03	12.33
Net Electrical Efficiency (HCV) of Steam Turbine + Boiler	[%]	9.32	7.58

Total CHP System Efficiency at MCR	Heat/Power Ratio
76.6 % (HCV)	7.22 (at MCR)

Steam Flow For 1 MWe Turbine (at MCR) 22000 [lbs/hr]

Table 26

PLANT	STEAM TURBINE	TYPE X

Net Electrical Output Range (MCR) 600-5920 [kW]

Boiler Outlet Conditions	46 [bar abs]	650 [psig]
	454 [deg C]	850 [deg F]
Steam Turbine Back Pressure	11.3 [bar abs]	150 [psig]
Minimum Turbine Output [% of MCR]	10 [%]	
Turbine Efficiency (mech + elec + allowance for boiler auxillaries)	81.2 [%]	
Useful Heat in Turbine Exhaust (at MCR)	2838 [kJ/kg]	

Comments:
Multi Stage Steam Turbine

Electrical Output (at 1 MWe)		MCR	50% MCR
Fuel Consumption (HCV)	[kW/kWe]	16.82	21.87
Steam Mass Flow Rate	[kg/kWhe]	15.19	19.75
Net Electrical Efficiency (HCV) of Steam Turbine + Boiler	[%]	5.95	4.57

Total CHP System Efficiency at MCR	Heat/Power Ratio
77.2 % (HCV)	12.0 (at MCR)

Steam Flow For 1 MWe Turbine (at MCR) 33000 [lbs/hr]

Table 27

PLANT	STEAM TURBINE	TYPE XI

Net Electrical Output Range (MCR) 600-7430 [kW]

Boiler Outlet Conditions	46 [bar abs]	650 [psig]
	454 [deg C]	850 [deg F]
Steam Turbine Back Pressure	7.9 [bar abs]	100 [psig]
Minimum Turbine Output [% of MCR]	10 [%]	
Turbine Efficiency (mech + elec + allowance for boiler auxillaries)	82.5 [%]	
Useful Heat in Turbine Exhaust (at MCR)	2770 [kJ/kg]	

Comments:
Multi Stage Steam Turbine

Electrical Output (at 1 MWe)		MCR	50% MCR
Fuel Consumption (HCV)	[kW/kWe]	13.42	16.91
Steam Mass Flow Rate	[kg/kWhe]	12.12	15.27
Net Electrical Efficiency (HCV) of Steam Turbine + Boiler	[%]	7.45	5.91

Total CHP System Efficiency at MCR	Heat/Power Ratio
77.0 % (HCV)	9.33 (at MCR)

Steam Flow For 1 MWe Turbine (at MCR) 26000 [lbs/hr]

Table 28

PLANT	STEAM TURBINE	TYPE XII	
Net Electrical Output Range (MCR)	600-8130 [kW]		
Boiler Outlet Conditions		46 [bar abs]	650 [psig]
		454 [deg C]	850 [deg F]
Steam Turbine Back Pressure		6.5 [bar abs]	80 [psig]
Minimum Turbine Output [% of MCR]		10 [%]	
Turbine Efficiency (mech + elec + allowance for boiler auxillaries)		82.9 [%]	
Useful Heat in Turbine Exhaust (at MCR)		2738 [kJ/kg]	

Comments:
Multi Stage Steam Turbine

Electrical Output (at 1 MWe)		MCR	50% MCR
Fuel Consumption (HCV)	[kW/kWe]	12.26	15.33
Steam Mass Flow Rate	[kg/kWhe]	11.07	13.84
Net Electrical Efficiency (HCV) of Steam Turbine + Boiler	[%]	8.16	6.53

Total CHP System Efficiency at MCR	Heat/Power Ratio
76.8 % (HCV)	8.42 (at MCR)

Steam Flow For 1 MWe Turbine (at MCR) 24000 [lbs/hr]

Table 29

PLANT	STEAM TURBINE	TYPE XIII

Net Electrical Output Range (MCR) 600-9150 [kW]

Boiler Outlet Conditions	46 [bar abs]	650 [psig]
	454 [deg C]	850 [deg F]
Steam Turbine Back Pressure	5.1 [bar abs]	60 [psig]
Minimum Turbine Output [% of MCR]	10 [%]	
Turbine Efficiency (mech + elec + allowance for boiler auxillaries)	83.5 [%]	
Useful Heat in Turbine Exhaust (at MCR)	2693 [kJ/kg]	

Comments:
Multi Stage Steam Turbine

Electrical Output (at 1 MWe)		MCR	50% MCR
Fuel Consumption (HCV)	[kW/kWe]	10.92	13.54
Steam Mass Flow Rate	[kg/kWhe]	9.86	12.23
Net Electrical Efficiency (HCV) of Steam Turbine + Boiler	[%]	9.16	7.39

Total CHP System Efficiency at MCR	Heat/Power Ratio
76.7 % (HCV)	7.38 (at MCR)

Steam Flow For 1 MWe Turbine (at MCR) 21000 [lbs/hr]

Table 30

PLANT	STEAM TURBINE	TYPE XIV

Net Electrical Output Range (MCR) 600-10000 [kW]

Boiler Outlet Conditions	46 [bar abs]	650 [psig]
	454 [deg C]	850 [deg F]
Steam Turbine Back Pressure	3.0 [bar abs]	30 [psig]
Minimum Turbine Output [% of MCR]	10 [%]	
Turbine Efficiency (mech + elec + allowance for boiler auxillaries)	84.2 [%]	
Useful Heat in Turbine Exhaust (at MCR)	2605 [kJ/kg]	

Comments:
Multi Stage Steam Turbine

Electrical Output (at 1 MWe)		MCR	50% MCR
Fuel Consumption (HCV)	[kW/kWe]	9.02	11.00
Steam Mass Flow Rate	[kg/kWhe]	8.15	9.94
Net Electrical Efficiency (HCV) of Steam Turbine + Boiler	[%]	11.09	9.09

Total CHP System Efficiency at MCR	Heat/Power Ratio
76.4 % (HCV)	5.90 (at MCR)

Steam Flow For 1 MWe Turbine (at MCR) 18000 [lbs/hr]

Table 31

PLANT	STEAM TURBINE	TYPE XV	
Net Electrical Output Range (MCR)	600-7490 [kW]		
Boiler Outlet Conditions		64 [bar abs] 482 [deg C]	900 [psig] 900 [deg F]
Steam Turbine Back Pressure		11.3 [bar abs]	150 [psig]
Minimum Turbine Output [% of MCR]		10 [%]	
Turbine Efficiency (mech + elec + allowance for boiler auxillaries)		82.4 [%]	
Useful Heat in Turbine Exhaust (at MCR)		2807 [kJ/kg]	

Comments:
Multi Stage Steam Turbine

Electrical Output (at 1 MWe)		MCR	50% MCR
Fuel Consumption (HCV)	[kW/kWe]	13.50	17.01
Steam Mass Flow Rate	[kg/kWhe]	12.02	15.15
Net Electrical Efficiency (HCV) of Steam Turbine + Boiler	[%]	7.41	5.88

Total CHP System Efficiency at MCR	Heat/Power Ratio
76.9 % (HCV)	9.38 (at MCR)

Steam Flow For 1 MWe Turbine (at MCR) 26000 [lbs/hr]

Table 32

PLANT	STEAM TURBINE	TYPE XVI

Net Electrical Output Range (MCR) 600-8830 [kW]

Boiler Outlet Conditions	64 [bar abs]	900 [psig]
	482 [deg C]	900 [deg F]
Steam Turbine Back Pressure	7.9 [bar abs]	100 [psig]
Minimum Turbine Output [% of MCR]	10 [%]	
Turbine Efficiency (mech + elec + allowance for boiler auxillaries)	83.3 [%]	
Useful Heat in Turbine Exhaust (at MCR)	2746 [kJ/kg]	

Comments:
Multi Stage Steam Turbine

Electrical Output (at 1 MWe)		MCR	50% MCR
Fuel Consumption (HCV)	[kW/kWe]	11.44	14.19
Steam Mass Flow Rate	[kg/kWhe]	10.19	12.63
Net Electrical Efficiency (HCV) of Steam Turbine + Boiler	[%]	8.74	7.05

Total CHP System Efficiency at MCR	Heat/Power Ratio
76.7 % (HCV)	7.77 (at MCR)

Steam Flow For 1 MWe Turbine (at MCR) 22000 [lbs/hr]

Table 33

PLANT	STEAM TURBINE	TYPE XVII

Net Electrical Output Range (MCR) 600-8830 [kW]

Boiler Outlet Conditions	64 [bar abs]	900 [psig]
	482 [deg C]	900 [deg F]
Steam Turbine Back Pressure	6.5 [bar abs]	80 [psig]
Minimum Turbine Output [% of MCR]	10 [%]	
Turbine Efficiency (mech + elec + allowance for boiler auxillaries)	83.6 [%]	
Useful Heat in Turbine Exhaust (at MCR)	2706 [kJ/kg]	

Comments:
Multi Stage Steam Turbine

Electrical Output (at 1 MWe)		MCR	50% MCR
Fuel Consumption (HCV)	[kW/kWe]	10.40	12.84
Steam Mass Flow Rate	[kg/kWhe]	9.29	11.47
Net Electrical Efficiency (HCV) of Steam Turbine + Boiler	[%]	9.62	7.79

Total CHP System Efficiency at MCR	Heat/Power Ratio
76.7 % (HCV)	6.98 (at MCR)

Steam Flow For 1 MWe Turbine (at MCR) 20000 [lbs/hr]

Table 34

PLANT	STEAM TURBINE	TYPE XVIII

Net Electrical Output Range (MCR) 600-10000 [kW]

Boiler Outlet Conditions	64 [bar abs]	900 [psig]
	482 [deg C]	900 [deg F]
Steam Turbine Back Pressure	5.1 [bar abs]	60 [psig]
Minimum Turbine Output [% of MCR]	10 [%]	
Turbine Efficiency (mech + elec + allowance for boiler auxillaries)	84.1 [%]	
Useful Heat in Turbine Exhaust (at MCR)	2668 [kJ/kg]	

Comments:
Multi Stage Steam Turbine

Electrical Output (at 1 MWe)		MCR	50% MCR
Fuel Consumption (HCV)	[kW/kWe]	9.56	11.76
Steam Mass Flow Rate	[kg/kWhe]	8.53	10.49
Net Electrical Efficiency (HCV) of Steam Turbine + Boiler	[%]	10.46	8.50

Total CHP System Efficiency at MCR	Heat/Power Ratio
76.6 % (HCV)	6.32 (at MCR)

Steam Flow For 1 MWe Turbine (at MCR) 18000 [lbs/hr]

Table 35

PLANT	STEAM TURBINE	TYPE XIX

Net Electrical Output Range (MCR) 600-10000 [kW]

Boiler Outlet Conditions	64 [bar abs]	900 [psig]
	482 [deg C]	900 [deg F]
Steam Turbine Back Pressure	3.0 [bar abs]	30 [psig]
Minimum Turbine Output [% of MCR]	10 [%]	
Turbine Efficiency (mech + elec + allowance for boiler auxillaries)	84.6 [%]	
Useful Heat in Turbine Exhaust (at MCR)	2580 [kJ/kg]	

Comments:
Multi Stage Steam Turbine

Electrical Output (at 1 MWe)		MCR	50% MCR
Fuel Consumption (HCV)	[kW/kWe]	8.09	9.79
Steam Mass Flow Rate	[kg/kWhe]	7.20	8.71
Net Electrical Efficiency (HCV) of Steam Turbine + Boiler	[%]	12.36	10.22

Total CHP System Efficiency at MCR	Heat/Power Ratio
76.2 % (HCV)	5.16 (at MCR)

Steam Flow For 1 MWe Turbine (at MCR) 16000 [lbs/hr]

2.7 Combined Cycle Plant

94 The combined cycle plant analysed for this study was a combination of a gas turbine generator with an after-fired waste heat boiler and back pressure steam turbine. The exhaust steam from the steam turbine provided the source of process heat. A diagram of a combined cycle plant is shown in Fig. 4.

95 Size Range - Four separate combined cycle plants were considered; two based on aero-derived gas turbines, rated at about 3MWe and 5MWe, and two based on industrial gas turbines also rated at about 3 MWe and 5MWe. In each case steam was generated from waste heat boilers after-fired to 800°C.

96 The steam outlet conditions from the waste heat boilers were 900°F/900 psig (482°C/63 bars abs) in each case and turbine back pressures of 150, 100 and 80 psig were considered. In each case a multi-stage steam turbine of size appropriate to the steam conditions was selected. In total 12 separate combined cycle plants were considered.

97 Only high pressure (900 psig) waste heat boilers were considered. The use of lower pressure boilers reduces the electrical output from the steam turbine and this was found to reduce the cost savings. The payback period also increased.

98 The gas turbine and steam turbines used in each case have been described in Sections 2.2 and 2.6 above.

99 Maximum Continuous Rating - The maximum electrical and heat output from a combined cycle plant occurs when the gas turbine is operating at full load with full after-firing. At reduced after-firing the total electrical output and the heat output are both reduced. Thus the electrical output depends on both the output of the gas turbine and the degree of after-firing. In view of the more complex behaviour of combined cycle plant, Tables 36-47 show only the performance at MCR. The performances at other conditions are shown in Figs 7 - 18.

100 Net Electrical Output - This corresponds to the gas turbine net electrical output plus the steam turbine net electrical output.

101 Useful Heat Output - This represents the useful heat in the exhaust steam from the steam turbine. It is expressed per unit of total electrical output from the whole plant. It is based on the enthalpy of the steam relative to saturated water at 50 ^{o}C.

102 Fuel Consumption - The fuel consumptions for the gas turbine and the after-burner are expressed separately per unit of total CHP plant net electrical output.

103 Net Electrical Efficiency This is defined as:

$$\frac{\text{Total Electrical Output}}{\text{Total fuel input (gas turbine and after-firing)}}$$

104 CHP System Efficiency - This is defined as:

$$\frac{\text{Total Electrical Output + Heat Output}}{\text{Total Fuel Input}}$$

105 A design flue gas temperature from the waste heat boiler of 150°C was assumed at MCR. At part loads the variation of flue gas temperatures was calculated based on the variation in exhaust gas flow rate and temperature.

106 Heat/Power ratio - This is defined as:

$$\frac{\text{Total Heat Output}}{\text{Total Electrical Output}}$$

107 Auxiliaries - The auxiliaries described previously for gas turbines and steam turbines were also used for combined cycle plant.

108 Part Load - Part Load performance is shown in Figs 7-18.

109 Fuel - Natural gas fuel was used for both the gas turbine and the after-fired boiler. Other fuels could be used for the after-burner and these would naturally affect the economics.

110 Minimum After firing - As high pressure waste heat boilers were used a minimum after-firing level of 550 C was specified otherwise the required superheat could not be achieved at part load conditions.

Table 36

PLANT	COMBINED CYCLE PLANT	TYPE I

Net Electrical Output (MCR)	4820 [kW]	
Gas Turbine Output (MCR)	2970 [kW]	(Plant Type III)
Steam Turbine Output (MCR)	1850 [kW]	(Plant Type XVII)
Waste Heat Boiler Outlet Conditions	64.0 [Bar abs]	900 [psig]
	482.0 [deg C]	900 [deg F]
Steam Turbine Back Pressure	6.5 [bar abs]	80 [psig]
Maximum After Firing Temperature	800 [deg C]	
Useful Heat Output at MCR	2.67 [kW/kWe]	
Gas Turbine Fuel Consumption (HCV)	3.24 [kW/kWe]	
After Firing Fuel Consumption (HCV)	1.80 [kW/kWe]	
Net Electrical Efficiency (HCV)	19.84 [%]	

Total CHP System Efficiency at MCR	Heat/Power Ratio at MCR
72.9 % (HCV)	2.67

Steam Flow from Plant at MCR	37000[lbs/hr]

Table 37

PLANT	COMBINED CYCLE PLANT	TYPE II
Net Electrical Output (MCR)	4652 [kW]	
Gas Turbine Output (MCR)	2970 [kW]	(Plant Type III)
Steam Turbine Output (MCR)	1682 [kW]	(Plant Type XVI)
Waste Heat Boiler Outlet Conditions	64.0 [Bar abs]	900 [psig]
	482.0 [deg C]	900 [deg F]
Steam Turbine Back Pressure	7.9 [bar abs]	100 [psig]
Maximum After Firing Temperature	800 [deg C]	
Useful Heat Output at MCR	2.81 [kW/kWe]	
Gas Turbine Fuel Consumption (HCV)	3.36 [kW/kWe]	
After Firing Fuel Consumption (HCV)	1.87 [kW/kWe]	
Net Electrical Efficiency (HCV)	19.15 [%]	

Total CHP System Efficiency at MCR	Heat/Power Ratio at MCR
72.9 % (HCV)	2.81

Steam Flow from Plant at MCR	37000 [lbs/hr]

Table 38

PLANT	COMBINED CYCLE PLANT	TYPE III
Net Electrical Output (MCR)	4400 [kW]	
Gas Turbine Output (MCR)	2970 [kW]	(Plant Type III)
Steam Turbine Output (MCR)	1430 [kW]	(Plant Type XV)
Waste Heat Boiler Outlet Conditions	64.0 [Bar abs]	900 [psig]
	482.0 [deg C]	900 [deg F]
Steam Turbine Back Pressure	11.3 [bar abs]	150 [psig]
Maximum After Firing Temperature	800 [deg C]	
Useful Heat Output at MCR	3.04 [kW/kWe]	
Gas Turbine Fuel Consumption (HCV)	3.55 [kW/kWe]	
After Firing Fuel Consumption (HCV)	1.97 [kW/kWe]	
Net Electrical Efficiency (HCV)	18.11 [%]	

Total CHP System Efficiency at MCR	Heat/Power Ratio at MCR
73.1 % (HCV)	3.04

Steam Flow from Plant at MCR	38000[lbs/hr]

Table 39

PLANT	COMBINED CYCLE PLANT	TYPE IV
Net Electrical Output (MCR)	7624 [kW]	
Gas Turbine Output (MCR)	5150 [kW]	(Plant Type IV)
Steam Turbine Output (MCR)	2474 [kW]	(Plant Type XVII)
Waste Heat Boiler Outlet Conditions	64.0 [Bar abs]	900 [psig]
	482.0 [deg C]	900 [deg F]
Steam Turbine Back Pressure	6.5 [bar abs]	80 [psig]
Maximum After Firing Temperature	800 [deg C]	
Useful Heat Output at MCR	2.27 [kW/kWe]	
Gas Turbine Fuel Consumption (HCV)	2.88 [kW/kWe]	
After Firing Fuel Consumption (HCV)	1.59 [kW/kWe]	
Net Electrical Efficiency (HCV)	22.39 [%]	

Total CHP System Efficiency at MCR	Heat/Power Ratio at MCR
73.3 % (HCV)	2.27

Steam Flow from Plant at MCR	50000[lbs/hr]

Table 40

PLANT	COMBINED CYCLE PLANT	TYPE V

Net Electrical Output (MCR)	7400 [kW]	
Gas Turbine Output (MCR)	5150 [kW]	(Plant Type IV)
Steam Turbine Output (MCR)	2250 [kW]	(Plant Type XVI)
Waste Heat Boiler Outlet Conditions	64.0 [Bar abs]	900 [psig]
	482.0 [deg C]	900 [deg F]
Steam Turbine Back Pressure	7.9 [bar abs]	100 [psig]
Maximum After Firing Temperature	800 [deg C]	
Useful Heat Output at MCR	2.37 [kW/kWe]	
Gas Turbine Fuel Consumption (HCV)	2.97 [kW/kWe]	
After Firing Fuel Consumption (HCV)	1.63 [kW/kWe]	
Net Electrical Efficiency (HCV)	21.74 [%]	

Total CHP System Efficiency at MCR	Heat/Power Ratio at MCR
73.3 % (HCV)	2.37

Steam Flow from Plant at MCR 50000[lbs/hr]

Table 41

PLANT	COMBINED CYCLE PLANT	TYPE VI	
Net Electrical Output (MCR)	7060 [kW]		
Gas Turbine Output (MCR)	5150 [kW]	(Plant Type IV)	
Steam Turbine Output (MCR)	1910 [kW]	(Plant Type XV)	
Waste Heat Boiler Outlet Conditions	64.0 [Bar abs]	900 [psig]	
	482.0 [deg C]	900 [deg F]	
Steam Turbine Back Pressure	11.3 [bar abs]	150 [psig]	
Maximum After Firing Temperature	800 [deg C]		
Useful Heat Output at MCR	2.54 [kW/kWe]		
Gas Turbine Fuel Consumption (HCV)	3.11 [kW/kWe]		
After Firing Fuel Consumption (HCV)	1.71 [kW/kWe]		
Net Electrical Efficiency (HCV)	20.74 [%]		

Total CHP System Efficiency at MCR	Heat/Power Ratio at MCR
73.5 % (HCV)	2.54

Steam Flow from Plant at MCR	51000 [lbs/hr]

Table 42

PLANT	COMBINED CYCLE PLANT	TYPE VII	
Net Electrical Output (MCR)	4549 [kW]		
Gas Turbine Output (MCR)	3140 [kW]	(Plant Type VI)	
Steam Turbine Output (MCR)	1409 [kW]	(Plant Type XVII)	
Waste Heat Boiler Outlet Conditions	64.0 [Bar abs]	900 [psig]	
	482.0 [deg C]	900 [deg F]	
Steam Turbine Back Pressure	6.5 [bar abs]	80 [psig]	
Maximum After Firing Temperature	800 [deg C]		
Useful Heat Output at MCR	2.16 [kW/kWe]		
Gas Turbine Fuel Consumption (HCV)	3.11 [kW/kWe]		
After Firing Fuel Consumption (HCV)	1.26 [kW/kWe]		
Net Electrical Efficiency (HCV)	22.88 [%]		

Total CHP System Efficiency at MCR	Heat/Power Ratio at MCR
72.3 % (HCV)	2.16

Steam Flow from Plant at MCR	28000 [lbs/hr]

Table 43

PLANT	COMBINED CYCLE PLANT	TYPE VIII
Net Electrical Output (MCR)	4422 [kW]	
Gas Turbine Output (MCR)	3140 [kW]	(Plant Type VI)
Steam Turbine Output (MCR)	1282 [kW]	(Plant Type XVI)
Waste Heat Boiler Outlet Conditions	64.0 [Bar abs]	900 [psig]
	482.0 [deg C]	900 [deg F]
Steam Turbine Back Pressure	7.9 [bar abs]	100 [psig]
Maximum After Firing Temperature	800 [deg C]	
Useful Heat Output at MCR	2.25 [kW/kWe]	
Gas Turbine Fuel Consumption (HCV)	3.20 [kW/kWe]	
After Firing Fuel Consumption (HCV)	1.30 [kW/kWe]	
Net Electrical Efficiency (HCV)	22.24 [%]	

Total CHP System Efficiency at MCR	Heat/Power Ratio at MCR
72.4 % (HCV)	2.25

Steam Flow from Plant at MCR	28000 [lbs/hr]

Table 44

PLANT	COMBINED CYCLE PLANT	TYPE IX	
Net Electrical Output (MCR)	4226 [kW]		
Gas Turbine Output (MCR)	3140 [kW]	(Plant Type VI)	
Steam Turbine Output (MCR)	1086 [kW]	(Plant Type XV)	
Waste Heat Boiler Outlet Conditions	64.0 [Bar abs]	900 [psig]	
	482.0 [deg C]	900 [deg F]	
Steam Turbine Back Pressure	11.3 [bar abs]	150 [psig]	
Maximum After Firing Temperature	800 [deg C]		
Useful Heat Output at MCR	2.41 [kW/kWe]		
Gas Turbine Fuel Consumption (HCV)	3.35 [kW/kWe]		
After Firing Fuel Consumption (HCV)	1.36 [kW/kWe]		
Net Electrical Efficiency (HCV)	21.26 [%]		

Total CHP System Efficiency at MCR	Heat/Power Ratio at MCR
72.5 % (HCV)	2.41

Steam Flow from Plant at MCR 29000 [lbs/hr]

Table 45

PLANT	COMBINED CYCLE PLANT	TYPE X

Net Electrical Output (MCR)	6580 [kW]		
Gas Turbine Output (MCR)	4830 [kW]	(Plant Type VIII)	
Steam Turbine Output (MCR)	1750 [kW]	(Plant Type XVII)	
Waste Heat Boiler Outlet Conditions	64.0 [Bar abs]	900 [psig]	
	482.0 [deg C]	900 [deg F]	
Steam Turbine Back Pressure	6.5 [bar abs]	80 [psig]	
Maximum After Firing Temperature	800 [deg C]		
Useful Heat Output at MCR	1.85 [kW/kWe]		
Gas Turbine Fuel Consumption (HCV)	2.81 [kW/kWe]		
After Firing Fuel Consumption (HCV)	1.05 [kW/kWe]		
Net Electrical Efficiency (HCV)	25.91 [%]		

Total CHP System Efficiency at MCR	Heat/Power Ratio at MCR
74.0 % (HCV)	1.85

Steam Flow from Plant at MCR	35000 [lbs/hr]

Table 46

PLANT	COMBINED CYCLE PLANT		TYPE XI
Net Electrical Output (MCR)	6420 [kW]		
Gas Turbine Output (MCR)	4830 [kW]	(Plant Type VIII)	
Steam Turbine Output (MCR)	1590 [kW]	(Plant Type XVI)	
Waste Heat Boiler Outlet Conditions	64.0 [Bar abs]		900 [psig]
	482.0 [deg C]		900 [deg F]
Steam Turbine Back Pressure	7.9 [bar abs]		100 [psig]
Maximum After Firing Temperature	800 [deg C]		
Useful Heat Output at MCR	1.93 [kW/kWe]		
Gas Turbine Fuel Consumption (HCV)	2.88 [kW/kWe]		
After Firing Fuel Consumption (HCV)	1.07 [kW/kWe]		
Net Electrical Efficiency (HCV)	25.28 [%]		

Total CHP System Efficiency at MCR	Heat/Power Ratio at MCR
74.0 % (HCV)	1.93

Steam Flow from Plant at MCR	35000 [lbs/hr]

Table 47

PLANT	COMBINED CYCLE PLANT		TYPE XII
Net Electrical Output (MCR)		6176 [kW]	
Gas Turbine Output (MCR)	4830 [kW]	(Plant Type VIII)	
Steam Turbine Output (MCR)	1346 [kW]	(Plant Type XV)	
Waste Heat Boiler Outlet Conditions		64.0 [Bar abs]	900 [psig]
		482.0 [deg C]	900 [deg F]
Steam Turbine Back Pressure		11.3 [bar abs]	150 [psig]
Maximum After Firing Temperature		800 [deg C]	
Useful Heat Output at MCR		2.05 [kW/kWe]	
Gas Turbine Fuel Consumption (HCV)		3.00 [kW/kWe]	
After Firing Fuel Consumption (HCV)		1.12 [kW/kWe]	
Net Electrical Efficiency (HCV)		24.32 [%]	

Total CHP System Efficiency at MCR	Heat/Power Ratio at MCR
74.2 % (HCV)	2.05

Steam Flow from Plant at MCR	36000 [lbs/hr]

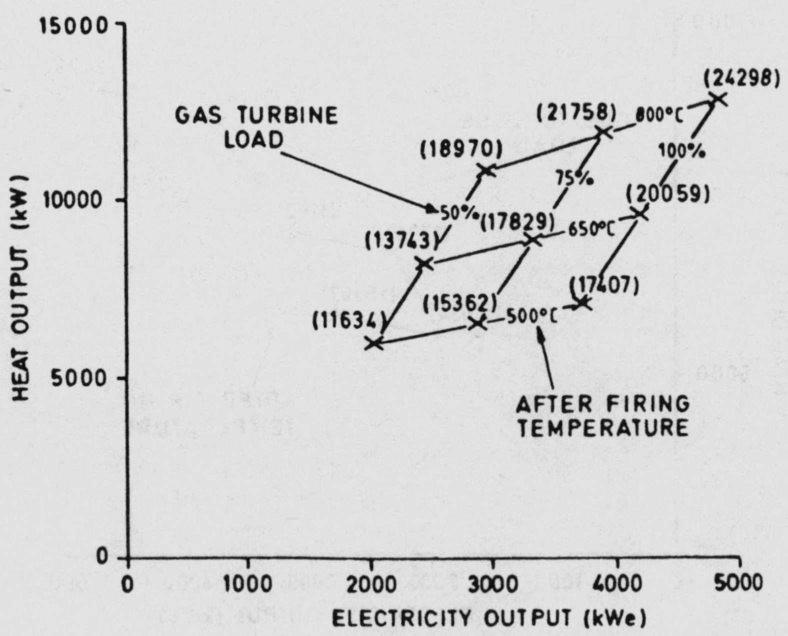

FUEL CONSUMPTION (kW) BRACKETED

Combined Cycle Plant Type I Turbine back pressure: 80 psig.

Fig. 7

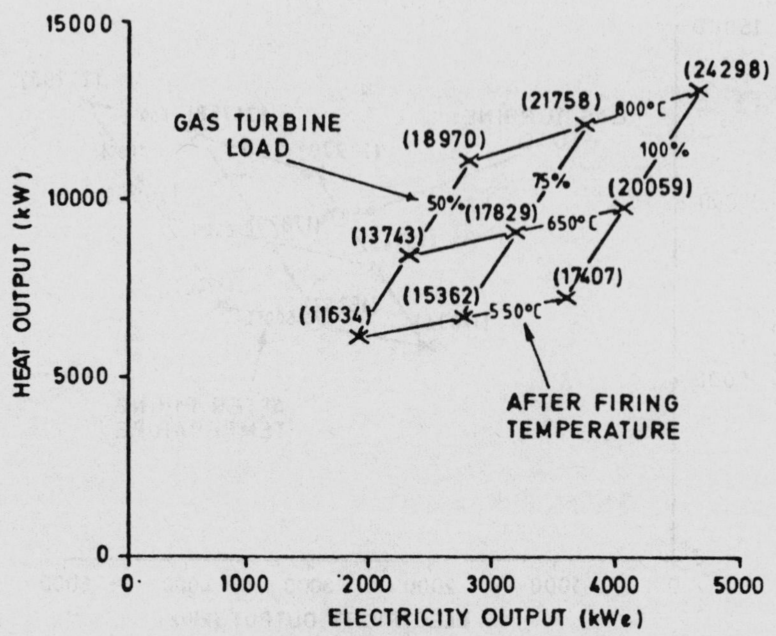

FUEL CONSUMPTION (kW) BRACKETED

Combined Cycle Plant Type II Turbine back pressure: 100 psig.

Fig. 8

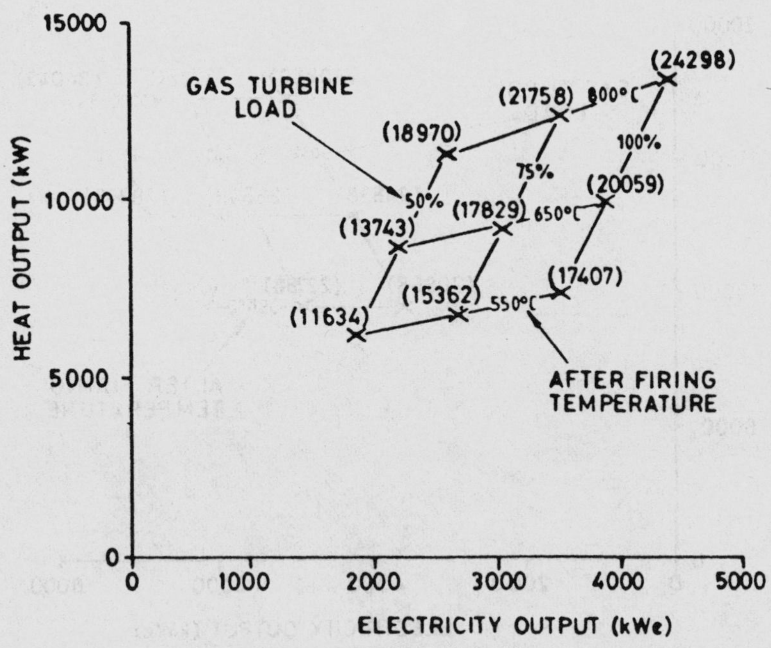

Combined Cycle Plant Type III Turbine back pressure: 150 psig.

Fig. 9

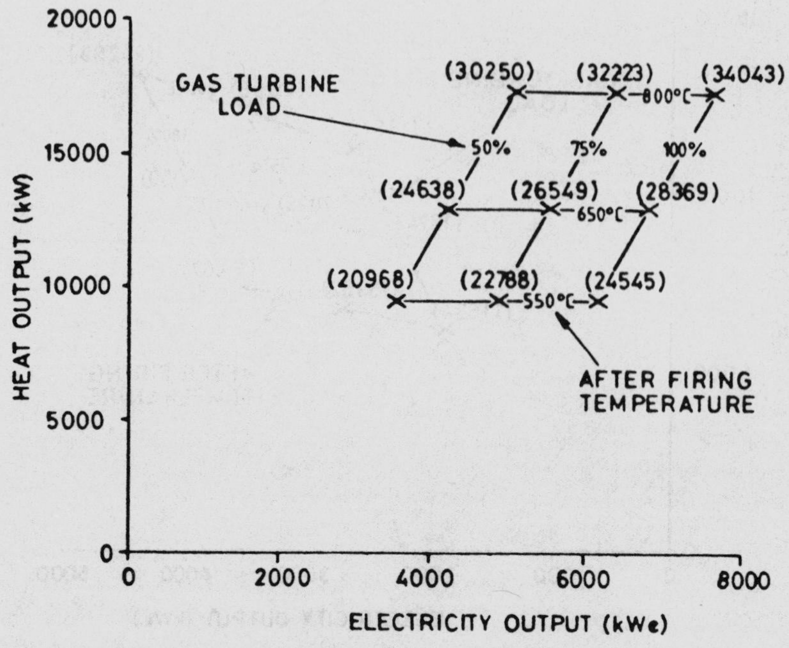

FUEL CONSUMPTION (kW) BRACKETED

Combined Cycle Plant Type IV Turbine back pressure: 80 psig.

Fig. 10

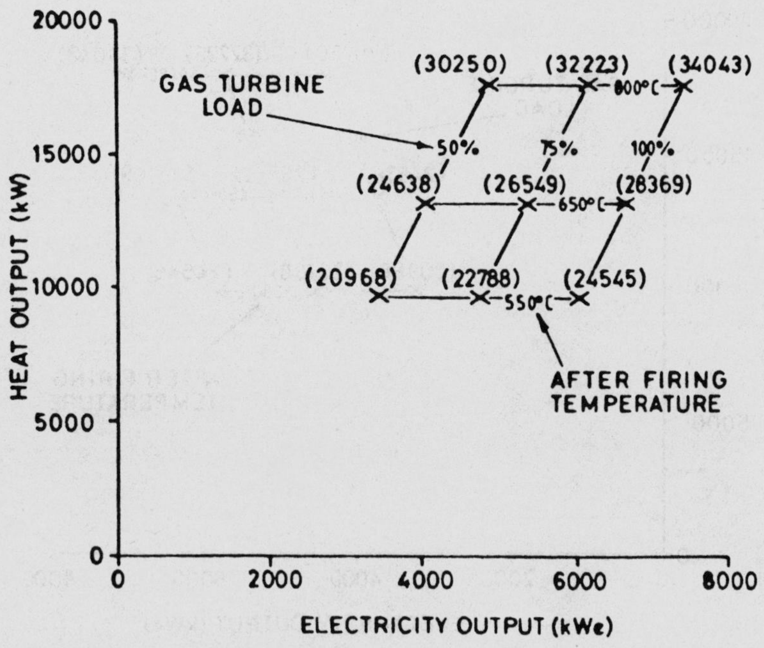

FUEL CONSUMPTION (kW) BRACKETED

Combined Cycle Plant Type V Turbine back pressure: 100 psig.

Fig. 11

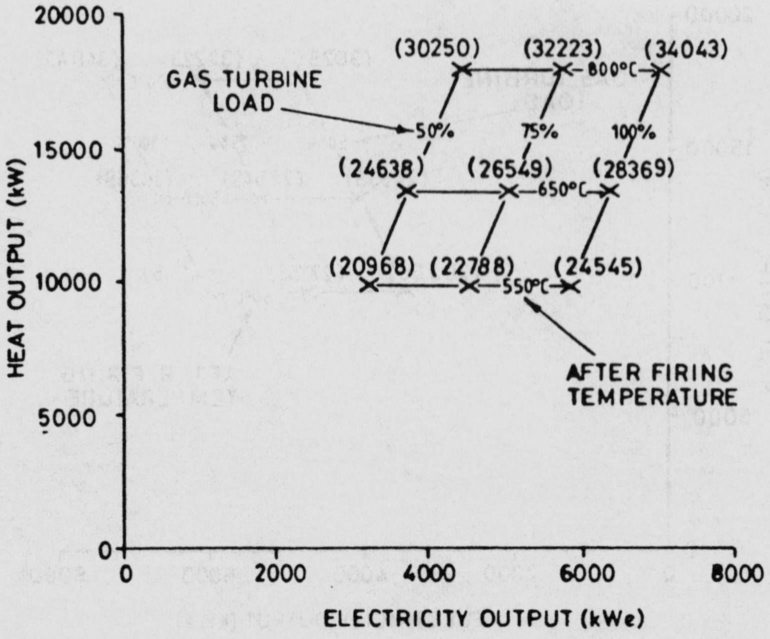

FUEL CONSUMPTION (kW) BRACKETED

Combined Cycle Plant Type VI Turbine back pressure: 150 psig.

Fig. 12

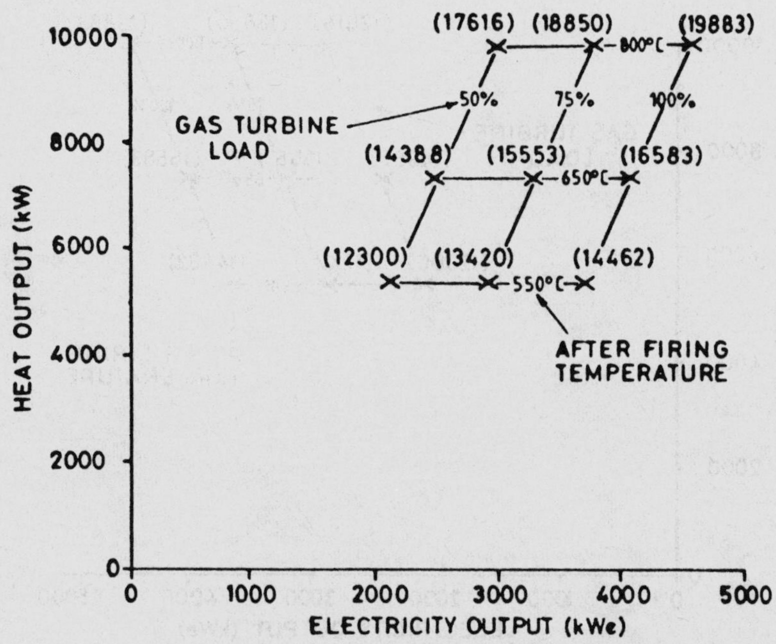

FUEL CONSUMPTION (kW) BRACKETED

Combined Cycle Plant Type VII Turbine back pressure: 80 psig.

Fig. 13

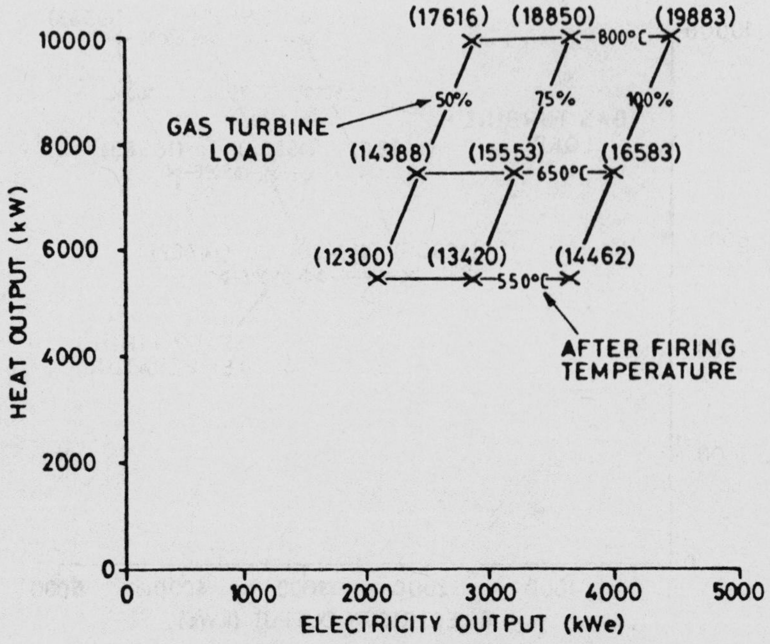

FUEL CONSUMPTION (kW) BRACKETED

Combined Cycle Plant Type VIII Turbine back pressure: 100 psig.

Fig. 14

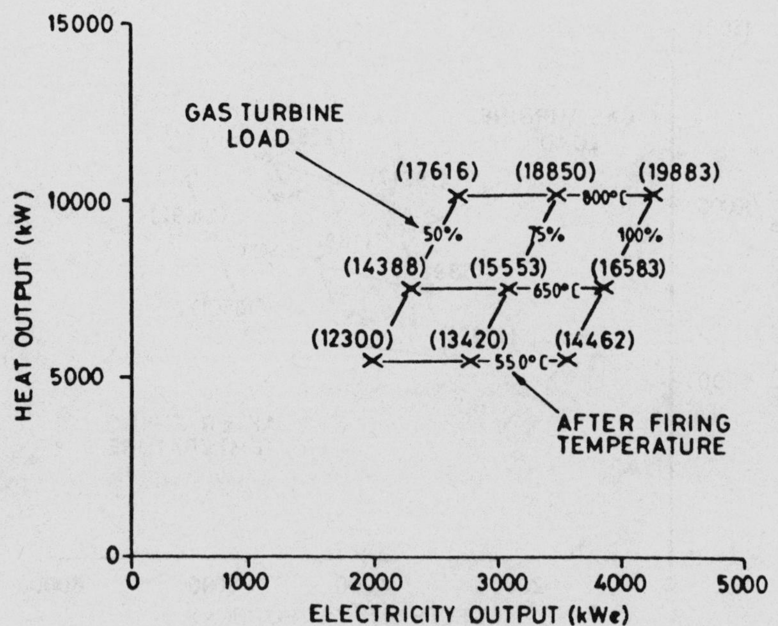

FUEL CONSUMPTION (kW) BRACKETED

Combined Cycle Plant Type IX Turbine back pressure: 150 psig.

Fig. 15

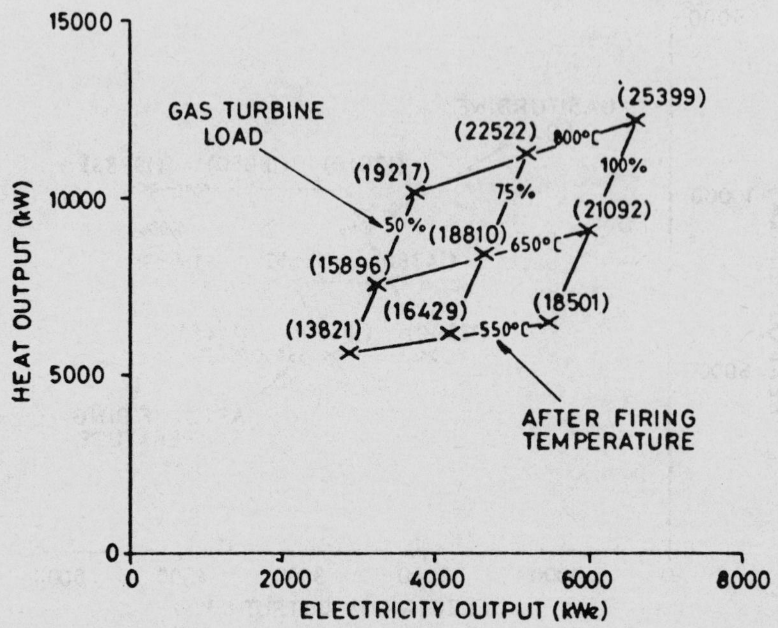

Combined Cycle Plant Type X Turbine back pressure: 80 psig.

Fig. 16

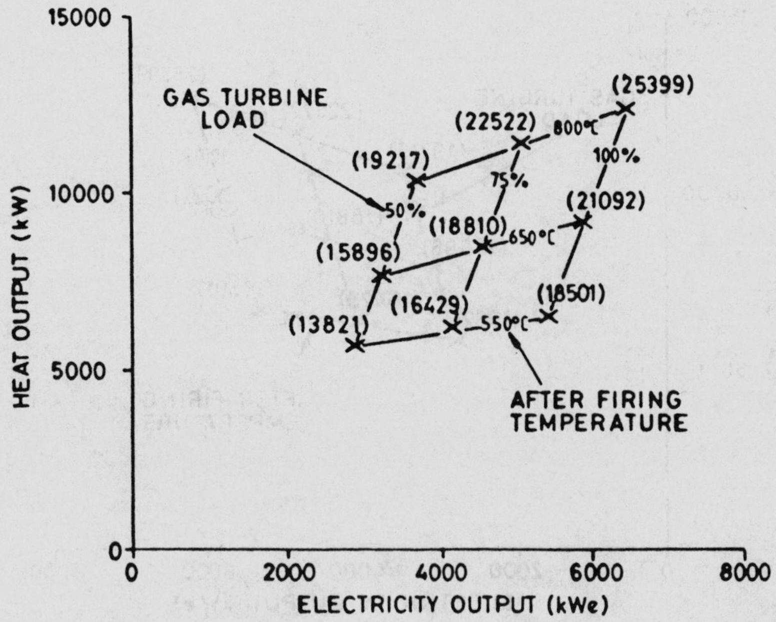

FUEL CONSUMPTION (kW) BRACKETED

Combined Cycle Plant Type XI Turbine back pressure: 100 psig.

Fig. 17

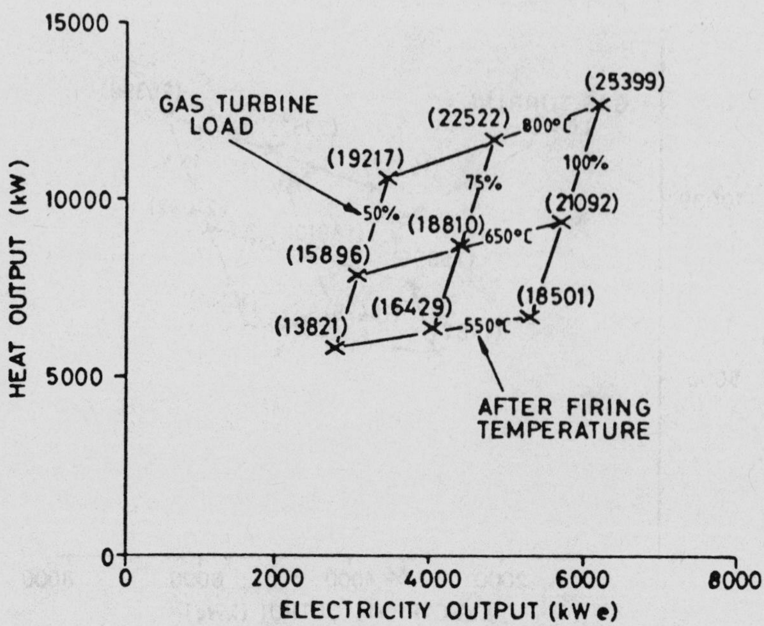

FUEL CONSUMPTION (kW) BRACKETED

Combined Cycle Plant Type XII Turbine back pressure: 150 psig.

Fig. 18

2.8 Retrofitted Low Pressure Steam Turbines

111 The retrofitting of single stage back pressure steam turbines to existing low pressure shell and tube type steam boilers were considered not in the main study but as a special case and are described in Appendix 4.

112 Steam conditions at outlets from the boiler were assumed to be 18.2 bar absolute (250 psig) and saturated. The back pressure chosen depends on the steam distribution pressure on the particular site, however for the purposes of the examples described in Appendix 4 a back pressure of 6.5 bar absolute (80 psig) was considered.

113 The performance of the steam turbine was calculated in a similar manner to that described in Section 2.6, except that boiler auxiliaries were not included. They are required in any case for the existing boiler installation. The performance is detailed in Table 48.

Table 48

PLANT	STEAM TURBINE	TYPE - LOW PRESSURE for DIRECT DRIVE

Net Electrical Output Range (MCR)	100-1000 [kW]	
Boiler Outlet Conditions	18 [bar abs] Saturated	250 [psig] Saturated
Steam Turbine Back Pressure	6.5 [bar abs]	80 [psig]
Minimum Turbine Output [% of MCR]	10 [%]	
Turbine Efficiency (mechanical)	93.0 [%]	
Useful Heat in Turbine Exhaust (at MCR)	2466 [kJ/kg]	

Comments:
Single Stage Steam Turbine for Direct Drive Applications

Electrical Output (at 500 kW)		MCR
Fuel Consumption (HCV)	[kW/kW]	30.0
Steam Mass Flow Rate	[kg/kWh]	31.3
Net Shaft Efficiency (HCV) of Steam Turbine + Boiler	[%]	3.3

Total CHP System Efficiency at MCR	Heat/Power Ratio
74.8 % (HCV)	21.4 (at MCR)

Steam Flow For 1 MW Turbine (at MCR) 69000 [lbs/hr]

2.9 Boilers

114 In calculating the performance of boilers the following assumptions were made:

Coal Fired Steam Generating Boilers

115 Efficiency - An average thermal efficiency of 78.5% based on higher calorific value was used.

116 Feed Water - A feed water temperature of 50°C was used. This assumes that there is 50% condensate return at 80°C and the make-up water is supplied at 20°C. Feed heating to 80°C is

frequently employed using steam from the boiler. This does not significantly affect the overall energy balance.

Exhaust Gas Waste Heat Boilers
(for low or high pressure steam or hot water)

117 Heat Recovery - the heat recovered from a waste heat boiler was calculated using the equation:

$$Q_h = m.Cp.T.(1 - P)$$

Q_h = Heat recovered
m = Exhaust gas mass flow rate
Cp = Gas mean specific test capacity
T = Gas temperature drop
P = Heat losses

118 Heat Losses - The convective and radiative heat losses from a waste heat boiler were assumed to be 2%.

119 Mean specific heat capacity - The mean specific heat capacities were determined from Figures 5 or 5a depending on the fuel used.

120 Gas Temperature Drop - This was defined as the difference in temperature of the exhaust gas between the the boiler inlet and outlet. It was assumed that in-duct after-firing would be used and therefore the combustion would be complete before the exhaust gases entered the boiler.

121 Boiler Outlet Flue Gas Temperature - The boiler outlet gas temperatures were calculated using the following assumptions:

a) Minimum boiler outlet gas temperatures at the design condition of 150°C for natural gas fuel and 220 C for heavy fuel oil.

b) A minimum pinch point of 30°C at the design condition.

c) A minimum approach temperature of 40°C at the boiler gas inlet.

d) A feed water temperature of 80°C for steam boilers.

e) A temperature difference of 20°C between the flow and return for hot water boilers.

The design conditions were assumed to be at the maximum continuous rating of the prime mover with maximum after-firing if appropriate.

122 These assumptions are somewhat conservative and better heat recovery may be obtained from some boilers which would result in improved economics.

123 Part Load - Part load boiler outlet flue gas temperatures were calculated assuming that the heat transfer coefficients in the boiler varied in proportion to (mass flow rate)

124 After firing fuel consumption - The fuel needed for after-firing was calculated from the equation:

$$Q_f = (mh_{fe} - mh_{fu}).E$$

Q_f = After-firing fuel consumption
m = Exhaust gas mass flow rate
h_{fe} = Enthalpy of fired exhaust gases
h_{fu} = Enthalpy of unfired exhaust gases
E = Efficiency of after-burner

An after-burner efficiency of 99% was assumed.

125 Steam or Hot Water Outlet Conditions - These were specified by the requirements of the heat distribution system at a site or the choice of a steam turbine.

126 Feed Water - Feed water was always assumed to be at 80°C on entry to the boiler. At sites with little or low temperature.

condensate return feed heating would be employed using steam or, in the case of internal combustion engine CHP plant, hot water recovered from the engine.

2.10 Maintenance

127 The following allowances are assumed for planned maintenance.

128 Gas turbine and steam turbine plant - 5% of operating hours
Internal combustion engine plant - 8% of operating hours

129 In addition it was assumed that there would be further unplanned breakdowns equal in duration to half of the planned maintenance periods.

3.0 CHP PLANT CAPITAL AND MAINTENANCE COSTS

130 The capital costs for CHP plant used in this study are those of complete installations. The costs include all expenditure that would be needed to complete a typical installation and include items such as new buildings, design fees etc. VAT is excluded.

131 The costs have been compiled from budget quotes obtained from manufacturers during December 1984. The prices of equipment varied between manufacturers by as much as 20%. The costs given here represent typical prices and in practice a spread of prices could be obtained for many of the items. In some cases the prices for equipment are very competitive and some manufacturers may offer substantial discounts. As such discounts are very variable they have not been considered in making up these costs. The budget quotes supplied by the manufacturers have also been compared and adjusted if necessary with cost estimates for actual projects. Sections 3.1 and 3.3 describe the capital costs used for various items of plant. Section 3.4 describes the total costs for complete CHP systems. It is these total costs that are used in the economic assessment of CHP installations.

132 Maintenance costs are described in Section 3.5. They have been derived from information supplied by equipment manufacturers and users.

3.1 Capital Costs of Prime Movers

Gas Turbines

133 The capital costs of the gas turbine generators described in Section 2.2 are shown in Table 49 for single unit installations. The costs are expressed per kW of net electrical output as described in Section 1.2. The gas turbine manufacturers quoted prices for the gas turbine generating sets only. These include the items shown below:

- i) Skid mounted gas turbine generating set (including gearbox and generator (11kV))
- ii) Air inlet filter, silencer and nominal ductwork
- iii) Exhaust diffuser and nominal ductwork to waste heat boiler
- iv) Electric starting
- v) Lubrication oil system and air blast coolers
- vi) Automatic starting and control system including synchronisation equipment
- vii) Delivery, installation and commissioning

TABLE 49

Capital Costs of Gas Turbine Generators (Single Installation)
(Scope as described in Section 3.1.1)

Gas Turbine Type	Capital Cost per kW of net electrical output
I	£420/kWe
II	£340/kWe
III	£210/kWe
IV	£210/kWe
V	£240/kWe
VI	£230/kWe
VII	£290/kWe
VIII	£240/kWe
IX	£300/KWe
X	£460/kWe

134 The gas turbine operates on natural gas and no extra allowance has been made for a liquid fuel standby, although some manufacturers supply a liquid fuel standby system as standard equipment.

135 The natural gas supply and compressor are described in Section 3.3.2.

136 Where applicable an exchange rate of £1.0 = $1.2 has been used.

137 Waste heat boilers for gas turbines are described in Section 3.2.

138 All the gas turbines utilise 11kV electricity generators except for Type X. This has an output less than 500kWe and so a 415V generator is used.

Heavy Fuel Oil Diesel Engines

139 The capital costs of the diesel engine systems described in Section 2.3 are shown in Fig 19, for single unit installations. Up to 4 MW the costs refer to engines operating at 750 rpm. Above 4 MW the costs refer to engines operating at 500 rpm. The lower speed engines cost proportionately more partly because the output for a given engine size is lower on account of the lower speed.

140 The costs are expressed per kW of net electrical output, as defined in Section 2.3. The performance of each of these types of engine is shown in Tables 11 and 12.

141 In the case of heavy fuel oil diesels the manufacturers will supply complete installations including engine generator and waste heat boiler. The costs shown in Figure 19 are thus for complete systems and include:

- i) Diesel engine plus generator (11kV)
- ii) Inlet ductwork plus air filter
- iii) Starting air compressor plus air receiver
- iv) Lubricating oil system
- v) Cooling water system including heat recovery system and heat dump air blast coolers
- vi) Fuel storage and treatment system for engine and boiler
- vii) Engine control system
- viii) Waste heat boiler with after-firing as appropriate
- ix) Boiler insulation
- x) Boiler feed pump, auxiliaries and control system
- xi) All exhaust flues including boiler bypass and chimney
- xii) Delivery installation and commissioning

142 Shell and tube type boilers are used generally in diesel engines CHP Systems. However, in the case of large engines with after-firing, water tube boilers are necessary in order to accommodate the large exhaust mass flow rates. The use of water tube boilers is indicated by the step increase in capital cost at 8MWe. After-firing of the exhaust gases to 1200°C is considered without the use of supplementary combustion air.

143 Heated oil storage tanks with a capacity sufficient for 3 weeks operation at full load have been allowed for.

144 Heat dump air blast coolers are included in the water cooling system to dump the heat not usefully recovered (e.g. second stage intercooler and oil cooler) and also to dump all the jacket cooling water heat at times when heat recovery is not required.

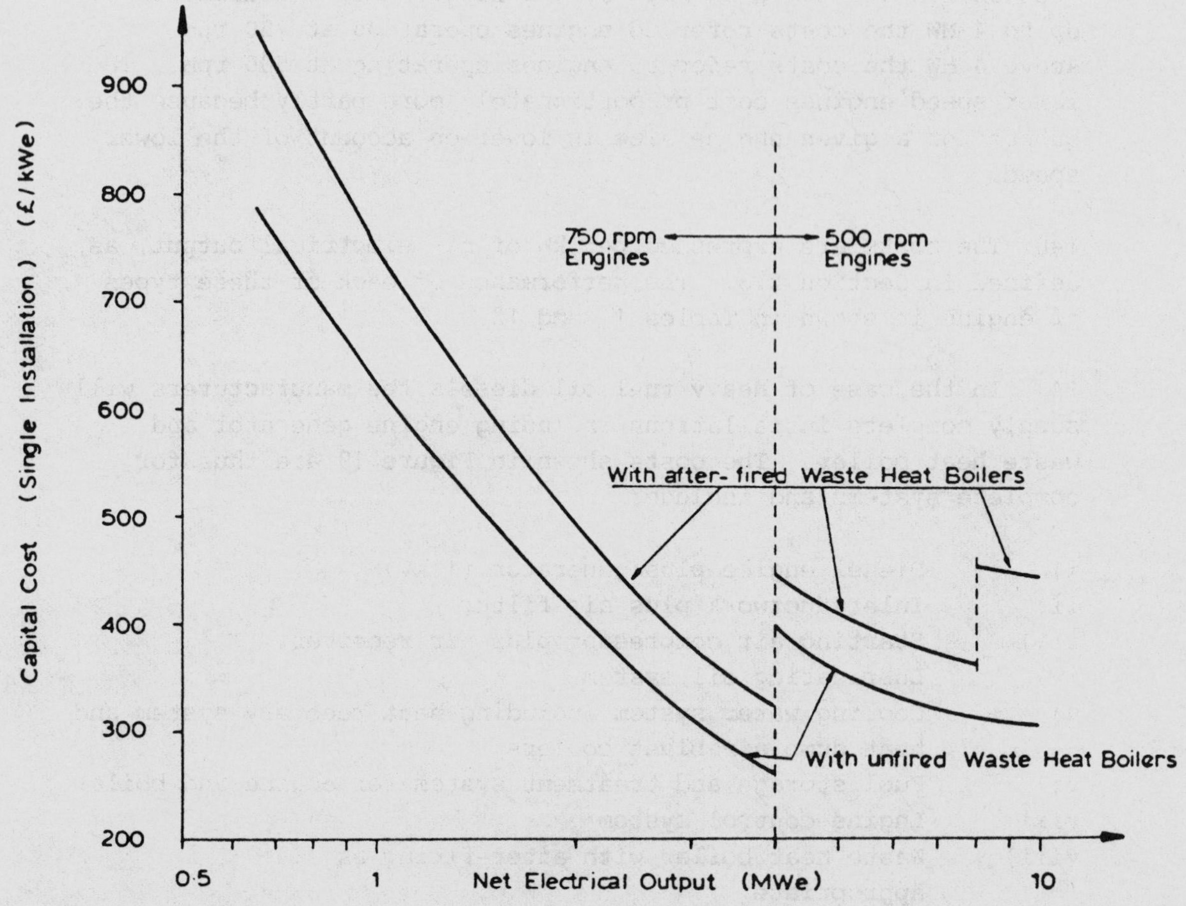

CAPITAL COSTS

of Heavy Fuel Oil Diesel Engines and Waste Heat Boilers

(Scope of Supply described in Section 3-1-2)

Fig. 19

Dual-Fuel Diesel Engines

145 The capital costs of dual-fuel diesel engine systems are shown in Figure 20 for single unit installations. Up to 2.1 MW the costs refer to engines operating at 600 rpm and above 2 MW the costs of engines operating at 500 rpm are shown. The performance of each of these types of engine is shown in Tables 13 and 14. The costs are expressed per kW of net electrical output as defined in Section 2.4.

146 The costs shown in Fig. 20 are for complete systems of engine plus waste heat boiler and include:

i) Dual-fuel engine plus generator (11kV)
ii) Inlet ductwork plus air filter
iii) Starting air compressor plus air receiver
iv) Lubricating oil system
v) Cooling water system including heat recovery system and heat dump air blast coolers
vi) Pilot fuel storage and supply system
vii) Engine Control system
viii) Waste heat boiler with after-firing as appropriate
ix) Boiler insulation
x) Boiler feed pump, auxiliaries and control system
xi) All exhaust flues including boiler bypass and chimney
xii) Delivery installation and commissioning.

147 Shell and tube type boilers are used in all cases and after-firing of the exhaust gases up to 1200°C without the use of supplementary combustion air is considered.

148 The cost of the natural gas supply including gas compressors is considered separately in Section 3.3.2.

149 Heat dump air blast coolers are included in the water cooling system to dump the heat not usefully recovered (e.g. second stage intercooler and oil cooler) and also to dump all the jacket cooling water heat at times when heat recovery is not required.

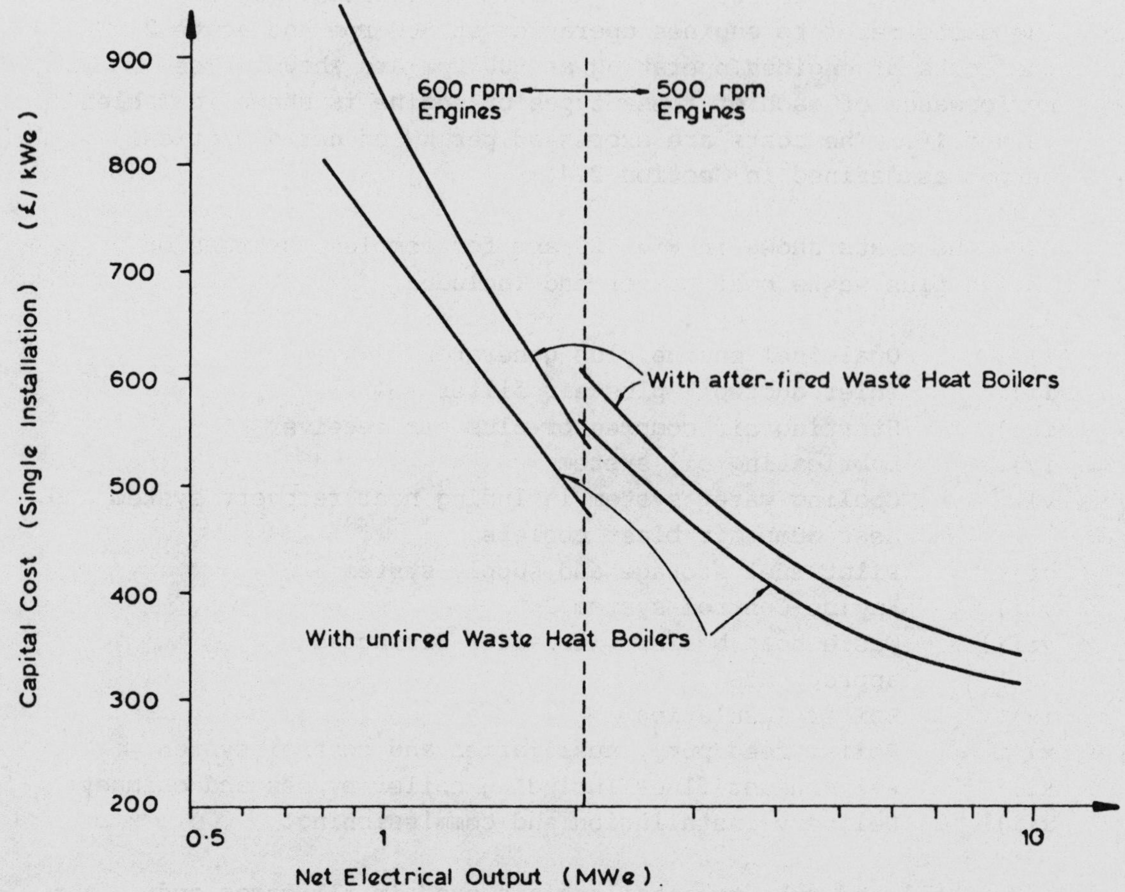

CAPITAL COSTS

of Dual-Fuel Diesel Engines and Waste Heat Boilers

(Scope of Supply described in Section 3-1-3)

Fig. 20

Industrial Spark Ignition Gas Engines

150 The capital costs of industrial spark ignition engines are shown in Fig. 21 for single unit installations. Two size ranges of engine are considered.

a) 100 kW - 550 kW - 4 stroke engines
b) 500 kW - 2000 kW - 2 stroke engines

151 For the 4 stroke engines 415V generators are used; the larger engines use 11kV generators. The manufacturers budget quotes for these engines include a heat recovery system. The costs shown in Fig 21 include the following items:

i) Spark ignition engine plus generator (415v or 11kV)
ii) Electric starting
iii) Lubricating oil system
iv) Cooling water system including heat recovery systems and heat dump air blast coolers
v) Engine control system including synchronisation equipment
vi) Exhaust gas waste heat boiler
vii) Boiler insulation
viii) Boiler feed pump, auxiliaries and control system
ix) All exhaust flues including boiler bypass and chimney
x) Delivery installation and commissioning

152 The cost of the natural gas supply system is described in Section 3.3.2.

153 In addition to generating sets, the use of 4 stroke spark ignition engines for the direct driving of plant is also considered. In this case the generator is omitted and the engine drives a reciprocating compressor or refrigeration plant directly. These costs are shown on Fig 21 in terms of net shaft output and include all the items detailed above excluding the generator and the electrical control equipment.

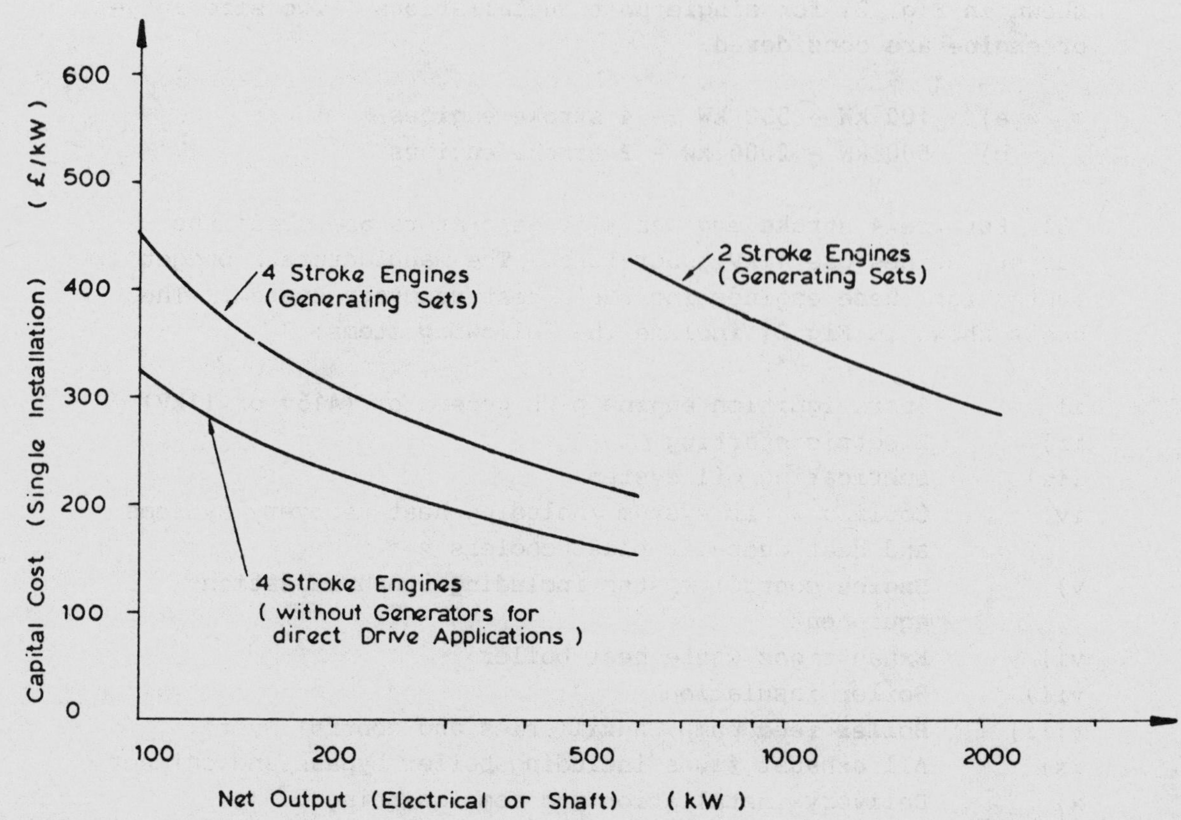

CAPITAL COSTS

of Industrial Spark Ignition Engines and Heat Recovery Plant

(Scope of Supply described in Section 3-1-4)

Fig. 21

High Pressure Boiler with Back Pressure
Steam Turbine Generating Sets

154 It was found that the capital costs for steam turbine generators varied with steam mass flow rate and electrical output. In particular the cost of the steam turbine depends largely on the frame size that must be used for particular steam conditions and flow rate.

155 It is most convenient therefore to express the capital costs for steam turbine generators in the form of equations which contain terms for steam mass flow rate and electrical output. These equations are given in Table 50 for single and multi-stage steam turbine generators, they are only applicable for the stated size range and steam conditions.
The costs shown in Table 50 include:-

i) Steam Turbine
ii) Gearbox
iii) Generator
iv) Generating set base frame
v) Lubrication system (including turbine driven pumps)
vi) Governor and control system (including synchronisation)
vii) Control panel
viii) Delivery, installation and commissioning

156 The performance of the back pressure steam turbines is given in Section 2.6.

157 Single stage steam turbines were considered for outputs up to 2 MW and for pressure ratios of less than 3:1. Multistage steam turbines in spite of the greater cost are found to be more attractive, at pressure ratios greater than 3:1, as they produce much more power for a given steam flow rate. Multi-stage steam turbines are generally not available for back pressures above 12 bar abs.

TABLE 50

Capital Costs for Back Pressure Steam Turbine Generating Sets

The general equation is of the form:

$$C = K + Am + Bw$$

C = Capital cost [£]
K,A and B = Constants
m = steam mass flow rate [kg/s]
w = Net electrical output [kWe]

The values of the constants for various size ranges and steam conditions are:

Type of Turbine	Single Stage	Multi-Stage	Multi-Stage	Multi-Stage
Electrical Output Range [kWe]	0-2000	600-10000	600-10000	600-10000
Steam Flow Range [kg/s]	0-15	0-9	9-15	15-25
Range of Inlet conditions:				
Temp. [C]	400-482	400-482	400-482	400-482
Pressure [bar abs]	30-64	30-64	30-64	30-64
Back Pressure Range [bar abs]	11-18	3-11	3-11	3-11
K	113300	432100	523300	660100
A	3810	0	0	0
B	19.1	19.1	19.1	19.1

Single Stage Back Pressure Steam
Turbines for Low Pressure Boilers

158 Retrofitting small single stage steam turbines onto low pressure boilers for the direct driving of plant or for generating electricity has been considered. The capital costs of these turbines are shown in Fig. 22. The capital costs are expressed in terms of net electrical output from the generator or net shaft output in the case of direct drives.

158 The steam pressures considered are 18.2 bar abs saturated at turbine inlet and 6.5 bar abs back pressure. Under these conditions the steam mass flow rate is proportional to the turbine output and there is no need for a relationship for capital cost which includes a term for steam flow and turbine output.

160 Turbines up to 1MW output are considered and costs are shown for generating sets and for direct drive applications. For direct drives the generator is omitted and a gearbox reducing the turbine speed to 1000 rpm, suitable for a reciprocating compressor, is included.
The costs shown in Fig 22 include:

i) Steam turbine
ii) Gearbox (as applicable)
iii) Generator (generating sets only (415V))
iv) Base frame
v) Turbine lubrication system
vi) Governor control system (and electrical control for generating sets)
vii) Control panel
viii) Delivery, installation and commissioning

161 In the case of steam turbines for direct drive application the cost of a driven machine (e.g. compressor) is not included.

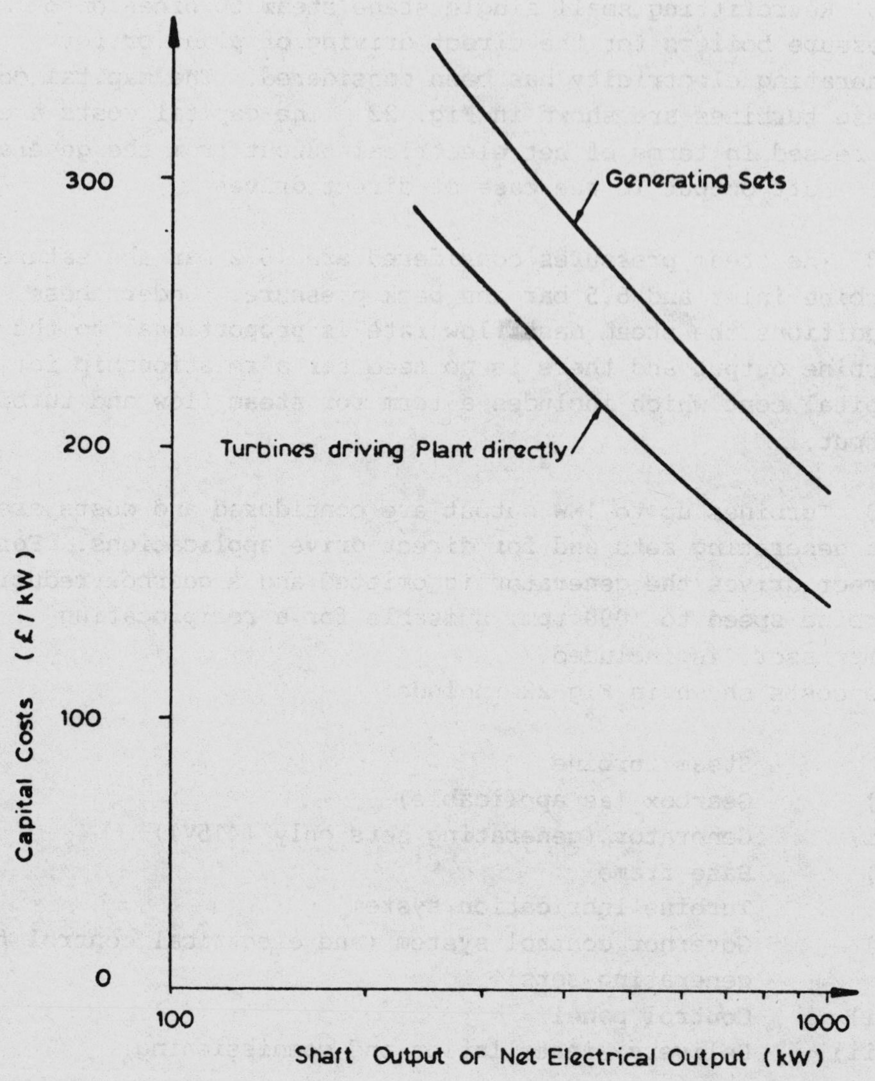

CAPITAL COSTS

of Single-Stage Back-Pressure Steam Turbines

for Low-Pressure Boilers

Fig. 22

Combined Cycle Plants

162 The items of plant required for a combined cycle system are gas turbines, waste heat boilers and steam turbines. The costs of these items are covered in other sections and need not be repeated here. The total costs for complete systems are given in Section 3.4.

3.2 Capital Costs of Boilers

Coal Fired Boilers

163 High pressure coal fired boilers have been considered in this study for providing steam for back pressure steam turbine CHP systems. The capital costs of these boilers are shown in Fig 23 and are expressed for single installations.

The costs include the following items:

i) Water tube boiler including superheater, steam drums and economiser
ii) Forced draught and induced draught fans
iii) Grit arrester, flues and chimney
iv) Feed pumps
v) Water treatment system
vi) Coal storage and handling system
vii) Ash handling system
viii) All feedwater and steam pipework and insulation to connect to turbines and existing systems.
ix) All structural steelwork and access ladders etc.
x) Boiler, control system including panel
xi) Delivery, installation and commissioning

164 The boiler steam outlet pressures and temperatures used for this study were:-

i) 32 bar abs/400°C (450 psig/650°F)

ii) 45 bar abs/454°C (750 psig/850°F)

iii) 64 bar abs/482 C (900 psig/900 F)

The boilers were of a standard stoker fired design.

165 A water treatment plant is considered necessary for boilers producing steam above 32 bar abs pressure. Existing water treatment plant for use with existing low pressure steam boilers would not be suitable for the higher pressure boilers. The water treatment plant has been sized assuming a 20% make up requirement.

166 A coal storage facility holding sufficient coal for 3 weeks operation of the boiler at full load has been allowed for. The coal would be stacked in a heap and a front end loader would take the coal into a short term storage hopper, which would feed into the plant. This form of coal storage is believed to be more economical than silos for the large quantities of coal necessary for CHP plants in the size range considered.

167 The most economic form of boiler construction is dependant on the size of the boiler required. For boilers of up to 9 kg/s of steam a factory assembled boiler is generally the most appropriate. For sizes of between 9 and 15 kg/s output of steam it is generally most economic to install two factory assembled boilers. For steam outputs greater than 15 kg/s a site assembled boiler is the cheapest. These three size ranges are shown in Fig. 23.

168 An allowance has been made for steam and water pipework to connect with a steam turbine and the existing works distribution system. This is likely to vary significantly from site to site, however a typical allowance of £24,000 per kg/s (£3 per lb/hr) of steam flow has been made based on experience at other installations.

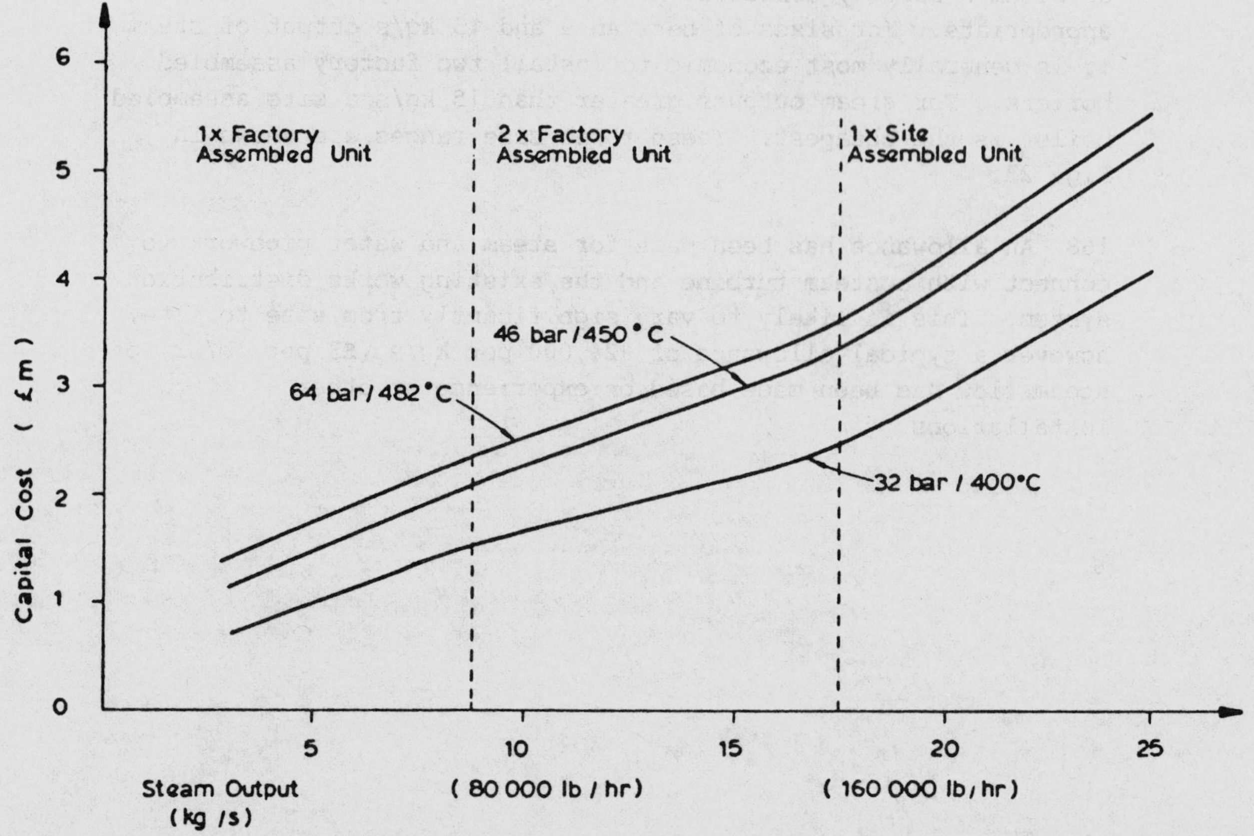

CAPITAL COSTS

of High Pressure Coal-Fired Boiler Installations

(Scope of Supply described in Section 3-2-1)

Fig. 23

Waste Heat Boilers for Gas Turbines

169 The costs of waste heat boilers for internal combustion engine plant were supplied by the engine manufacturers. However in the case of gas turbine plant boiler manufacturers were approached for prices of suitable waste heat boilers. These costs are shown on Fig 24 for single boiler installations.

170 For single cycle gas turbine based CHP plants waste heat boilers were used to generate low pressure steam for process heating purposes. However for combined cycle plant high pressure boilers, operating at steam outlet conditions of 64 bar abs, 482 ^{o}C, were used. They were found to give more economic systems than lower pressure boilers.

The costs shown in Fig 24 include the following items:

i)	Water tube boiler plus economiser and steam drums
ii)	Feed pumps and circulation pumps
iii)	Nominal flue gas inlet duct
iv)	Boiler insulation
v)	After-firing burner and duct (where applicable)
vi)	Flue gas bypass
vii)	Flue gas outlet duct and chimney
viii)	All feed water and steam pipework and insulation to connect to existing systems
ix)	All structural steelwork, access ladders etc.
x)	Boiler control system including panel
xi)	Water treatment plant (high pressure boilers only)
xii)	Delivery installation and commissioning

171 The low pressure boilers generate saturated steam at up to 18 bar abs. The high pressure boilers generate superheated steam at 64 bar abs, 482°

172 In all cases water tube boilers are used for gas turbines as they can accept the large volumes of exhaust gas without imposing too much back pressure on the gas turbine.

173 After-firing up to 800°C has been considered. In general most manufacturers will use a relatively inexpensive form of boiler construction for gas temperatures up to about 800°C. However with after-firing above 800-1000°C (depending on the manufacturer) a much more expensive boiler construction is required. It was found that the costs of waste heat boilers with after-firing up to 800°C were not significantly greater, per unit of steam output, than unfired boilers.

174 The cost of a gas supply to the after-firing burner is considered in Section 3.3.

175 Low pressure boilers can utilise the feedwater that would otherwise have been supplied to existing boilers at a site. However in the case of high pressure boilers additional water treatment would be required. The plant has been sized assuming a 20% feed water make up requirement.

176 An allowance of £12,000 per kg/s (£1.5 per lb/hr) of steam has been made for steam and water pipework to connect up with the existing site distribution systems. This cost will vary significantly depending on the site. The figure used is typical based on experience at other installations.

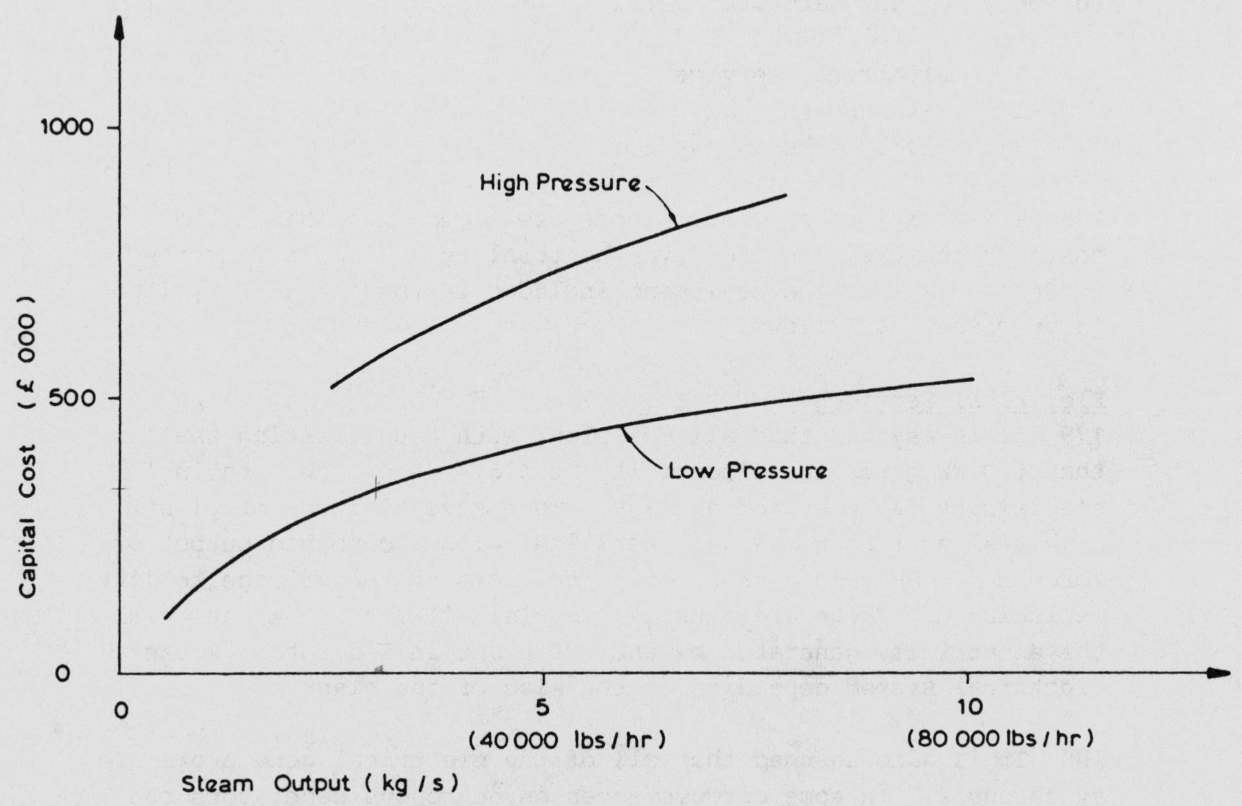

CAPITAL COSTS

of Waste Heat Boilers for Gas Turbines

(Scope of Supply described in Section 3-2-2)

Fig. 24

3.3 Capital Costs of Auxiliary Items

177 Allowances for the following auxiliary items have been added to the basic CHP hardware costs:

1) Electrical services
2) Gas supply
3) Civil Works

178 These capital cost allowances are added onto the capital costs of the CHP plant to give the total system costs described in Section 2.4. The equipment included in these auxiliary items is described as follows.

Electrical Services

179 It is assumed that all CHP plant with a unit rating smaller than 600 kW generates electricity at 415V. Above this rating electricity is generated at 11kV. When several items of plant each smaller than 600kW are installed, with a combined output of more than 1 MW then a transformer to increase the voltage to 11kV is included. There are thus two levels: 11kV or 415kV at which the electricity generated by the CHP plant is fed into a works electrical system depending on the size of the plant.

180 It is also assumed that all of the electrical generators are synchronous. In some circumstances asynchronous generators can be appropriate as they are cheaper and simpler to control, however the benefits of using the CHP plant for standby generation are then lost. The use of synchronous generators therefore provides a more flexible CHP system which can be used in 'stand alone' applications and has been assumed here.

The costs of the electrical services include the following items:

a) All 11 kV and/or 415V switchgear and cabling for the CHP generator and 415V switchgear and cabling for the auxiliaries and building services.
b) All metering of imported and exported electricity
c) All electrical building services, lighting, ventilation etc.

181 It is assumed that 100m of cabling is required to the nearest 11 kV substation in the case of high voltage generation and that the generation at low voltage can be fed into a local distribution system. As described in Section 3.3.3 it is assumed that where small spark ignition engines or gas turbines are installed with not more than two units each of rating less than 600kW then they can be installed in an existing boilerhouse and the existing electrical installations are of sufficient capacity to add on the CHP plant. Above this size range a new boilerhouse must be provided.

182 No provision is made for a new substation or any costs incurred by the local electricity board.

183 Typical costs of the electrical services range from £18,000 for a 200kW spark ignition engine generating at 415V to £75,000 for a 5MW gas turbine generating at 11kV.

Gas Supplies

184 It is assumed that gas is supplied to a site at 26" W.G pressure by the gas authority. Thus it will require compression before it can be used in any turbocharged internal combustion engines or gas turbines. After-fired boilers will also need gas booster blowers.

185 The gas pressures required by turbocharged internal combustion engines are in the order of 2-4 bar abs. For industrial gas turbines, pressures of 6-13 bar abs are required and 18-20 bar abs for aero derived gas turbines. Gas pressures of 30-40" W.G. are required for after-fired boilers.

The costs allowed for gas supplies include:

a) 60m of pipeline to new CHP installation.
b) Separate meters for the prime mover and after fired boiler (where applicable).
c) Gas compressors and blowers of appropriate size.
d) All interconnecting pipework.

186 The power required for compressing gas to the pressures required by gas turbines can be considerable; in some cases as much as 10% of the output from the gas turbine. There are thus very substantial savings both in terms of power saved and reduced capital cost of compression plant if the gas authority can supply gas at a higher pressure. For instance an increase of gas supply pressure to 3 bar gauge (45 psig) will halve the power needed to compress gas to 18 bar abs for a gas turbine. This could increase the net output power from the gas turbine by 5% at no increase in fuel consumption.

187 Typical costs of gas supply installations are £100,000 for a gas supply of 15 MW thermal (10 MW to a gas turbine at 6 bar abs and 5 MW to an after-fired boiler at 40" W.G.) increasing to £240,000 for a gas supply of 25 MW thermal (18 MW to a gas turbine at 20 bar abs and 7 MW to an after-fired boiler at 40" W.G.). If a high pressure gas supply is available these costs can be substantially reduced as a smaller compressor would be required.

Civil Works

188 It has been assumed that a new plant house will be required for all CHP plant, except for gas turbines and spark ignition engines where the unit size rating is less than 600kW and two or less units are installed. In the case of these small spark ignition and gas turbine engines and retrofitted turbines on low pressure steam boilers it is assumed that they are small enough and simple enough to be installed into an existing boiler house, in which case the cost of civil work is very small.

The civil works for a new plant house include:-

a) All excavations and foundations (in the case of large diesel engines these can be substantial)
b) Supply and erection of a building to house the prime movers and heat recovery boilers
c) Additional building extensions for gas meters and compressors and the CHP plant control room
d) Maintenance cranes (for internal combustion engines)

189 Typical costs used for the civil works were in the range of £400/m² floor area for gas turbines to £500/m² floor area for diesel engines. The buildings were assumed to be of steel frame construction with blockwork walls and sheet steel roof.

190 Typical costs of EOT maintenance cranes were in the order of 15-20% of the cost of a building.

3.4 Total System Costs

191 The total costs for CHP systems are made up from the costs described in sections 3.1 to 3.3 with additional allowances for contingencies and engineering design.

Engineering Design and Contingencies

192 Any CHP plant installation will incur some engineering design costs and these will typically be up to 5% of the total system cost. An allowance of 5% of total system cost has been added for each type of system except for gas turbine single cycle systems where 3% has been allowed. Gas turbine systems are generally somewhat simpler than the other types.

193 After considering the costs of several CHP plants that have been installed it was decided to allow for contingencies of 10% of the total system cost. This was because most CHP plants considered had one very large cost such as:-

Very high civil costs
New electricity substation required
New gas supply main needed to site
High costs for interconnecting pipework

194 These large contingencies generally added about 5% to the system cost and so, including the contingencies for other small items, a contingency allowance of 10% was considered to be appropriate.

Total CHP System Costs (December 1984)

195 The costs for single unit CHP plant installations are shown in Figs. 25 and 26. Figure 26 shows the detailed costs for gas turbine, internal combustion engine and combined cycle plant All of the costs are expressed in terms of net electrical output (or shaft output in the case of direct drive applications) rather than the gross rating of the prime mover.

The make up of each total system cost is described below:

a) Gas Turbine Plant - The total costs for systems with unfired and after-fired waste heat boilers include:-

i) Gas turbine costs (from Section 3.1.1, Table 49)
ii) Low pressure heat recovery boiler costs (from Section 3.2.2, Fig. 24)
iii) Electrical services, gas supply, civil works (from Section 3.3)
iv) Contingencies 10%
v) Design fees 3%

b) <u>Heavy Fuel Oil Diesel Engine Plant</u> - The total costs for systems with unfired and after-fired waste heat boilers include:-

i) Engine and boiler costs (from Section 3.1.2, Fig. 19)
ii) Electrical services, civil works (from Section 3.3)
iii) Contingencies 10%
iv) Design Fees 5%

c) <u>Dual-Fuel Diesel Engine Plant</u> - The total costs for systems with unfired and after-fired waste heat boilers include:-

i) Engine and boiler costs (from section 3.1.3 Fig. 20)
ii) Electrical services, gas supply, civil works (from section 3.3)
iii) Contingencies 10%
iv) Design Fees 5%

d) <u>Industrial Spark Ignition Engine Plant</u> - The total costs for electricity generating or direct drive applications include:

i) Engine and heat recovery costs (from Section 3.1.4, Fig. 21)
ii) Electrical services, gas supply, civil works (from Section 3.3)
iii) Contingencies 10%
iv) Design fees 5%

e) <u>High Pressure Boiler and Back-Pressure Steam Turbine Plant</u> - The total costs for high pressure steam boilers and back-pressure steam turbines include:

i) High pressure coal fired costs (from Section 3.2.1, Fig. 23)
ii) High pressure steam turbine costs (from Section 3.1.5, Table 50)
iii) Electrical services, civil works (from Section 3.3)
iv) Contingencies 10%
v) Design fees 5%

196 The costs for all 19 types of steam turbine configuration used have not been shown. Instead 6 examples have been given which indicate how the costs vary with the type of turbine, the boiler pressure and the back-pressure.

f) <u>Back-Pressure Steam Turbine Plant for Low Pressure Boilers</u> - The total costs for retrofitted steam turbines onto low pressure steam boilers for electricity generation and direct drive applications include:

i) Steam turbine costs (from Section 3.1.6, Fig. 22)
ii) Electrical services, civil works
iii) Contingencies 10%
iv) Design fees 5%

g) <u>Combined Cycle Plant</u> - The total costs for combined cycle plant consisting of gas turbine, after-fired high pressure waste heat recovery boiler and back pressure steam turbine include:

i) Gas turbine costs (from Section 3.1.1, Table 49)
ii) High pressure heat recovery boiler costs (from Section 3.2.2, Fig. 24)
iii) Back pressure steam turbine costs (from Section 3.1.5, Table 50)
iv) Electrical services, gas supply, civil works (from Section 3.3)
v) Contingencies 10%
vi) Design fees 5%

<u>Multiple Unit Installations</u>

197 In the analysis of CHP systems on industrial sites the installation of up to four prime movers of equal size was considered. The total system cost per prime mover will decrease when multiple units are installed as the costs of providing all the auxiliary items do not increase proportionally when multiple units are installed. Cases of multiple units were considered and it was found that the following multipliers can be used on the total system costs for one unit to obtain the costs for multiple units.

a) 1 unit installation - Multiply x 1
b) 2 unit installation - Multiply x 0.93
c) 3 unit installation - Multiply x 0.91
d) 4 unit installation - Multiply x 0.89

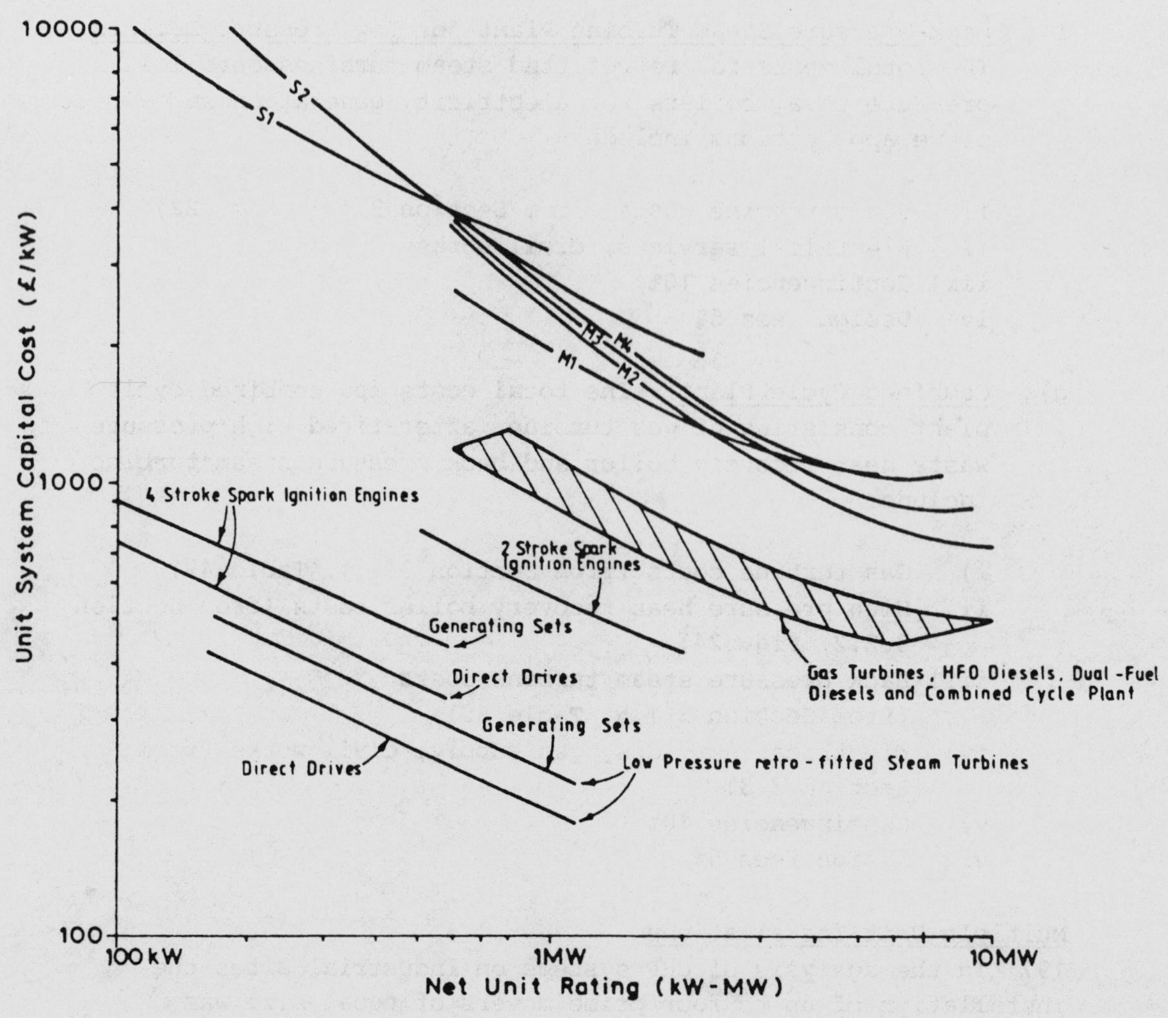

S1 - Single Stage Steam Turbine - 64 bar, 482°C Boiler - 18 bar back pressure.
S2 - Single Stage Steam Turbine - 32 bar, 400°C Boiler - 18 bar back pressure.
M1 - Multi Stage Steam Turbine - 32 bar, 400°C Boiler - 7 bar back pressure.
M2 - Multi Stage Steam Turbine - 64 bar, 482°C Boiler - 5 bar back pressure.
M3 - Multi Stage Steam Turbine - 64 bar, 482°C Boiler - 7 bar back pressure.
M4 - Multi Stage Steam Turbine - 64 bar, 482°C Boiler - 11 bar back pressure.

C.H.P. Total System Costs

Fig. 25

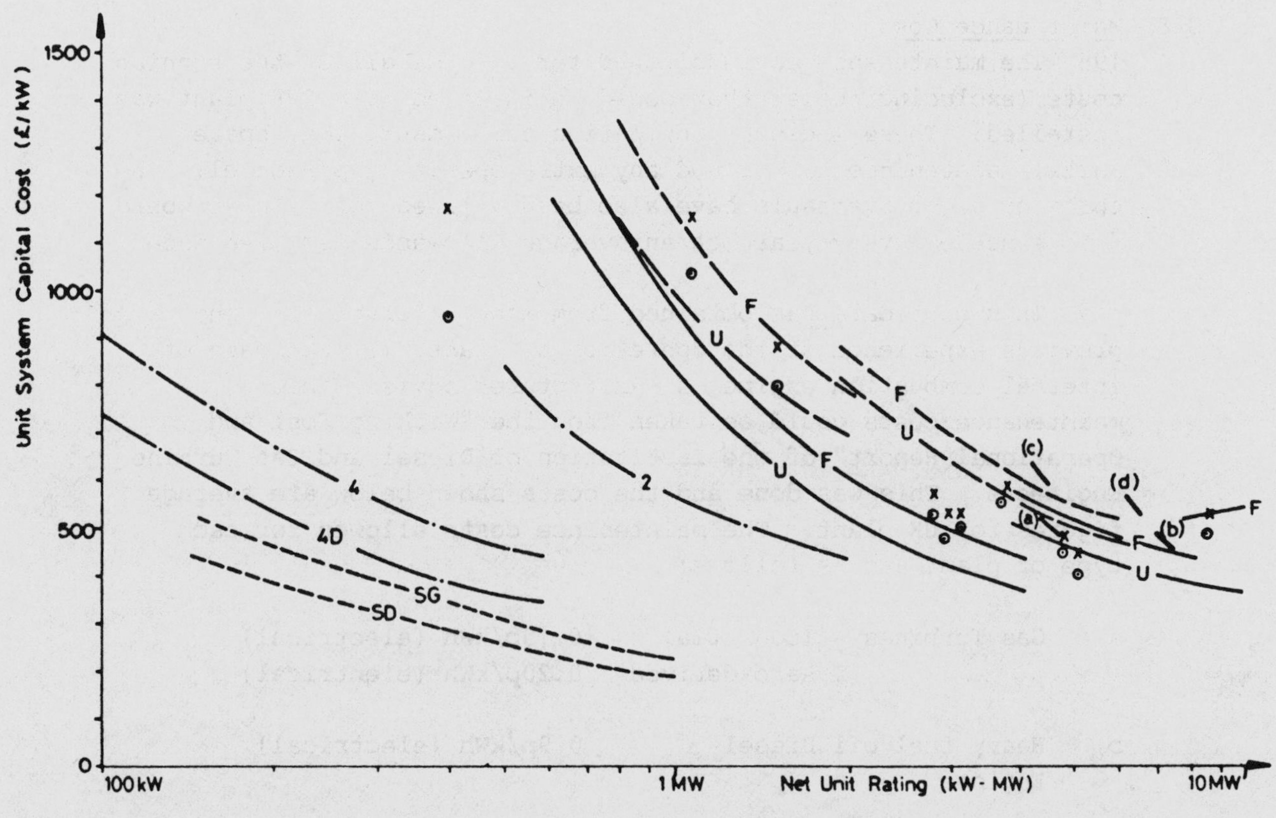

Gas Turbine - Unfired Heat Recovery Boiler	⊙
Gas Turbine - Fired Heat Recovery Boiler	x
H.F.O. Diesel - Unfired Heat Recovery Boiler	—U—
H.F.O. Diesel - Fired Heat Recovery Boiler	—F—
Dual Fired Diesel - Unfired Heat Recovery Boiler	--U--
Duel Fired Diesel - Fired Heat Recovery Boiler	--F--
Spark Ignition Engine - 2 Stroke	—·—2—·—
Spark Ignition Engine - 4 Stroke	—·—4—·—
Spark Ignition Engine - 4 Stroke - Direct Drive	—·—4D—·
Low Pressure Steam Turbine - Generating Set	---SG--
Low Pressure Steam Turbine - Direct Drive	---SD--
(a) Combined Cycle Plant Types I-III	—(a)—
(b) Combined Cycle Plant Types IV-VI	—(b)—
(c) Combined Cycle Plant Types VII-IX	—(c)—
(d) Combined Cycle Plant Types X-XII	—(d)—

Details of C.H.P. Total System Costs (from Fig. 25)

Fig. 26

3.5 Maintenance Costs

198 The maintenance costs allowed for include all of the running costs (excluding fuels) that would be incurred if a CHP plant was installed. These include lubrication oil consumption, spare parts, maintenance labour and any extra operating personnel. The costs of major overhauls have also been allowed for. These would not be needed every year but an average allowance has been made.

199 This cost data was obtained from manufacturers data and previous experience of the operation of plant. In the case of internal combustion engines a manufacturer advised that maintenance costs could be taken from the "Working Cost and Operational Report" of the Institution of Diesel and Gas Turbine Engineers. This was done and the costs shown below are average figures for UK plant. The maintenance costs allowed for each type of plant are as follows:

a) Gas Turbines - Industrial 0.15p/kWh (electrical)
Aero-derived 0.20p/kWh (electrical)

b) Heavy fuel oil Diesel Engines 0.9p/kWh (electrical)

c) Dual fuel Diesel Engines 0.9p/kWh (electrical)

d) Industrial Spark Ignition Engines 0.5p/kWh (electrical)

e) Coal fired Boilers and Steam Turbines

Boiler Pressure	Maintenance Allowance
32 bar absolute	2% of capital cost/year + 0.07p/kWh of boiler thermal output
46 & 64 bar absolute	2% of capital cost/year + 0.1p/kWh of boiler thermal output

f) Low pressure retrofitted steam turbines 0.15p/kWh (electrical)

g) Combined cycle plant 0.3p/kWh (electrical)

200 These costs are somewhat conservative and in practice lower maintenance costs may be achieved, which would enhance the cost effectiveness of a CHP scheme. In particular some modern diesel engines may give considerably lower maintenance costs than the average figures presented here.

4.0 FOSSIL FUEL PRICES

4.1 Comments on Fuel Prices

201 Four fossil fuels were considered in the study:

a) Natural Gas
b) Coal
c) Heavy Fuel Oil (3500 secs)
d) Light Oil (35 secs)

Natural Gas

202 Prices for interruptible natural gas were used in the study. At times of interruption it was assumed that the CHP plant would shut down and the existing boilers would be used with a standby fuel. Typically 10 days of interruption per year can be expected.

203 Gas is often supplied to an industrial site at a pressure of 1-2 psig. Gas turbines require high pressure gas, up to 18 bar absolute in some cases. The power needed for gas compression can absorb up to 10% of the output of a gas turbine. The British Gas Corporation can, in some cases, supply gas at a high pressure to gas turbine CHP plant and it is in the consumers interest to request this if it is appropriate.

Coal

204 Prices for washed singles supplied to a large industrial consumer were used in the study.

Heavy Fuel Oil

205 Prices for 3500 second oil as supplied to a large industrial consumer were used in the study.

Light Oil

206 Light 35 second oil is used as a pilot fuel for dual-fuel diesel engines. It provides 5-10% of the total fuel input. Consequently the economics of running a dual fuel engine are not particularly sensitive to the price of light oil.

4.2 Changes in Fuel Price

207 The price of heavy fuel oil changed considerably during the course of the study and this had an effect on the price of other

fuels. In December 1984 typical prices were:

Natural Gas	28.3p/therm	£2.68/GJ	28.3p/therm
Coal	£62.5/tonne	£2.20/GJ	23.2p/therm
Heavy Fuel Oil	15.6p/litre	£3.80/GJ	40.0p/therm
Light Oil	18.0p/litre	£4.70/GJ	50.0p/therm

Prices are based on higher (or gross) calorific value of the fuel.These prices were used in the initial assessment of the potential for CHP.

208 During the spring of 1985 the price of heavy fuel oil rose to 20p/litre and then fell to under 13p/litre during the summer. Towards the end of 1985 heavy fuel oil had fallen in price to 11p/litre and the potential for CHP was reassessed at the following prices:

Natural Gas	28.3p/therm	£2.68/GJ
Coal	£62.5/tonne	£2.20/GJ
Heavy Fuel Oil	11.0p/litre	£2.68/GJ

209 At the beginning of 1986 the price of heavy fuel oil fell sharply again and this time the price of interruptible natural gas also fell. The prices were at their lowest during the summer and rose again slightly in the Autumn. The graph shown in Figure 27 shows the trends over the last two years.

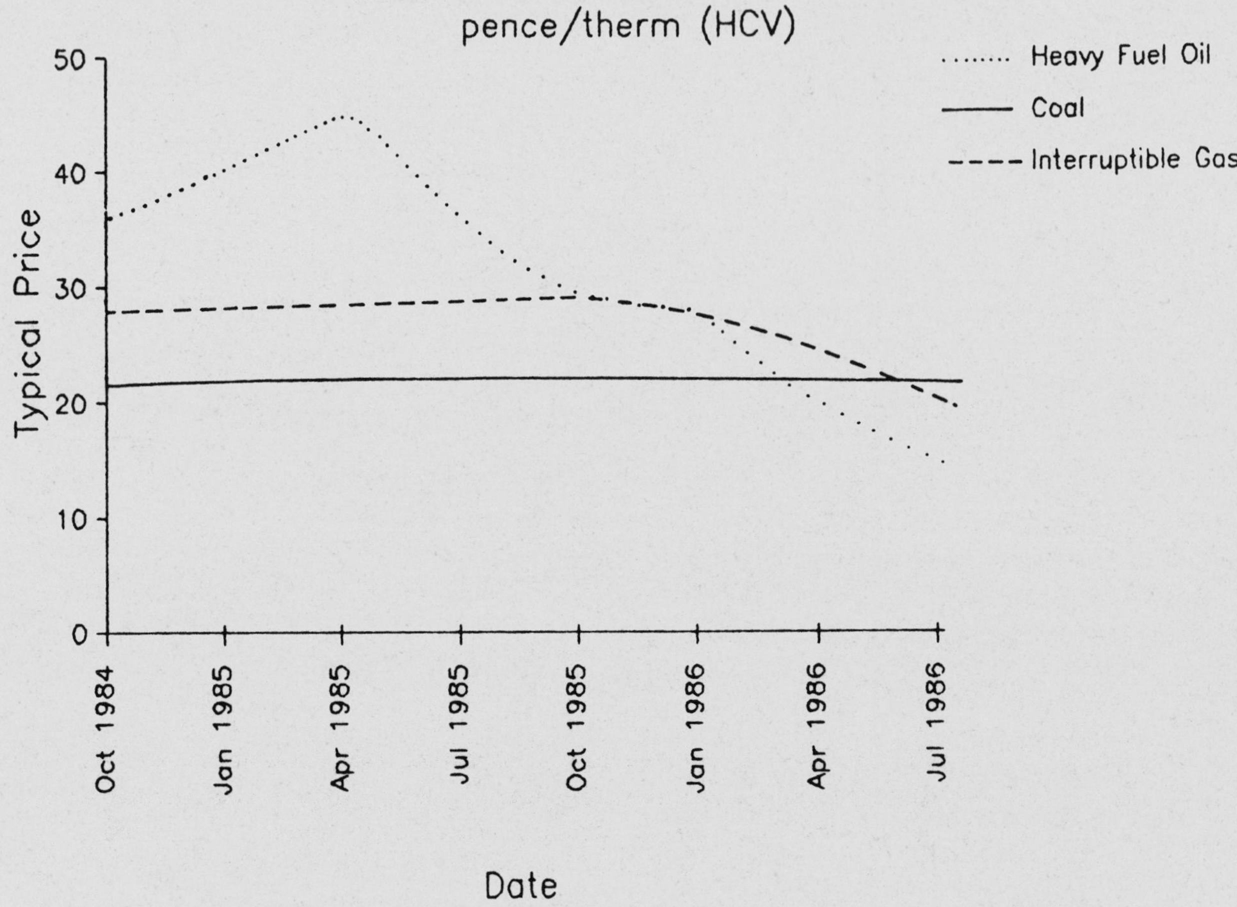

Fuel Price Changes (Oct. 1984 - Jul. 1986)

Fig. 27

4.3 Calculation of T.C.E. Savings

210 The overall primary energy savings resulting from the use of CHP plant were expressed in terms of tonnes of coal equivalent. It was assumed that the calorific value of coal was 26 GJ/tonne for the purposes of these calculations and that the overall efficiency of electricity generation and distribution in the U.K. was 30.5%.

The primary energy saving (T.C.E.) was calculated from the formula:

$$T.C.E. = (Q_f + E/0.305)/26$$

T.C.E. - Tonnes of Coal Equivalent energy saved per year

Q_f - Fossil fuel saving per year as a result of operating CHP plant (GJ) (this could be negative).

E - Electricity generated by CHP plant per year (GJ)

5.0 ELECTRICITY TARIFFS

5.1 Electricity Tariffs for Industrial Consumers

i) Non CHP Users

211 There are three electricity tariffs in use for non-CHP industrial consumers. These are:

a) Maximum demand based tariffs
b) Unit based tariffs
c) Load management tariffs

212 Maximum demand tariffs are the most popular for small and medium sized sites. The tariff is based on the maximum electrical demand drawn in addition to the number of units consumed.

213 Unit based tariffs are rare and only provided by a few electricity boards. They do not contain a maximum demand charge; all of the tariff is taken on unit charges. As this tariff is rare it was not considered in the study.

214 Load management tariffs are used for very large consumers who are prepared to shed load, when required, in order to reduce the overall demand on the electricity network at times of peak demand. As it is only the very large consumers who use the load management tariff, it was not considered in the study.

ii) CHP Users

215 Since the 1983 Energy Act the electricity boards have been obliged to publish tariffs for private generators of electricity. These tariffs cover the purchase of electricity from private generators and the use of an electricity board's distribution system for a private generator to supply private consumers. Only the tariffs concerned with the purchase of electricity from a private generator are used in the study.

216 There are two types of tariff for private generators:

a) Maximum demand based tariff
b) Unit based tariff

The unit based tariff is offered only by a few electricity boards, thus it was not considered in the study. All of the other electricity boards offer maximum demand based tariffs.

5.2 Components of the Electrical Tariff

217 For industrial consumers considered in this study the maximum demand tariffs are by far the most common and are therefore the only ones considered. The tariff comprises six components.

i) A standing charge
ii) A capacity or availability charge
iii) A maximum demand charge
iv) Charges for each unit of electricity
v) Charges for reactive power consumed
vi) An adjustment to account for the variation of fuel costs.

218 These are described as follows:

i) Standing Charge - This is a fixed monthly charge payable for the provision of an electrical connection to the local electricity board.

ii) Capacity Charge - This is a monthly charge in accordance with the size of, the demand of the site. It is based on the 'Chargeable Service Capacity' which is the electrical capacity made available to the site by the electricity board and is negotiated when the supply is installed. It can be re-negotiated if the demand required by the site changes. For the purposes of this study the capacity changes are based on the maximum electrical demand of a site. In the case of a private generator the charge may correspond to the maximum export capacity from the CHP system if this is larger than the site demand.

iii) Maximum Demand Charge - This is a monthly charge payable for the maximum consumption during any half hour period in that month. The size of the charge will vary from month to month being largest during the winter months. In some cases the charge can be restricted to consumption taken only during the day, typically between 8am and 8pm, on weekdays.

iv) Unit Charge - The charges for each unit of consumption may be constant or two rates, one for night and one for day consumption may be used. In the case of units exported to the electricity board, the payment made by the board varies with time of year in addition to time of day.

v) Reactive Power - The reactive power charge is usually small and can be zero if the overall power factor at the site is high or if the supply is metered at high voltage.

vi) Fuel Cost Adjustment - This is a supplement to the unit charge dependant on the average cost of the fuel purchased by the CEGB.

219 The charges may vary depending whether the electrical supply is metered at high voltage (>1kV) or low voltage (<1kV).

220 In general, over a year, a typical maximum demand tariff could be made up as follows, depending on load factor:

Standing and capacity charges	3- 5%
Maximum Demand charges	9-11%
Unit charges	84-88%

221 The components described above apply to both conventional maximum demand tariffs and those for private generators.

Change in Tariff on Installation of CHP

222 When an industrial consumer installs a CHP plant the electricity board will change the tariff from a conventional maximum demand tariff to a tariff for a private generator. The demand supplied by the electricity board will normally be reduced and the private generator may negotiate a reduced capacity charge. However if the private generator wishes to use the electricity board to supply the full site demand when the CHP breaks down then the full capacity charge must be paid for the privilege of the standby provided by the electricity board.

223 In assessing the economics of CHP systems comparison is made of the electrical charges without a CHP system, based on a conventional maximum demand tariff, and the electrical charges with CHP, based on a tariff for a private generator.

5.3 Tariff Used for the Study

224 There are 15 electricity boards in the United Kingdom and each offer slightly different tariffs. The variation is about 7% between the cheapest and the most expensive maximum demand tariffs.

225 The Midlands Electricity Board tariffs are close to the national average in overall cost. However as the Midlands tariff for private generators is unusual, in that it is a unit based

tariff with no maximum demand component, the London Electricity Board tariffs were chosen for use in the study.

226 It is assumed that the electrical supply to a site will be metered at high voltage (i.e. >1kV) and that the power factor is greater than 0.9 and therefore there will be no charge for reactive power.

Tariff for Non-CHP site (based on LEB 1984/85 tariff)

Monthly standing charge	£51.00
Monthly capacity charge	£0.75 per kVA of capacity
Monthly maximum demand charges	£3.85 per kW during November and March. £5.45 per kW during December, January and February. Zero at other times.
Unit Charges	1.54p/kWh for units consumed between midnight and 07.00 hours. 3.21p/kWh for units consumed at other times.
Reactive power charge	zero
Fuel Cost adjustment	+0.286p/kWh (based on a fuel cost of £51.5/tonne which applied at the end of 1984)

Tariff for CHP Site (based on LEB 1984/85 tariff)

Monthly standing charge	£66.5
Monthly capacity charge	£0.75 per kVA of site demand or export capacity whichever is the larger
Monthly maximum demand charges (for imported electricity	£3.85 per kW during for November and March £5.45 per kW during December, January and February. Zero at other times
Unit charges for imported electricity	1.54p/kWh for units consumed between midnight and 07.00 hours 3.21p/kWh for units consumed at other times
Unit payments for exported electricity	1.37p/kWh for units exported between 00.30 hours and 07.30 hours. 3.46p/kWh for units exported between 07.30 hours and 20.00 hours, on Mondays to Fridays during November and March. 6.05p/kWh for units exported between 07.30 hours and 20.00 hours, on Mondays to Fridays during December, January and February. 2.43p/kWh for units exported at other times.
Reactive power charge	zero
Fuel cost adjustment	+0.286p/kWh for imported electricity +0.26p/kWh for exported electricity (with a fuel cost of £51.50/ tonne)

6.0 COMPUTER MODEL FOR ASSESSMENT OF CHP PLANT

227 The economic assessment of CHP plants at industrial sites was carried out using a computer model to simulate the operation of the CHP plant. The operation of the computer model, the assumptions contained and the checking of the results is described as follows.

6.1 Description of the Computer Model

Operation of the Model

228 The CHP computer model contains, as its core, a method for determining the optimum mode of operating a CHP plant to obtain the greatest cost saving, and a method of establishing the quantity of those savings.

229 The optimum mode of operating the CHP plant is determined for each hour of the day of the year by considering the CHP plant operating in an electricity matching or a heat matching (with exported electricity) mode. The hourly running costs of operating in either mode are compared with the cost of switching the plant off. The optimum mode is considered to be the one with the lowest running cost. Having established the optimum mode of operation of the CHP plant for each hour of each day of the year, the computer model sums the hourly running costs for the whole year and compares them with annual running costs without a CHP plant to obtain an annual saving in running costs. The running cost saving can then be compared with the capital cost to obtain a simple payback for the installation of the CHP plant. The cost analysis takes into consideration all of the following costs:

a) Fuel cost for the CHP plant
b) Fuel cost for existing boiler plant
(to top up the CHP plant)
c) Maintenance cost of CHP plant
(including allowances for plant breakdown and the effect on maximum demand)
d) Cost of imported electricity
e) Cost benefit of exporting electricity

230 Considering the operation of the CHP plant at each hour of the day allows the effect of hourly variations in site heat and power demands to be taken into account as well as the variations in electricity tariff at different times of the day. When determining the optimum mode of operation of the CHP plant it is not considered practical to choose an optimum mode for each hour as a situation may occur where the plant switches on and off every alternate hour. The program thus sums the hourly running

costs over 3 periods of the day corresponding to the 3 electricity tariff periods, i.e.:

a) midnight to 07.00
b) 07.00 to 20.00
c) 20.00 to midnight

231 The optimum mode of operating the CHP plant in each of these 3 periods is calculated by comparing the running costs in each mode. This limits the maximum number of changes in the operating mode to 3 per day, which is considered to be more reasonable.

<u>Data Input to the Model</u>

232 The computer model requires input data for

a) CHP plant performance and capital cost
b) Fuel tariffs
c) Heat and power demands of a site

The CHP plant performance data, capital costs and fuel tariffs have been presented elsewhere in this appendix.
The heat and power demand data required is as follows:

a) Total annual energy consumption of electricity; fuel for boilers and fuel for other processes.
b) Pressures and temperatures at which steam and/or hot water is generated in boilers
c) Proportion of the steam and/or hot water used at a temperature of <80°C
d) Temperature of any hot air used for other processes and whether the process is suitable for the direct utilisation of exhaust gas from a CHP plant.
e) The efficiency of the existing boilers at the site and the fuel used.
f) The proportion of condensate returned to the boilers from the site steam system
g) The number of non-working days at the site
h) Electricity demands, boiler fuel demands and fuel demands for any direct fired hot air processes, (expressed as a percentage of the maximum demand) for each hour of a typical summer working, winter working and non-working day.

<u>Matching of CHP Plant Output to Site Demands</u>

233 The heat and power demands of a site for typical summer working, winter working and non-working days are input to the computer model. From these three day types the program calculates a further 2 day types to represent days in between summer and winter. It is assumed that the differences between

the summer and winter demands are on account of weather changes and that the variation follows the changes in degree days.

234 The model considers the utilisation of the waste heat in the exhaust from the CHP plant and includes a procedure for calculating the quantity of waste heat that can be recovered from a waste heat boiler at both full and part loads taking into account the quality of the steam or hot water generated. When an internal combustion engine is used as the CHP prime mover utilisation of the jacket cooling water is considered provided that there is a demand for heat at <80°C. This low temperature heat demand could be either for low temperature process heat or preheating boiler feedwater. In the case of steam turbines and combined cycle plant the steam from the turbine exhaust is utilised.

235 The computer model has the facility to analyse sites where the heat demand is in the form of steam (up to 250 psig) and/or hot water up to 200°C. Additionally the model can analyse a case where the exhaust gas from the CHP plant is used directly in a plant; such as the use of gas turbine exhaust in a drying plant.

236 Further detailed assumptions contained in the computer model are described in Section 6.3.

6.2 Use of the Computer Model

237 The computer model was used in two versions to analyse CHP installations at a particular site. A diagram indicating this arrangement is shown in Figure 28.

Version I of the Computer Model

238 Version I of the computer model was set up to analyse each of the six different types of CHP plant to the site. For each type of CHP plant several sizes of plant were considered by incrementing across the size ranges or, in the case of gas turbines and combined cycle plant, analysing in turn each of the plants selected for the study. For each plant and size analysed the following alternatives were also considered:

a) Installation from 1 to 4 units of plant (each unit of equal size).

b) Use of unfired or after-fired waste heat boilers (when applicable).

239 Consequently there could be up to 8 CHP plants analysed of each size of each type. Thus altogether over 600 different CHP plant configurations could be analysed at any site. Version I of the computer model was set up to print out the results of each

analysis giving the energy savings and simple payback. An example of the printout from Version I is shown in Fig. 29. It can be seen that the printout enables the plant with the lowest payback or greatest running cost savings to be identified easily.

240 The program does not printout those cases where the analysis results is a payback greater than 10.2 years.

Version II of the Computer Model

241 Version II of the computer model performed the same analysis process as Version I but only for one specified CHP plant configuration. Version II was set up to provide a detailed printout of this analysis, as shown in the example in Figure 30. This printout shows a detailed breakdown of the running costs and fuel consumptions at the site both with CHP and without CHP using the existing boiler system. The printout also shows the optimum mode of operating the plant during each period of the day and for each day type.

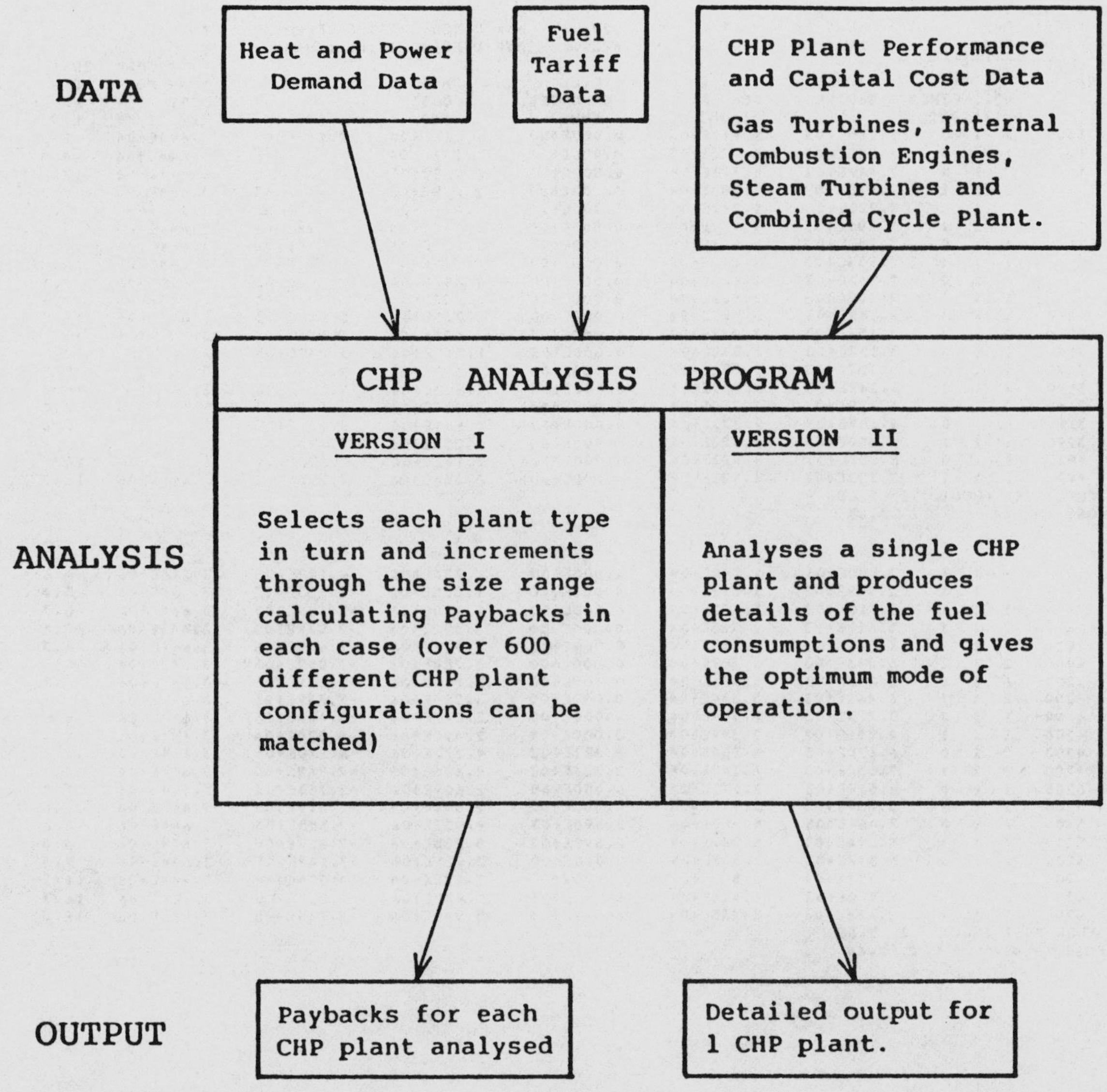

C.H.P. Computer Model

Fig. 28

THE EXISTING FUEL IS 1
MAX ELEC DEMAND= 9426. MAX HEAT DEMAND (STEAM)= 14108. MAX DIRECT FIRING DEMAND= 0.
AVE ELEC DEMAND= 7739. AVE HEAT DEMAND (STEAM)= 11622. AVE DIRECT FIRING DEMAND= 0.

PLANT TYPE	SIZE BAND	UNIT RATING	NO. UNITS	ECON	FIRED BOIL	TCE SAVING	ELEC GENERATED [MWh]	ELEC EXPORTED [MWh]	CAPITAL COST [£]	NPV 5% [£]	EXISTING RUNNING COST [£/YR]	RUNNING COST SAVING [%]	PAY-BACK [YRS]
1	2	1520	1	1	0	1.235E+03	8.915E+03	0.000E+00	1.213E+06	-2.190E+05	3.644E+06	3.5	9.42
1	2	1520	1	1	1	1.763E+03	8.915E+03	0.000E+00	1.346E+06	-6.838E+04	3.644E+06	4.5	8.13
1	2	1520	2	1	0	2.449E+03	1.774E+04	0.000E+00	2.255E+06	-2.801E+05	3.644E+06	7.0	8.81
1	2	1520	2	1	1	2.575E+03	1.783E+04	0.000E+00	2.503E+06	-4.587E+05	3.644E+06	7.3	9.45
1	3	2970	1	1	0	2.922E+03	1.742E+04	0.000E+00	1.407E+06	8.090E+05	3.644E+06	7.9	4.90
1	3	2970	1	1	1	3.381E+03	1.742E+04	0.000E+00	1.583E+06	8.798E+05	3.644E+06	8.8	4.96
1	4	5145	1	1	0	5.752E+03	3.013E+04	0.000E+00	2.056E+06	2.154E+06	3.644E+06	15.0	3.77
1	4	5145	1	1	1	5.933E+03	3.018E+04	0.000E+00	2.286E+06	2.022E+06	3.644E+06	15.3	4.10
1	5	2840	1	1	0	2.567E+03	1.666E+04	0.000E+00	1.481E+06	4.508E+05	3.644E+06	6.9	5.92
1	5	2840	1	1	1	3.108E+03	1.666E+04	0.000E+00	1.626E+06	5.972E+05	3.644E+06	7.9	5.65
1	5	2840	2	1	0	4.187E+03	2.987E+04	0.000E+00	2.755E+06	5.656E+05	3.644E+06	11.8	6.41
1	5	2840	2	1	1	4.156E+03	3.041E+04	0.000E+00	3.025E+06	2.856E+05	3.644E+06	11.8	7.05
1	6	3140	1	1	0	3.557E+03	1.842E+04	0.000E+00	1.549E+06	9.742E+05	3.644E+06	9.0	4.74
1	6	3140	1	1	1	5.157E+03	2.331E+04	0.000E+00	1.677E+06	1.106E+06	3.644E+06	9.9	4.65
1	6	3140	2	1	0	5.547E+03	3.171E+04	0.000E+00	2.882E+06	1.271E+06	3.644E+06	14.8	5.36
1	6	3140	2	1	1	5.498E+03	3.220E+04	0.000E+00	3.119E+06	1.012E+06	3.644E+06	14.7	5.83
1	7	3790	1	1	0	4.189E+03	2.223E+04	0.000E+00	2.101E+06	8.892E+05	3.644E+06	10.6	5.42
1	7	3790	1	1	1	4.580E+03	2.223E+04	0.000E+00	2.256E+06	9.445E+05	3.644E+06	11.4	5.44
1	8	4825	1	1	0	8.826E+03	3.933E+04	0.000E+00	2.192E+06	2.122E+06	3.644E+06	15.3	3.92
1	8	4825	1	1	1	9.322E+03	3.933E+04	0.000E+00	2.356E+06	2.225E+06	3.644E+06	16.3	3.97

EXISTING FUEL COST (HCV) £ 2.68
CHP FUEL COST (HCV) £ 2.68

PLANT TYPE	SIZE BAND	UNIT RATING	NO. UNITS	ECON	FIRED BOIL	TCE SAVING	ELEC GENERATED [MWh]	ELEC EXPORTED [MWh]	CAPITAL COST [£]	NPV 5% [£]	EXISTING RUNNING COST [£/YR]	RUNNING COST SAVING [%]	PAY-BACK [YRS]
3	2	2000	1	1	1	2.588E+03	1.140E+04	0.000E+00	1.720E+06	-4.147E+05	3.644E+06	4.6	10.17
3	2	2300	1	1	1	2.899E+03	1.311E+04	0.000E+00	1.825E+06	-3.653E+05	3.644E+06	5.2	9.65
3	2	2600	1	1	1	3.210E+03	1.482E+04	0.000E+00	1.930E+06	-3.159E+05	3.644E+06	5.7	9.23
3	2	2600	2	1	1	5.619E+03	2.965E+04	0.000E+00	3.592E+06	-7.939E+05	3.644E+06	9.9	9.91
3	2	2900	1	1	1	3.521E+03	1.653E+04	0.000E+00	2.035E+06	-2.665E+05	3.644E+06	6.3	8.88
3	2	2900	2	1	1	6.046E+03	3.307E+04	0.000E+00	3.788E+06	-7.853E+05	3.644E+06	10.7	9.74
3	2	3200	1	1	1	3.832E+03	1.824E+04	0.000E+00	2.140E+06	-2.171E+05	3.644E+06	6.8	8.59
3	2	3200	2	1	1	6.462E+03	3.640E+04	0.000E+00	3.984E+06	-7.789E+05	3.644E+06	11.4	9.60
3	2	4200	1	1	0	3.775E+03	2.394E+04	0.000E+00	2.297E+06	-4.480E+05	3.644E+06	6.6	9.59
3	2	4200	1	1	1	4.795E+03	2.394E+04	0.000E+00	2.490E+06	-9.220E+04	3.644E+06	8.5	8.02
3	2	4200	2	1	0	6.670E+03	4.505E+04	2.331E+02	4.272E+06	-8.970E+05	3.644E+06	12.0	9.77
3	2	4200	2	1	1	7.533E+03	4.505E+04	2.331E+02	4.636E+06	-7.962E+05	3.644E+06	13.6	9.32
3	2	5200	1	1	0	4.569E+03	2.975E+04	0.000E+00	2.609E+06	-3.783E+05	3.644E+06	7.9	9.03
3	2	5200	1	1	1	5.637E+03	2.975E+04	0.000E+00	2.840E+06	-3.429E+04	3.644E+06	10.0	7.81
3	2	5200	2	1	0	7.461E+03	5.067E+04	2.598E+03	4.852E+06	-7.555E+05	3.644E+06	14.6	9.14
3	2	5200	2	1	1	8.216E+03	5.067E+04	2.598E+03	5.288E+06	-7.853E+05	3.644E+06	16.0	9.07
3	2	6200	1	1	0	5.347E+03	3.547E+04	0.000E+00	2.921E+06	-3.147E+05	3.644E+06	9.3	8.65
3	2	6200	1	1	1	6.351E+03	3.547E+04	0.000E+00	3.190E+06	-4.329E+04	3.644E+06	11.2	7.83
3	2	6200	2	1	0	7.970E+03	5.435E+04	6.267E+03	5.432E+06	-7.222E+05	3.644E+06	16.7	8.90
3	2	6200	2	1	1	8.636E+03	5.435E+04	6.267E+03	5.940E+06	-8.718E+05	3.644E+06	18.0	9.05

EXISTING FUEL COST (HCV) £ 2.68
CHP FUEL COST (HCV) £ 2.68

C.H.P. Computer Model – Sample Printout from Version I

Fig. 29

```
DAILY AVE ELEC DEMAND =  4566.KW DAILY AVE HEAT DEMAND =18820.KW
DAILY AVE DIRECT FIRE =     0.KW
PLANT IS 1    NO OF UNITS = 2   UNIT RATING =  4825 KW
BAND IS  8        BOILER ECONOMISER = 1    FIRED BOILER = 1
MACHINE HOURS PER YEAR =  15423.  EXISTING FUEL TYPE 1
EXISTING FUEL COST (HCV)    £  2.68/GJ
CHP FUEL COST (HCV)         £  2.68/GJ
                                       CHP            EXISTING SYSTEM
TOTAL ANNUAL RUNNING COST          £ 2369822.            £ 3511319.
CHP FUEL COST                      £ 2826044.            £       0.
FOSSIL FUEL COST (EXISTING PLANT) £  199671.            £ 2120751.
ELEC IMPORT COST   UNIT CHARGES,   £   91750.            £ 1222102.
ELEC EXPORT CREDIT   UNIT CHARGE   £  986630.            £       0.
ANNUAL MAINTENANCE COST (PLANNED) £  126377.            £       0.
TOTAL ANNUAL CHP FUEL                   9995.  kTh              0.  kTh
TOTAL FOSSIL FUEL (EXISTING PLANT)       706.  kTh           7501.  kTh
TOTAL ELEC IMPORT                    2991522.  KWH       40000016.  KWH
TOTAL ELEC EXPORT                   26180176.  KWH              0.  KWH
TOTAL ELEC GENERATED                63188664.  KWH              0.  KWH
ELEC MAX DEMAND CHARGES,           £       0.            £   99483.
ELEC STANDING CHARGES              £   87648.            £   68984.
WORST MAX DEMAND PENALTY OWING TO BREAKDOWN   £ 24961.

     CAPITAL COST                  £ 4382346.
     NET COST SAVINGS              £ 1141497.
     SIMPLE PAYBACK                       3.839 YEARS

     NET NATIONAL ENERGY SAVING       15691.   TCE/YR
     NET PRESENT VALUE (30%RR),       £ -852837.
     NET PRESENT VALUE (10%RR),       £ 2626447.
     NET PRESENT VALUE (5%RR),        £ 4430013.

                    MATCHING        (1=ELEC 2=HEAT 3=SWITCH OFF)
                    ********
                                         *** PERIOD ***
                                       NIGHT       DAY        EVENING
       DAY TYPE WINTER                  1.          2.            2.
                WINTER/SPRING           1.          2.            2.
                SPRING/SUMMER           1.          2.            2.
                SUMMER                  1.          2.            2.
                NONWORKING,             3.          3.            3.
```

C.H.P. Computer Model – Sample Printout from Version II

Fig. 30

6.3 Assumptions contained within the Computer Model

242 The following is a list of the main assumptions contained within the computer model. These are in addition to the assumptions on plant performance and capital cost described earlier in this appendix and those on the data on industrial sites described in Appendix 3.

Assumptions concerning the site demand data

243 Most of the assumptions were made as a consequence of the quality of the data available from the industrial data base.

i) Heat and power loads were expressed as a proportion of the maximum site load for each hour of the day during typical summer and winter working days and a typical non-working day. It was assumed that the variation of the heat and power demand data from summer to winter working days would be in proportion to the variation in degree days.

ii) It was assumed that four working day types and one non-working day type would satisfactorily represent the operation at a site. The day types represent the following periods of the year.

Day Type	Working days in:
1	December, January and February
2	November and March
3	October, April and May
4	June to September

Where non-working days were specified they were assumed to be spread evenly throughout the year (i.e. weekends).

iii) The demand profiles were specified for electricity, boilers (steam and/or hot water) and other fuels. For boilers, three temperatures and pressures for steam and one temperature for hot water and the quantity of heat (as a proportion of the total heat demand) required in each case could be specified. It was assumed that the heat demand at each temperature level was a constant proportion of the site demand at any time of the day or year.

Similarly with other fuels two temperatures of heat required for direct fired plant could be specified.

iv) Any demand for heat at a temperature less than 80°C was specified as a fixed proportion of the hourly site boiler heat demand. Only heat used for process applications at a temperature less than 80°C was included.

244 Assumptions concerning the matching of site demand and CHP plant

i) The boiler heat demand was matched to the heat produced from an exhaust gas waste heat boiler, in the case of gas turbine and internal combustion engine CHP plant, and matched to the exhaust steam in the case of steam turbines and combined cycle plant.

It was assumed that the temperature of the exhaust gas from the CHP plant engine must be at least 40°C higher than the temperature of the heat demand matched, in the case of gas turbine and internal combustion engines. Any heat demanded at a temperature too high for the CHP engine exhaust must be supplied by existing boilers at the site.

ii) When a site heat demand was matched to the CHP plant, the CHP plant waste boiler was assumed to provide all of the heat at the highest steam or hot water temperature required. In the case of steam turbine and combined cycle plant the steam turbine corresponding to the highest site steam or hot water temperature was considered.

iii) In the case of internal combustion engine plant the proportion of the site heat demand less than 80°C was matched if possible to the heat recovered from the jacket cooling water. In addition the jacket cooling water could be used to preheat to 80°C any boiler make up feed water required. If there was no demand for heat less than 80°C then the jacket cooling water was not utilised and was dumped. If the demand for heat less than 80° C was too large to be met by the jacket cooling water then the demand was made up with heat from the exhaust gas waste heat boiler.

iv) Any heat that could not be matched by the CHP plant was assumed to be provided by the existing boiler installations or direct fired plant.

v) Electricity matching - It was assumed that the CHP plant was run to match the site electrical demand. If the demand was higher than the maximum continuous rating (MCR) of the plant then electricity must be imported. If the demand was lower(than the MCR of the plant then the CHP plant was operated at part load. If the electricity demand was lower than the minimum electrical output of one unit of the CHP plant then the CHP plant was turned off. The heat recovered from the CHP plant was either matched by the site demand or dumped. After-firing of waste heat boilers (if applicable) was used to match the site heat demand if required.

vi) Heat matching - It was assumed that the CHP plant was run to match the site heat demand. If the heat demand was too low for the minimum output of the CHP plant then the plant was switched off and existing boilers were used. If the CHP plant had the capability for after-firing then the minimum after-firing was used. The CHP plant would operate at the maximum load necessary to match the heat demand without after-firing and only if the heat demand could not be met at MCR of the CHP plant was after-firing used. If the corresponding electricity output from the CHP plant was greater than the site demand then electricity was exported.

vii) The optimum mode of operating the CHP plant was chosen by comparing the hourly running costs for electricity matching, heat matching or switching the CHP plant off and using existing boilers. The optimum was that which gave the lowest hourly running cost.

viii) Where more than 1 unit of CHP plant was installed it was assumed that all units would always run at the same load and that the minimum number of units would be run to match the load. This would, under most circumstances, be the most efficient way of operating the plant.

245 <u>Assumptions concerning maintenance and breakdown of the CHP plant</u>

i) It was assumed that all planned maintenance of the CHP plant would take place on non-working days or, if these were insufficient, summer days. It was assumed that maintenance would take place under 3 shift working.

ii) It was assumed that there would be unplanned stoppages, attributable to plant breakdown, equal to half of the planned maintenance but distributed uniformly throughout the plant operating periods. In addition it was assumed that there would be one stoppage every year during the period when the largest maximum demand penalty could be incurred.

iii) A gas interruption of 10 days per year was found to decrease the annual CHP cost saving by 2-3%. This small effect is less than the performance tolerance of a CHP plant and has thus been omitted from the program for simplicity.

In any case a maximum demand penalty is included to account for the CHP plant shutting down during the time of heaviest demand and this often forms the major part of the penalty attributable to gas interruptions.

6.4 Comparison of the Computer Model with other CHP data

246 In order to test the computer model a number of runs were carried out on sites where detailed CHP feasibility studies had been undertaken. The results of some of these analyses are described below. In general the computer model produced results similar to those produced by the possibility studies.

Gas Turbine CHP plant

247 Data was available for a paper mill which was previously operating on heavy fuel oil and was to be converted to a gas turbine CHP scheme.

248 The site previously consumed 20,000 MWh of electricity and 3870 kTherms of oil for the boilers per year.

249 A comparison of the costs (mid 1985 prices) based on the actual performance of the installation is shown below:-

	Actual Costs	Computer Cost Prediction
Annual Running Cost Saving.	£774,000	£730,000
Capital Cost of Installation.	£1.61m	£1.67m
Single Payback.	2.1 years	2.3 years

250 The computer model thus gives a slightly more conservative prediction. This is largely on account of the assumption in the model that the gas is supplied at a pressure of 26" W.G.

251 The site described was supplied with gas at 100 psig. This reduces the requirements for gas compression; giving additional savings in running cost and capital cost.

Coal Fired Boiler and Steam Turbine CHP Plant

252 Data was available for a food factory where processing and canning take place. The data was from a study in 1981 and consequently the costs could not be compared realistically with those used in the computer model. Nevertheless a useful comparison was made by comparing the energy consumption figures.

253 The site previously used 8500 kilotherms per year of fossil fuel (oil and gas) and consumed 37,000 MWh/year of electricity.

The predictions of the energy consumption of a coal fired boiler and steam turbine CHP plant are shown below:

	Predictions from Previous Study	Predictions from Computer Model
Electricity generated by CHP plant	17,000 MWh/year	16,500 MWh/year
Coal needed to run CHP plant	40,000 Tonnes/year	39,900 Tonnes/year

254 The boilers generate superheated steam at 650 psig and the back pressure steam turbines exhaust at 100 psig. There are two steam turbines each rated at 1800 kWe.

255 The computer model gives a lower output of electricity from the turbines than the previous study of the site. This is believed to arise because the previous study took insufficient account of the low load factor of the steam turbines. Each steam turbine is sized to match well over half the process steam load. This is to provide an adequate standby. However this reduces the amount of electricity that can be generated as the overall load factor is low which reduces the efficiency of the steam turbines. The computer model demonstrated this by showing that 1 steam turbine rated at 3000 kWe could generate 18500 MWh of electricity with a coal consumption of 40,200 tonnes.

<u>Payback</u>

256 The computer model predicts a capital cost of £5.0m for the plant. Based on 1985 fuel prices and, assuming that site previously ran entirely on heavy fuel oil, this gives annual running cost savings of £1.36m and hence a payback of 3.7 years.

257 As originally calculated in the study, the annual cost savings were -1.1m and the capital cost estimated at -5.6m thus giving a payback of 5.1 years.

258 The payback has decreased largely on account of the increase in heavy fuel oil prices and the lower capital cost estimate used by the computer model.

Heavy Fuel Oil Diesel Engine

259 Data was available for a site in the rubber products industry where a previous investigation of a heavy fuel oil diesel engine based CHP plant had been made. The site used heavy fuel oil to provide steam from the boilers and the annual energy consumptions were:

Electricity	39,500 MWh
Heavy Fuel Oil	9.5 million litres

260 The proposed CHP plant was a 3.3 MWe diesel engine with an after-fired waste heat boiler and heat recovery from jacket cooling water to preheat the boiler feed water.

261 The site was supplied with electricity under a load management tariff the form of which made it advantageous to operate the CHP plant at reduced load during certain periods of the day.

262 The computer model does not include the facility for simulating a load management tariff of this type, nevertheless the predictions obtained compare closely with those from the previous investigation as shown below:

	Estimation from previous investigation	Computer model prediction
Site electricity consumption with CHP	27,600 MWh	26,800 MWh
Site heavy fuel consumption with CHP	10.9m litres	11.1m litres

263 The computer model predicts a slightly higher load factor for the diesel engine as the peculiarities of the load management tariff are not considered.

Payback

264 The previous investigation used a lower price for heavy fuel oil than the computer model and an adjustment in fuel cost to match the computer model resulted in the following costs:

	Estimation from previous investigation	Computer model prediction
Net annual fuel savings (including breakdowns)	£225,000	£235,000
Annual maintenance cost	£33,000	£110,000
Capital cost	£1.35m	£1.58m
Simple payback	6.1 years	12 years

265 The net annual savings are of similar size. The capital cost estimated in the previous investigation is somewhat lower than the estimates used in the computer model. This is largely on account of a lower civil cost as the engine was to be located inside an existing building.

266 The annual maintenance cost estimated by the computer model is much higher than that estimated in the previous investigation. This is derived from a working cost report which has investigated many engine installations and it includes major overhauls as well as extra operating personnel (see Section 3.5). The difference in maintenance cost accounts for the large difference in payback.

<u>Gas Turbine for Direct Fired Plant</u>

267 Data was available for a site that operated a large hot air drier. The exhaust from a gas turbine was suitable for use directly in the drier. A study had already been done and a comparison of the running cost savings and capital cost is shown below:

268 The gas turbine selected had a net electrical output of 1,100 kW. The turbine exhaust was ducted directly into the drier.

	Estimation from previous study	Computer model prediction
Annual running cost savings	£200,000	£185,000
Capital cost	£600,000	£615,000
Simple payback	3 years	3.3 years

The results from the computer model show a close agreement (within 10%) of the estimate from the previous study.

Table 51

269 We acknowledge the help and assistance of the following companies who provided technical information.

Gas Turbines	Internal Combustion Engines
A.E.G. Kanis	N.E.I. APE Ltd. - Crossley Engines
Centrax Ltd.	N.E.I. APE Ltd. - W.H. Allen
GEC Ruston Gas Turbines	Sulzer
Kongsberg	Harland and Wolff Ltd.
Sulzer	Stork Diesel
Dale Electric	Dale Electric
Ingersoll Rand	Deutz Engines
Noel Penny Turbines	Bowmaker
	R.A. Lister
Steam Turbines	Mirlees Blackstone
N.E.I. APE W.H. Allen	Applied Energy Systems
KKK Ltd.	Petbow Ltd.
Turbodyne Dresser	Rolls Royce Perkins Engines
A.E.G. Kanis	Clark Kincaid
R.G. Joliffe & Co. Ltd.	Auto Diesels (Jenbacher)
Peter Brotherhood	GEC Dorman Diesels
	MAN-GHH
Boiler Plant	
Foster Wheeler Power Products	
Senior Green	
Babcock Power Ltd.	
APV Spiro gills	
Stone-Platt Crawley	
Wanson Co. Ltd.	

INDUSTRIAL COMBINED HEAT AND POWER
THE POTENTIAL FOR NEW USERS

APPENDIX 3
INDUSTRIAL DATA BASE

CONTENTS

1.0 Introduction

1 The objective of this aspect of the project was to create a database on the patterns of energy consumption in those industries which have the greatest economic potential for the installation of CHP plants (and which are subsequently referred to as the 'selected industries'). Sites which operate CHP already and those whose annual energy consumption was less than approximately 250,000 therms have been excluded. The following sections describe the methodology which was employed to generate the Industrial Database.

2.0 Selection of Priority Industries

2 The Department of Energy maintains a database containing information on the size distribution of manufacturing establishments by annual energy consumption. This was used to identify the most promising subsectors of industry. The data is the result of an energy purchases enquiry carried out in 1979 and it covers approximately 11,000 out of a total of some 100,000 manufacturing establishments. The survey coverage of large energy users was virtually 100%, of medium energy users around 85%, while small energy users were covered through only a small sample.

3 The Economics and Statistics Branch of the Department of Energy extracted non-confidential data from the DEn database tailored to suit the requirements of the selection exercise. This data subsequently formed part of the Industrial Database.

4 An initial screening of this and other data identified 50 industry groups (either individual Minimum List Headings of the Standard Industrial Classification, sub-divisions of Minimum List Headings, or groupings of more than one Minimum List Heading) which have a significant energy consumption and which were thought to have possible potential for new CHP.

5 The screening made use of rough estimates for both the energy use characteristics of the different subsectors of industry and the range of sizes and heat to power ratios appropriate to the different types of prime mover. An algorithm was then devised and applied to the industries covered by the Department of Energy data. Those remaining subsectors which contain existing CHP sites were modified or deleted. This procedure identified 50 industries representing the following sectors:

- food, drink and tobacco
- chemicals and allied industries
- rubber and plastics processing
- paper and board
- textiles
- land vehicles, ships and aircraft industries
- mechanical, electrical, electronic and instrument engineering.

6 The oil refining, iron and steel, dyestuffs, sugar and a large proportion of the inorganic and organic chemicals industries were not included in the industrial data base as they already operate CHP plant.

7 The cement, glass, bricks and pottery industries were not included as the temperature of the process heat requirement in these industries is too high to be met by conventional CHP plant.

The industries covered are listed below:

FOOD INDUSTRIES & TOBACCO

Profile No.	SIC Code(s)	Product/Activity
1	4116	Edible Oils
2	4122	Frozen Meat and Meat Products
3	4122	Canned Meat Products
4	4122	Meat Products (not frozen or canned)
5	4123	Poultry
6	4130	Liquid Milk
7	4130	Milk and Milk Products - Creameries
8	4130	Milk Products
9	4147	Frozen Fruit and Vegetables
10	4147	Canned Fruit and Vegetables
11	4150	Fish
12	4196	Bread and Flour Confectionery
13	4197	Biscuits and Crispbread
14	4214	Cocoa, Chocolate and Sugar Confectionery
15	4221	Compound Animal Feeds
16	4222	Petfoods
17	4239	Miscellaneous Foods
18	4240	Malt Whisky
19	4240	Potable Spirits (excl. malt whisky)
20	4270	Beer and Other Brewing Products
21	4270	Malt and Malt Products
22	4290	Tobacco

CHEMICAL INDUSTRIES

Profile No.	SIC Code(s)	Produce/Activity
23	2511	Inorganic Chemicals
24	2512	Organic Chemicals
25	2513	Fertilisers
26	2514	Synthetic Resins and Plastics
27	2515	Synthetic Rubber
28	2551	Paints
29	2562	Formulated Adhesives and Sealants
30	2563	Chemical Treatment of Oils and Fats
31	2564	Essential Oils and Flavourings
32	2567	Miscellaneous Chemicals
33	2568	Formulated Pesticides
34	2570	Pharmaceutical Products - Antibiotics
35	2570	Pharmaceutical Products (excl. antibiotics)
36	2581	Soap and Synthetic Detergents
37	2582	Perfumes and Cosmetics
38	2591	Photographic Materials

GENERAL ENGINEERING

Profile No.	SIC Code(s)	Product/Activity
39	32	Mechanical Engineering
40	33/34/37	Electrical, Electronic & Inst. Engineering
41	35/36/ 3120-69	Vehicles, Aircraft, Ships & Metal Goods

TEXTILES

Profile No.	SIC Code(s)	Product/Activity
42	43 (excl.4370)	General Textiles (excl. textile finishing)
43	4370	Textile Finishing
44	44/45	Leather (tanneries)
45	44/45	Leather Goods

PAPER		
Profile No.	**SIC Code(s)**	**Product/Activity**
46	47 (excl.4710)	Paper Conversion (excl. Paper & Board)
47	4710	Paper and Board

PLASTIC & RUBBER PROCESSING		
Profile No.	**SIC Code(s)**	**Product/Activity**
48	481/482	Rubber Tyres
49	481/482	Rubber Products(other than tyres)
50	483	Plastics Conversion

3.0 Site Description

8 The sites in each of the 50 selected industries were split into size bands on an energy consumption basis. Up to three bands were identified for each industry and they were derived from the data supplied by the DEn. A 'typical' site was then developed for each size band in each industry and the descriptions of these sites formed the industrial data base.

9 A standard format was adopted for the site description. This is shown in Figure 1. Three energy inputs are defined:

- electricity
- fossil fuels used in boilers for raising steam and hot water
- fossil fuels used for firing plant directly.

10 The figure lists the items which are included in each site description. The profiles of demand for each energy input employ three day types:

- a summer working day(s)
- a winder working day (w)
- a non-working day (n)

11 The site descriptions developed for each energy consumption band in each of the 50 industry groups represented 'typical' sites within the group rather than any particular site. The data for each typical site was developed as follows:

- The site energy consumption was the average consumption of all of the sites within each of the size range bands considered for each industry group.

- The heat quality defined from available data on the most typical steam pressure, hot water temperature, boiler efficiency, proportion of condensate return and split between process and space heating demand in each industrial group.

- The profiles of electricity and heat demand derived from available data on the most typical operation of a site within each industrial group.

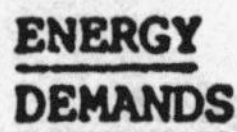

ENERGY DEMANDS	CONSUMPTION DETAILS	DEMAND PROFILES
		S = SUMMER DAY W = WINTER DAY N = NON WORKING DAY
ELECTRICITY	(i) Total annual consumption	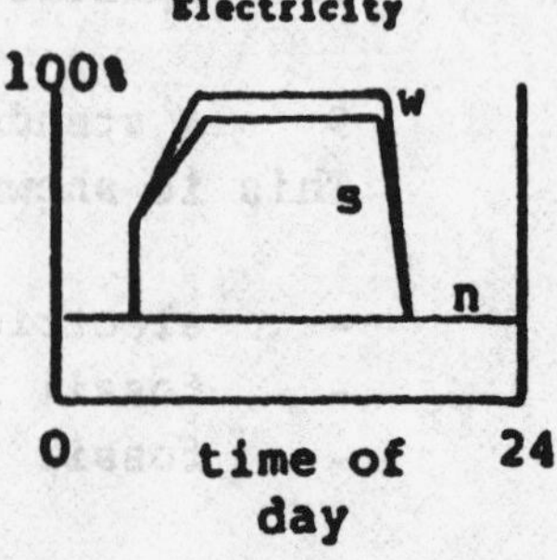
HEAT (Fossil fuel for existing boilers)	(i) Total annual consumption (ii) Steam/Hot water condition (iii) Fraction of demand which can be met by water at less than 85° C (iv) Percentage steam condensate return (v) Efficiency of existing boilers (based on HCV)	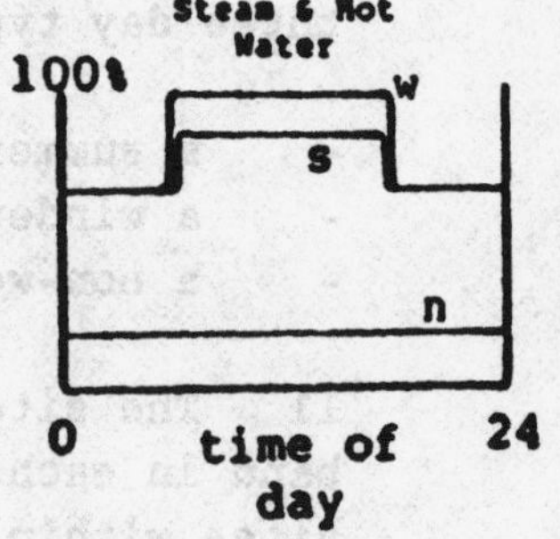
HEAT (Fuel for directly fired plant suitable for gas turbine exhaust)	(i) Total annual consumption (ii) Temperature required	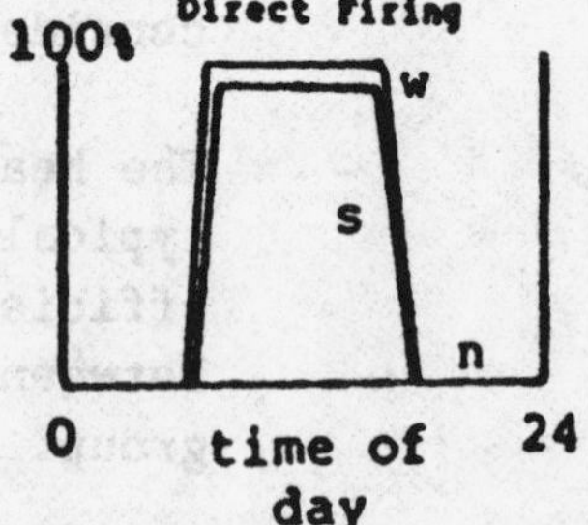

DATA DEFINING AN INDUSTRIAL SITE **Fig. 1**

4.0 Information Sources

12 The energy use details of the selected industries were obtained from the following sources:

Government Statistics

- Department of Energy Database on the energy consumption and size distribution of British industry. This data results from the 1979 Energy Purchases Enquiry. The database covers virtually 100% of the large energy users, around 85% of the medium energy users and a small sample of small energy users. To take account of the fact that the heat to power ratios of many industries have changed (usually as a result of reductions in heat demand), heat scaling factors were applied which were calculated from the trends identified in the 1984 Digest of UK Energy Statistics.

- Disaggregated Model of Energy Consumption in UK Industry by Bush, R.P. and Chadwick, A.T. of the Energy Technology Support Unit.

- The Pattern of Energy Use in the UK, 1976 by Bush, R.P. and Matthews, B.J. of the Energy Technology Support Unit, R7, Harwell, 1979.

- 1984 Digest of UK Energy Statistics, Department of Energy, HMSO, London.

- 1979 Census of Production and Purchases Inquiry. Business Statistics Office, HMSO London.

- Energy Use and Energy Efficiency in UK Manufacturing Industry up to the Year 2000. Volume 1, Chief Scientist's Group, Energy Technology Division, AERE Harwell, 1984.

- Standard Industrial Classification, Revised 1980, HMSO, London.

- Inquiry into Private Generation of Electricity in Great Britain 1977. Published by Department of Energy.

- Combined Heat and Power and Electricity Generation in British Industry 1983-1988, Energy Efficiency Series No 5, Energy Efficiency Office, HMSO, 1986

Industrial Energy Use Details

- Energy Use and Energy Efficiency in UK Manufacturing

Industry up to the Year 2000. Volume II, Sector Reports containing the Detailed Analyses of the Industries, their Energy Use and Potential Energy Savings, Chief Scientists Group, Energy Technology Division, AERE Harwell, 1984.

- Energy Audit Series issued jointly by the Department of Energy and Department of Industry.

- Industrial Energy Thrift Scheme Reports, 1 to 54, Department of Industry.

- 'Application of the Concept of Heat/Process Integration in the Non-Chemical Manufacturing Industries'. An unpublished market research study by the Energy Technology Support Unit (1983).

- 'Breakdown of Energy Use within the UK Chemicals Industry'. An unpublished report by the Energy Technology Support Unit (1984).

- The results of detailed CHP feasibility studies carried out on seven sites.

- Information provided directly by the Energy Technology Support Unit, Harwell.

- British Paper and Board Industry Federation Report on Energy Consumption and Costs for 1984. BPBIF, 1983.

- Energy Use in the Food Manufacturing Industry. Crawford, A.G. and Elson, C.R, Scientific and Technical Surveys No. 128, January 1982, Leatherhead Food Research Association.

- Private Communications with the Shirley Institute of the textile industry.

- In-house knowledge of the Consultants, Ove Arup and Partners, London.

- References listed in 'Energy Use and Energy Efficiency, in UK Manufacturing Industry up to the Year 2000' (indicated above).

- CHP Assessment Forms completed by pre-selected industrial establishments. These are the subject of the section 5.0.

13 Data extracted from the Department of Energy Database by The Economics and Statistics Branch provided the basis of the energy consumption data of the Industrial Database. This data

quantified the energy consumption of individual industries in 1979. It contained three energy consumption size bands and for each band specified the number of sites involved and their total energy consumption sub-divided into electricity and fossil fuel consumption. Average energy consumption figures were therefore obtained for each band by dividing the totals by the number of sites.

5.0 CHP Assessment Forms - Data on Individual Sites

14 In order to fill important gaps in the available data on energy use in industry, forty industrial establishments were contacted and asked to complete a specially designed assessment form. An example of the form is included at the end of this Appendix. Twenty two completed forms were obtained and relevant information was included into the industrial database profiles. In return for the information, these establishments were provided with outline economic assessments of the viability of CHP at their particular site. The forms were particularly beneficial for information on the daily profiles of electricity and heat demand. Table 1 lists the activities of these establishments. The organisations are listed in the acknowledgement in the main report.

6.0 Changing Industrial Heat Demands

15 The data on industrial energy consumption was taken primarily from Department of Energy assessment of data produced by the Business Statistics Office 1979 Energy Purchases Enquiry. As described in the introduction, there is a trend of declining heat demands in industry while electricity demands have remained fairly constant.

16 Figures 2 to 5 present electricity consumption as a percentage of the total energy consumption (heat supplied basis) for four different areas of industry. The data was extracted from Table 12 of the 1984 Digest of UK Energy Statistics. Extrapolating to 1985, the following table indicates the heat scaling factors used for construction of the Industrial Database.

Sector of Industry	Ratio=(Fossil Fuel)/(Electricity) 1979	1985*	Heat Scaling Factor
Food, Drink and Tobacco	7.95	6.41	0.806
Textiles, Leather and Clothing	5.44	4.88	0.897
Engineering and Metal Trades	3.96	3.18	0.803
Chemicals and Allied Industries	7.95	8.09	1.017£

* estimates obtained by extrapolation
£ a figure of 1.0 was in fact used.

17 In applying the heat scaling figures it was assumed that the electricity consumption at the 'average' sites had not changed since 1979.

Table 1

Industries from which Heat and Power demand Data on Actual Sites was obtained

Site Code Letter	INDUSTRY	No. of Employees @ site	Current Fuel	Energy Cost £ M
	Small energy users:			
A	PLASTICS PROCESSING	170	GAS	.2
B	MECHANICAL ENGINEERING	481	GAS	.3
C	FOOD	120	GAS	.6
D	OTHER	1000	GAS	.8
E	PLASTICS PROCESSING	400	GAS	.8
F	DRINKS	192	HFO	.8
G	CHEMICALS	400	GAS	.8
H	FOOD	200	GAS	.9
	Medium energy users:			
I	DRINKS	600	HFO	1
J	FOOD	1400	GAS	1.3
K	DRINKS	700	HFO/GAS	1.4
L	MECHANICAL ENGINEERING	1700	HFO	1.5
M	CHEMICALS	1180	COAL	1.8
N	CHEMICALS	425	COAL	2
O	OTHER	270	GAS	2
	Large energy users:			
P	FOOD	3000	GAS	2.4
Q	CHEMICALS	320	HFO	2.4
R	FOOD	1200	GAS	2.5
S	DRINKS	2000	GAS	3.5
T	FOOD	6370	COAL	3.7
U	OTHER	1000	HFO	4.5
V	CHEMICALS	670	GAS	4.8
	Average: small	370.38		.65
	: medium	896.43		1.57
	: large	2080		3.4
	: overall	1081.73		1.82

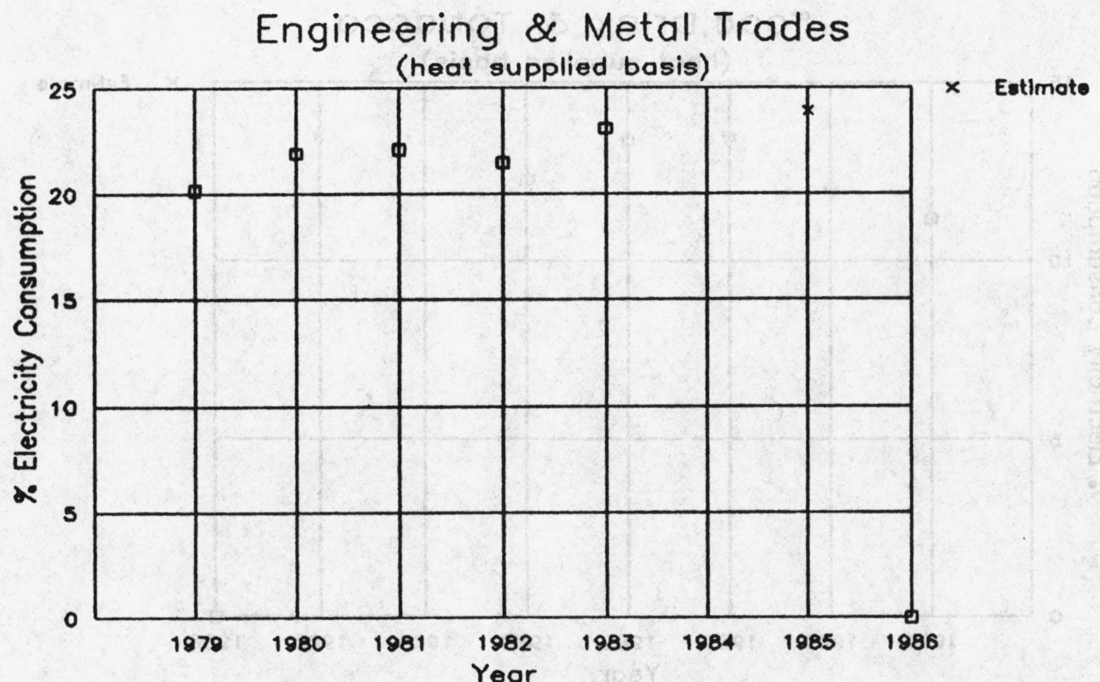

Heat to Power Ratio Changes in the Engineering and Metal Trades Industries.

Fig. 2

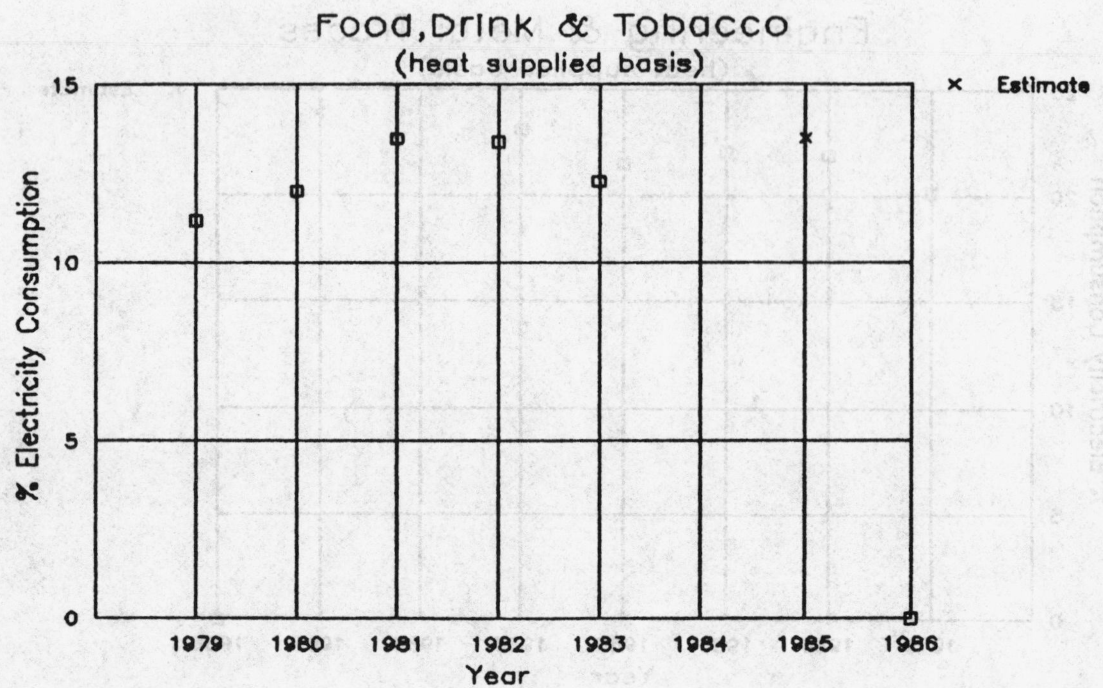

Heat to Power Ratio Changes in the Food, Drink and Tobacco Industries.

Fig. 3

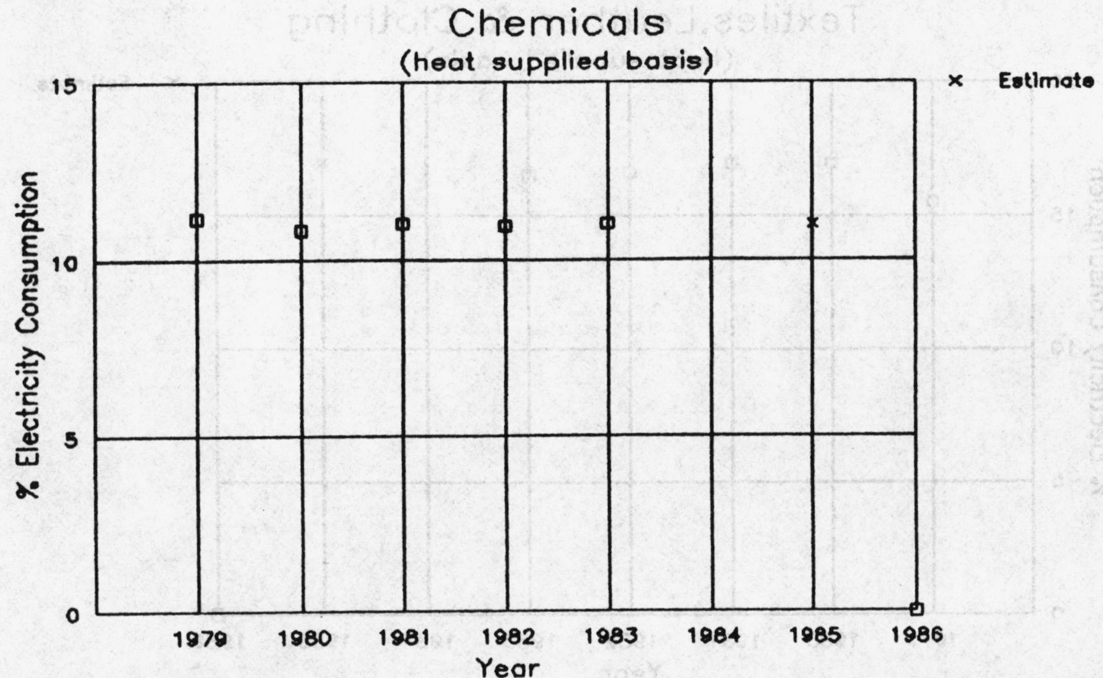

Heat to Power Ratio Changes in the Chemicals and Allied Industries.

Fig. 4

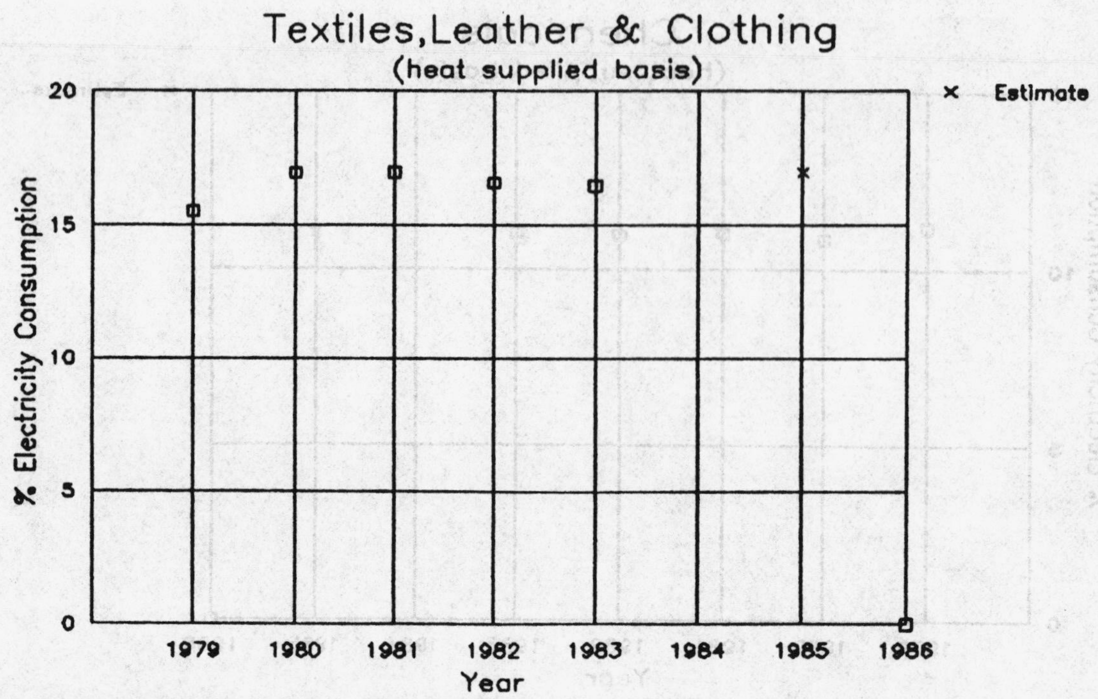

Heat to Power Ratio Changes in the Textiles, Leather and Clothing Industries.

Fig. 5

7.0 Number of Establishments in UK Industry

18 The number of establishments (sites) in UK industry has fallen since 1979. No correction for this was attempted. However, the Department of Energy Database had less than a 100% coverage of 1979 industry (about 100% of all large sites, 85% of medium sites, and a small coverage of small sites) and this offsets the errors in the number of establishments.

7.1 Conversions to Coal

19 Data held by the Department of Energy was used to modify the database to take account of known conversions from heavy fuel oil to coal which occurred over the period 1981 to 1985.

8.0 Industry Profiles

20 The following simplifications and limitations should be noted when considering the industry profiles:

- The splits between electricity, fuels for steam or hot water raising and other fuels are derived from industry averages with appropriate modifications made to eliminate existing CHP sites. Heat scaling factors were also applied to take account of the reductions in heat to power ratio which have occurred between 1979 (the year of the source data) and 1985.

- The 'number of sites' refers to the situation in 1979, except for SIC 4710 Paper and Board which refers to 1983.

- Insofar as information was available, the profiles of electricity and heat demand were selected to be typical approximations of those most frequently encountered in the industry. In practice, the sites in any particular industry may operate a variety of different shift patterns and the demand profiles may vary from day to day.

- Boiler efficiency is based on higher (gross) calorific value.

- 'Appropriate Direct Firing' refers to direct firing heat demands which, it was judged, might be suitable for fulfilment directly by prime mover exhaust. Generally, very little information was available on direct firing.

- In practice the proportion of heat used for space heating may vary widely between sites involved in the same industry. Where possible the industry average is given.

- Short term variations in shift patterns due to, for example, overtime at peak periods have not been included.

- Figures of percentage condensate return are based on actual site data when available, otherwise a default figure of 80% was used.

8.1 Quality of Data

21 The information assembled in the industrial database contains the data necessary to assess the cost effectiveness of CHP plants on 'typical' sites in 50 major industries. This data was not previously available and a wide variety of sources were tapped and a considerable amount of judgement and experience, on the part of Ove Arup and Partners and ETSU, was brought to bear in establishing the database. Major assumptions were made concerning changes in industry since the 1979 Energy Purchases Enquiry and also in defining the energy demand patterns of individual sites.

22 Recognising the limitations of the data, the quality was nonetheless deemed sufficient to fulfil the objectives of the study.

INDUSTRY PROFILES

1 EDIBLE OILS [SIC 4116]

Profile of the Industry

Vegetable oil extraction and refining is the most important activity of the subsector in energy terms. Oil is extracted from soya beans and a variety of seeds by mixing crushed seeds with a solvent. The mixture is filtered and the solvent boiled off to leave the oil. The seed residue is used in animal feedstocks.

Refining renders the oils colourless, odourless and free from fatty acids. The main heat requirements of the process are steam or hot water at 100-120°C for the neutralising process and steam at 180-250°C for hydrogenation and deodorisation.

Typical energy consumption and heat quality details for two different sizes of non-CHP site are given in Table 1.1 and typical profiles of electricity and heat demand are presented in Figure 1.1.

Energy Consumption and Heat Quality Details for the Edible Oil Processing Industry

AVERAGE ANNUAL ENERGY CONSUMPTION PER SITE	Size Band 1 kTh	Size Band 1 GWh	Size Band 1 %	Size Band 2 kTh	Size Band 2 GWh	Size Band 2 %	Size Band 3 kTh	Size Band 3 GWh	Size Band 3 %
1) ELECTRICITY	33.3	1.0	7.69	242.0	7.1	7.68			
2) FUELS for STEAM & HOT WATER GENERATION	366.1	10.7	84.47	2663.4	78.1	84.49			
3) FUELS for OTHER USE	34.0	1.0	7.84	247.1	7.2	7.84			
Totals	433.4	12.7	100	3152.5	92.4	100			

DETAILS OF STEAM AND HOT WATER GENERATION AND USE

Steam generation details: 250.0 psig
290.0 deg.C

Details of use: 88.0 % process heat
12.0 % space heat

Fraction of process heat demand which can be met by hot water at < or = 85 deg.C: 0 %

Boiler efficiency: 77.0 % Condensate return: 55.0 %

DETAILS OF APPROPRIATE DIRECT FIRED PLANT CONTAINED WITHIN 3)

Fraction of 3) used in the plant: 0 %

Operating temperature of plant: 0 deg.C

NUMBER OF SITES (all size bands): 11 NUMBER OF NON-WORKING DAYS PER YEAR: 28

Edible Oil Processing

Table 1

Typical Daily Profiles of Heat and Power Demand for the Edible Oil Processing Industry

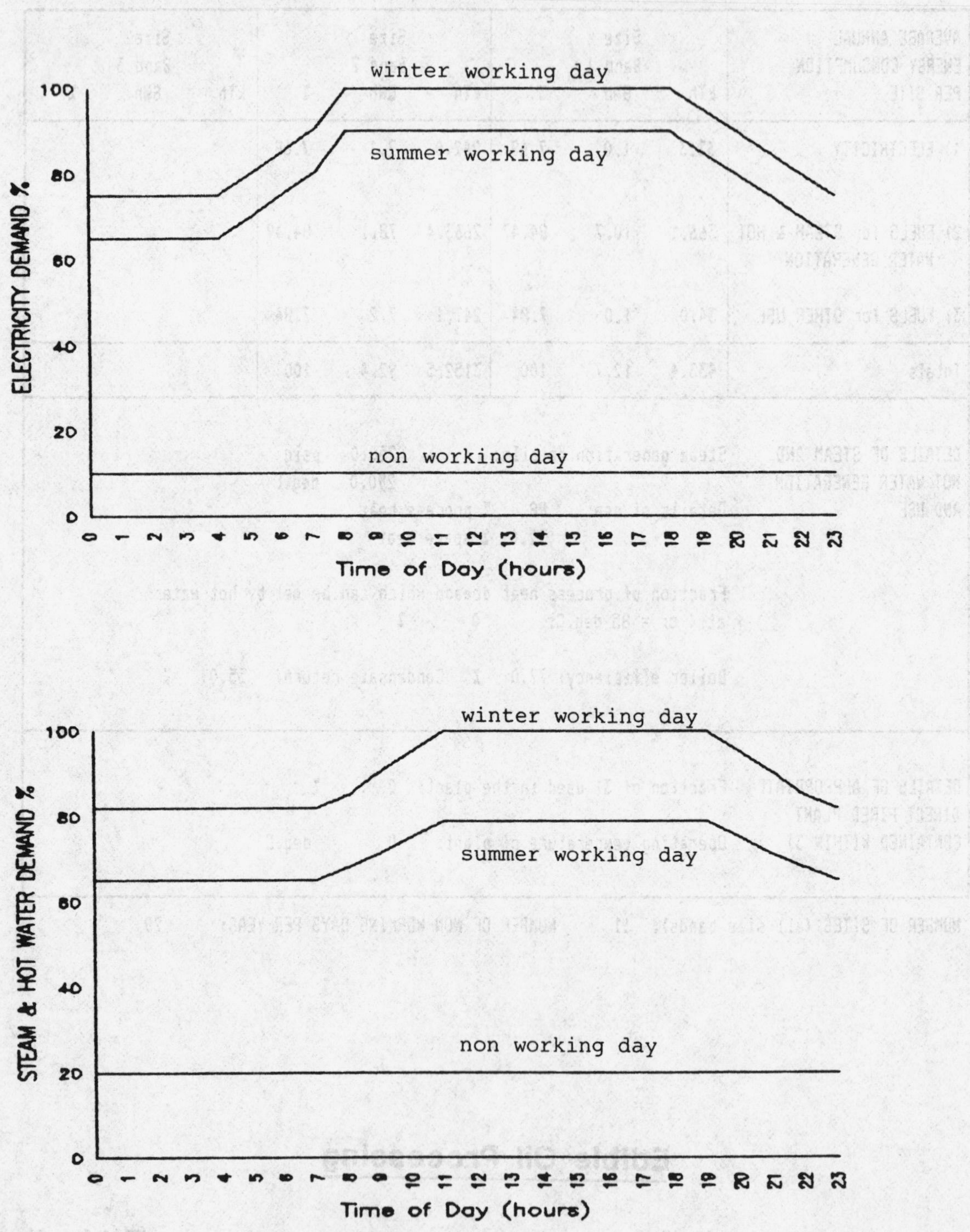

Edible Oil Processing **Fig. 1**

2. FROZEN MEAT AND MEAT PRODUCTS [part of SIC 4122]

Profile of the Industry

The preparation of meat products from carcasses involves cutting, trimming, grinding , mixing and extrusion. These processes require electricity for motive power. Refrigeration is a major user of electricity, and heat is required for cooking operations, space heating and to provide hot water for cleaning purposes.

Average energy consumption and heat quality details for two different sizes of site are given in Table 2.1 and the profiles of electricity and heat demand used for the analysis of this industry are presented in Figure 2.1.

Energy Consumption and Heat Quality Details for the Frozen Meat and Meat Products Industry

AVERAGE ANNUAL ENERGY CONSUMPTION PER SITE	Size Band 1 kTh	GWh	%	Size Band 2 kTh	GWh	%	Size Band 3 kTh	GWh	%
1) ELECTRICITY	105.1	3.1	48.15	459.3	13.5	45.16			
2) FUELS for STEAM & HOT WATER GENERATION	81.5	2.4	37.30	401.3	11.8	39.45			
3) FUELS for OTHER USE	31.8	.9	14.55	156.5	4.6	15.39			
Totals	218.4	6.4	100	1017.1	29.8	100			

DETAILS OF STEAM AND HOT WATER GENERATION AND USE

Steam generation details: 150.0 psig saturated

Details of use: 90.0 % process heat
10.0 % space heat

Fraction of process heat demand which can be met by hot water at < or = 85 deg.C: 10.0 %

Boiler efficiency: 77.5 % Condensate return: 80.0 %

DETAILS OF APPROPRIATE DIRECT FIRED PLANT CONTAINED WITHIN 3)

Fraction of 3) used in the plant: 0 %

Operating temperature of plant: 0 deg.C

NUMBER OF SITES (all size bands): 6 NUMBER OF NON-WORKING DAYS PER YEAR: 119

Frozen Meat and Meat Products

Table 2

Typical Daily Profiles of Heat and Power Demand for the Frozen Meat and Meat Products Industry

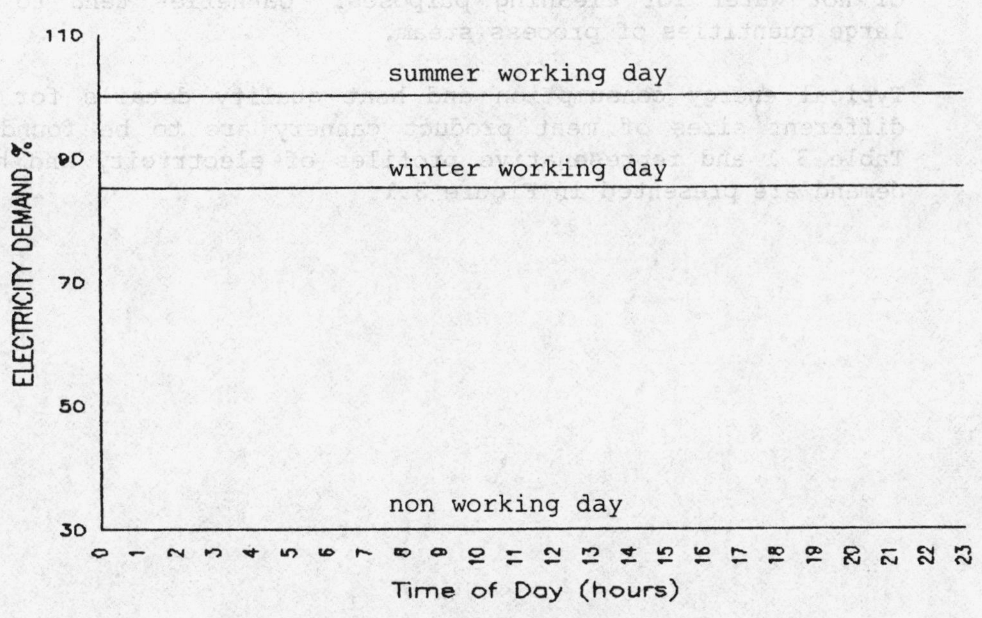

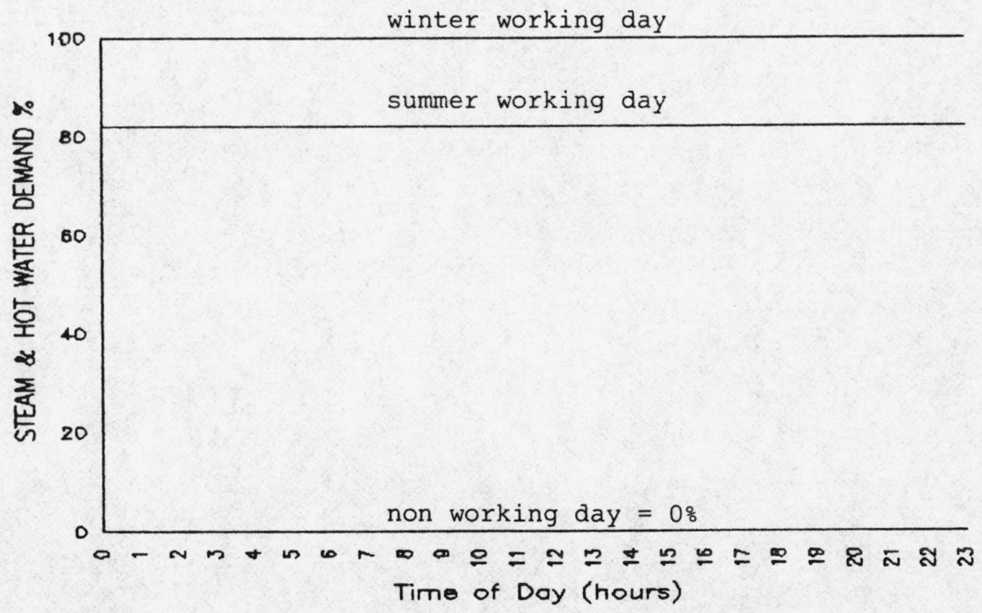

Frozen Meat and Meat Products **Fig. 2**

3. CANNED MEAT PRODUCTS [part of SIC 4122]

Profile of the Industry

The production of canned meat products involves the following processes : meat preparation, cooking, can filling, retorting (for sterilisation) cooling and packaging. Steam is used for cooking, heating retorts, space heating and for the production of hot water for cleaning purposes. Canneries tend to use large quantities of process steam.

Typical energy consumption and heat quality details for two different sizes of meat product cannery are to be found in Table 3.1 and represenative profiles of electricity and heat demand are presented in Figure 3.1.

Energy Consumption and Heat Quality Details for the Canned Meat Products Industry

AVERAGE ANNUAL ENERGY CONSUMPTION PER SITE	Size Band 1 kTh	Size Band 1 GWh	Size Band 1 %	Size Band 2 kTh	Size Band 2 GWh	Size Band 2 %	Size Band 3 kTh	Size Band 3 GWh	Size Band 3 %
1) ELECTRICITY	105.1	3.1	19.38	459.3	13.5	17.57			
2) FUELS for STEAM & HOT WATER GENERATION	365.4	10.7	67.34	1799.8	52.8	68.85			
3) FUELS for OTHER USE	72.1	2.1	13.29	355.2	10.4	13.59			
Totals	542.6	15.9	100	2614.3	76.6	100			

DETAILS OF STEAM AND HOT WATER GENERATION AND USE

Steam generation details: 120.0 psig saturated

Details of use: 93.0 % process heat
7.0 % space heat

Fraction of process heat demand which can be met by hot water at < or = 85 deg.C: 10.0 %

Boiler efficiency: 75.0 % Condensate return: 58.0 %

DETAILS OF APPROPRIATE DIRECT FIRED PLANT CONTAINED WITHIN 3)

Fraction of 3) used in the plant: 0 %

Operating temperature of plant: 0 deg.C

NUMBER OF SITES (all size bands): 8 NUMBER OF NON-WORKING DAYS PER YEAR: 119

Canned Meat Products

Table 3

Typical Daily Profiles of Heat and Power Demand for the Canned Meat Products Industry

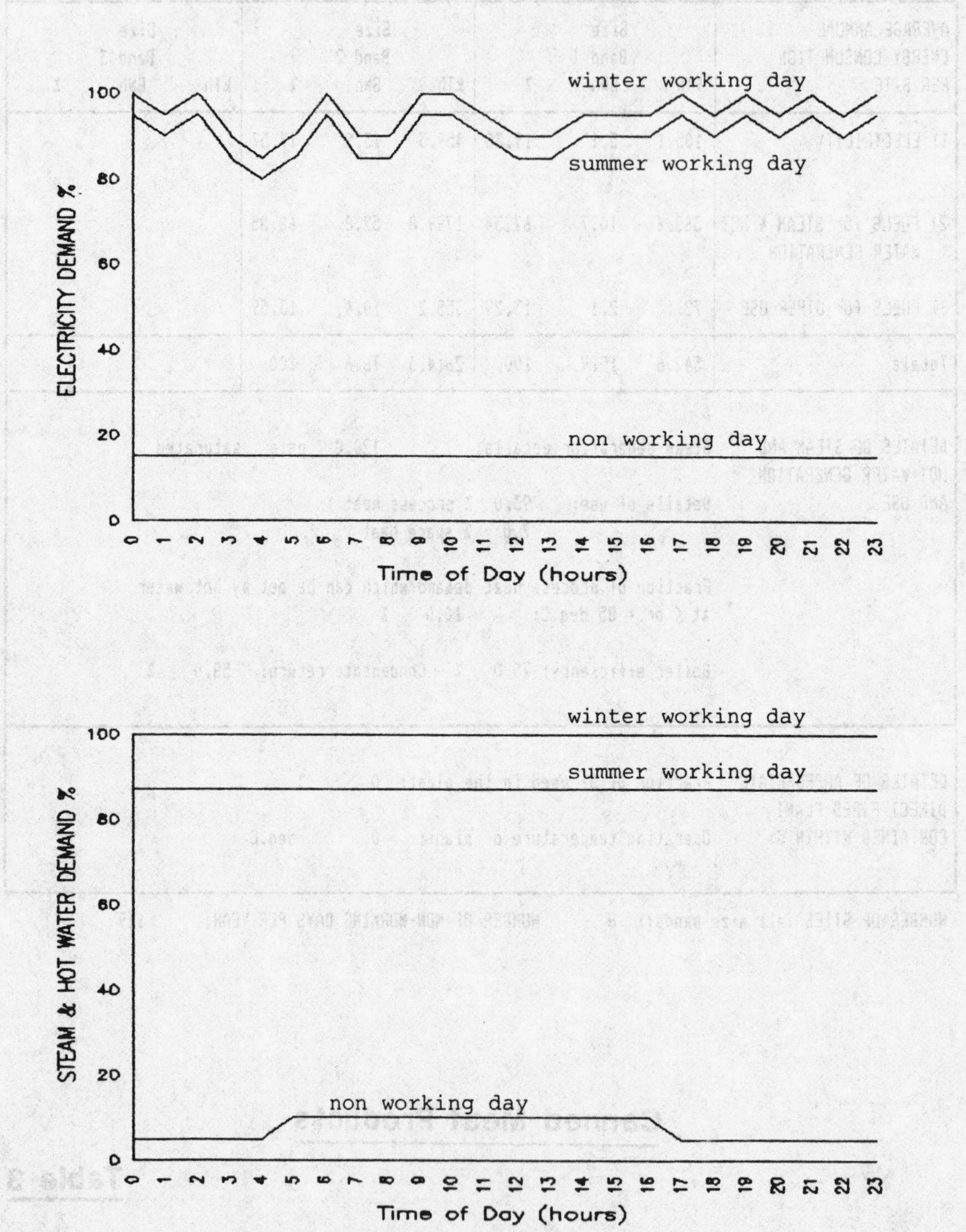

Canned Meat Products

Fig. 3

4. MEAT PRODUCTS (NOT FROZEN OR CANNED) [part of SIC 4122)

Profile of the Industry

This part of SIC 4122 concerns meat products which are preserved by cooking, curing, smoking and pickling and which are not canned or frozen.

Electricity provides the motive power for operations such as meat cutting, grinding, mixing and extrusion. Heat is required for cooking and curing operations, for space heating and for the production of hot water for cleaning purposes.

Average energy consumption and heat quality details for two different sizes of site are presented in Table 4.1 and the profiles of electricity and heat demand used for the analysis of this industry are presented in Figure 4.1.

Energy Consumption and Heat Quality Details for the Industry

AVERAGE ANNUAL ENERGY CONSUMPTION PER SITE	Size Band 1 kTh	GWh	%	Size Band 2 kTh	GWh	%	Size Band 3 kTh	GWh	%
1) ELECTRICITY	105.1	3.1	37.10	459.3	13.5	34.34			
2) FUELS for STEAM & HOT WATER GENERATION	136.8	4.0	48.26	673.7	19.7	50.38			
3) FUELS for OTHER USE	41.5	1.2	14.64	204.4	6.0	15.28			
Totals	283.4	8.3	100	1337.4	39.2	100			

DETAILS OF STEAM AND HOT WATER GENERATION AND USE	Steam generation details: 120.0 psig saturated Details of use: 92.0 % process heat 8.0 % space heat Fraction of process heat demand which can be met by hot water at < or = 85 deg.C: 10.0 % Boiler efficiency: 75.0 % Condensate return: 58.0 %
DETAILS OF APPROPRIATE DIRECT FIRED PLANT CONTAINED WITHIN 3)	Fraction of 3) used in the plant: 0 % Operating temperature of plant: 0 deg.C

NUMBER OF SITES (all size bands): 36 NUMBER OF NON-WORKING DAYS PER YEAR: 119

Other Meat Products

Table 4

Typical Daily Profiles of Heat and Power Demand for the Industry

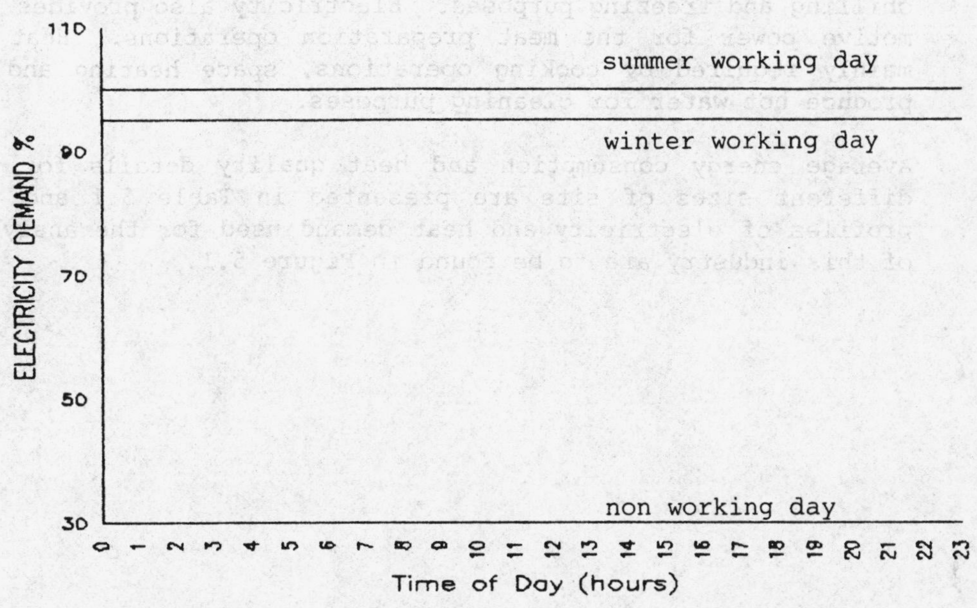

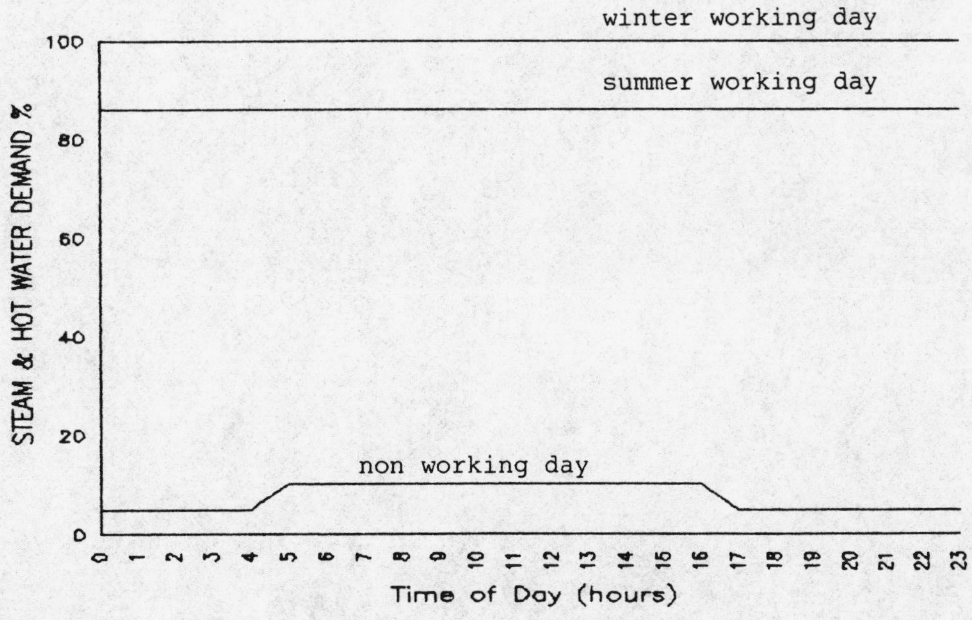

Other Meat Products

Fig. 4

5. POULTRY AND POULTRY MEAT PRODUCTS [SIC4123]

Profile of the Industry

The production of poultry products consists principally of meat preparation (cutting, trimming, grinding, mixing etc.), processes to preserve the food (cooking, freezing etc.) and packaging. Refrigeration is a major electricity user for both chilling and freezing purposes. Electricity also provides the motive power for the meat preparation operations. Heat is mainly required by cooking operations, space heating and to produce hot water for cleaning purposes.

Average energy consumption and heat quality details for two different sizes of site are presented in Table 5.1 and the profiles of electricity and heat demand used for the analysis of this industry are to be found in Figure 5.1.

Energy Consumption and Heat Quality Details for the Poultry and Poultry Meat Products Industry

AVERAGE ANNUAL ENERGY CONSUMPTION PER SITE	Size Band 1 kTh	Size Band 1 GWh	Size Band 1 %	Size Band 2 kTh	Size Band 2 GWh	Size Band 2 %	Size Band 3 kTh	Size Band 3 GWh	Size Band 3 %
1) ELECTRICITY	143.0	4.2	37.92	1450.3	42.5	47.12			
2) FUELS for STEAM & HOT WATER GENERATION	195.1	5.7	51.73	1356.4	39.8	44.07			
3) FUELS for OTHER USE	39.0	1.1	10.35	271.3	8.0	8.81			
Totals	377.1	11.1	100	3078.1	90.2	100			

DETAILS OF STEAM AND HOT WATER GENERATION AND USE	Steam generation details: 100.0 psig saturated Details of use: 90.0 % process heat 10.0 % space heat Fraction of process heat demand which can be met by hot water at < or = 85 deg.C: 5.0 % Boiler efficiency: 75.0 % Condensate return: 58.0 %
DETAILS OF APPROPRIATE DIRECT FIRED PLANT CONTAINED WITHIN 3)	Fraction of 3) used in the plant: 0 % Operating temperature of plant: 0 deg.C

NUMBER OF SITES (all size bands): 14 NUMBER OF NON-WORKING DAYS PER YEAR: 119

Poultry and Poultry Meat

Table 5

Typical Daily Profiles of Heat and Power Demand for the Poultry and Poultry Meat Products Industry

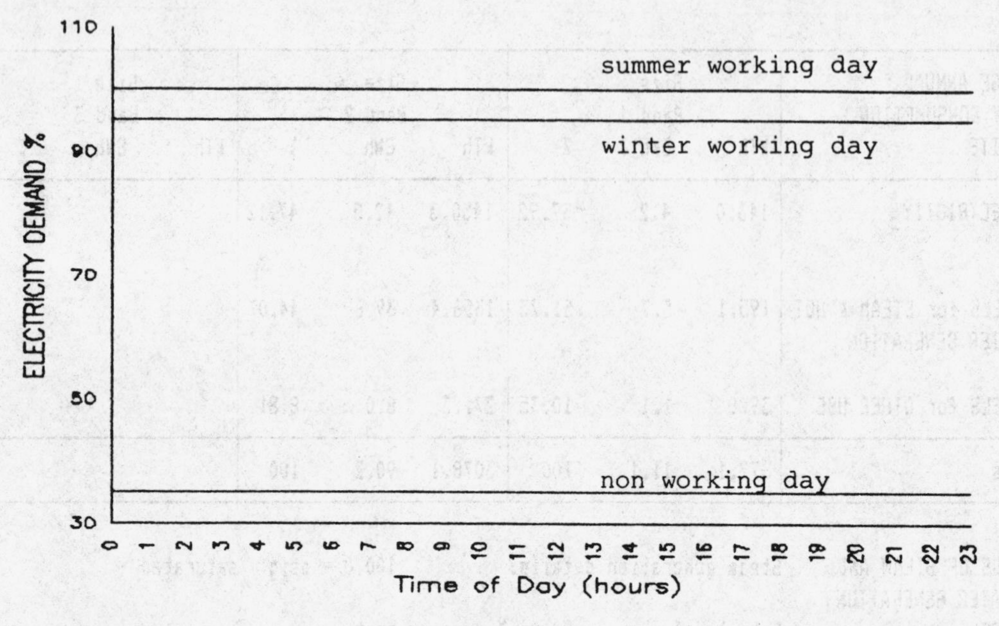

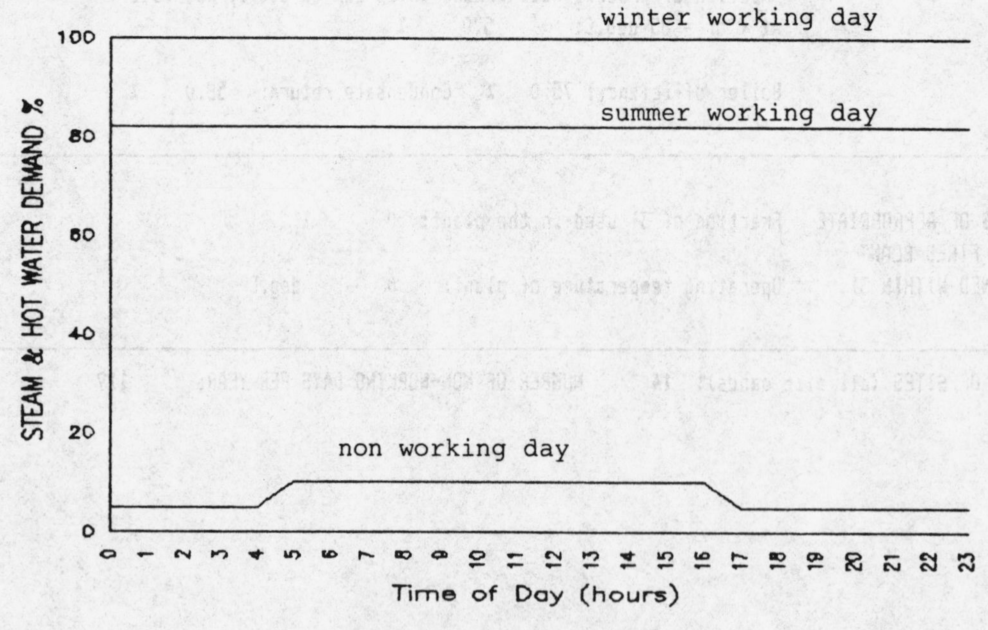

Poultry and Poultry Meat

Fig. 5

6. MILK AND MILK PRODUCTS - LIQUID MILK PROCESSING [part of SIC 4130]

Profile of the Industry

Liquid milk processing is the least energy intensive of the three sub-divisions of SIC 4130 and is mainly concerned with the heat treatment and bottling of pasteurised milk. Sterilised milk, UHT milk and homogenised milk use more energy per litre but account for only about 10% of the milk sales. The demand for liquid milk changes little throughout the year and any change in the milk supply is taken up by the milk product industry. Liquid milk processing plants tend to be small compared with milk product manufacturing sites.

The plants tend to operate on a regular daily cycle starting with the receipt of milk from the farms, and ending with the despatch of bottled milk for the next morning's delivery. The main process steps are refrigeration of the milk received, thermal treatment of the milk, cooling, and bottling. Significant amounts of hot water are used in cleaning. The main site utilities are boiler plant for process and space heating, and refrigeration plant.

Typical energy consumption and heat quality details for liquid milk processing sites are presented in Table 6.1 and typical profiles for electricity and heat demand are presented in Figure 6.1.

Energy Consumption and Heat Quality Details for the Liquid Milk Processing Industry

AVERAGE ANNUAL ENERGY CONSUMPTION PER SITE	kTh	Size Band 1 GWh	%	kTh	Size Band 2 GWh	%	kTh	Size Band 3 GWh	%
1) ELECTRICITY	71.0	2.1	22.90						
2) FUELS for STEAM & HOT WATER GENERATION	229.5	6.7	73.96						
3) FUELS for OTHER USE	9.7	.3	3.14						
Totals	310.2	9.1	100						

DETAILS OF STEAM AND HOT WATER GENERATION AND USE	Steam generation details: 100.0 psig saturated Details of use: 60.0 % process heat 40.0 % space heat Fraction of process heat demand which can be met by hot water at < or = 85 deg.C: 20.0 % Boiler efficiency: 77.5 % Condensate return: 80.0 %
DETAILS OF APPROPRIATE DIRECT FIRED PLANT CONTAINED WITHIN 3)	Fraction of 3) used in the plant: 0 % Operating temperature of plant: 0 deg.C

NUMBER OF SITES (all size bands): 23 NUMBER OF NON-WORKING DAYS PER YEAR: 64

Liquid Milk Processing

Table 6

Typical Daily Profiles of Heat and Power Demand for the Liquid Milk Processing Industry

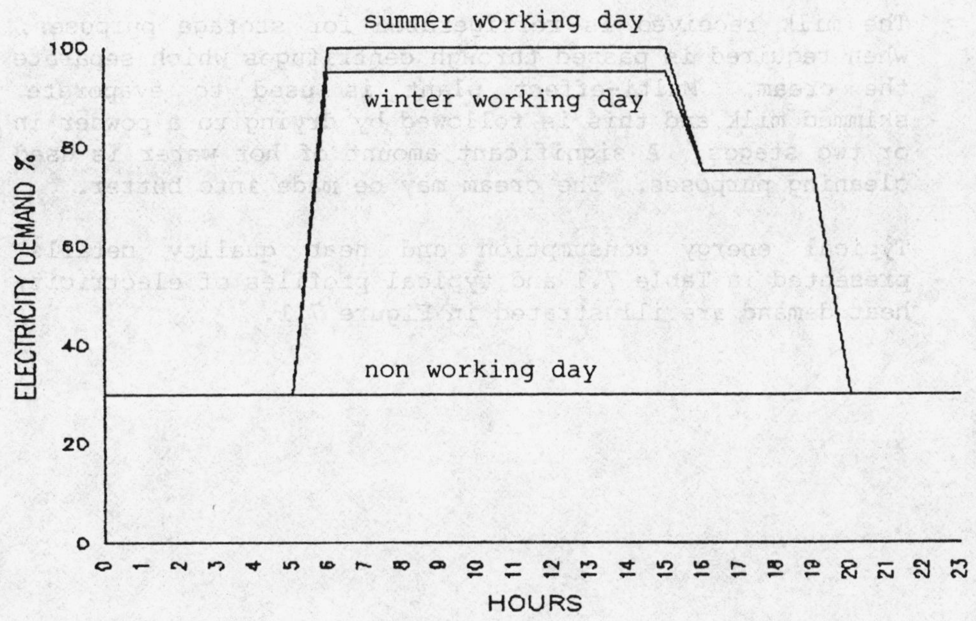

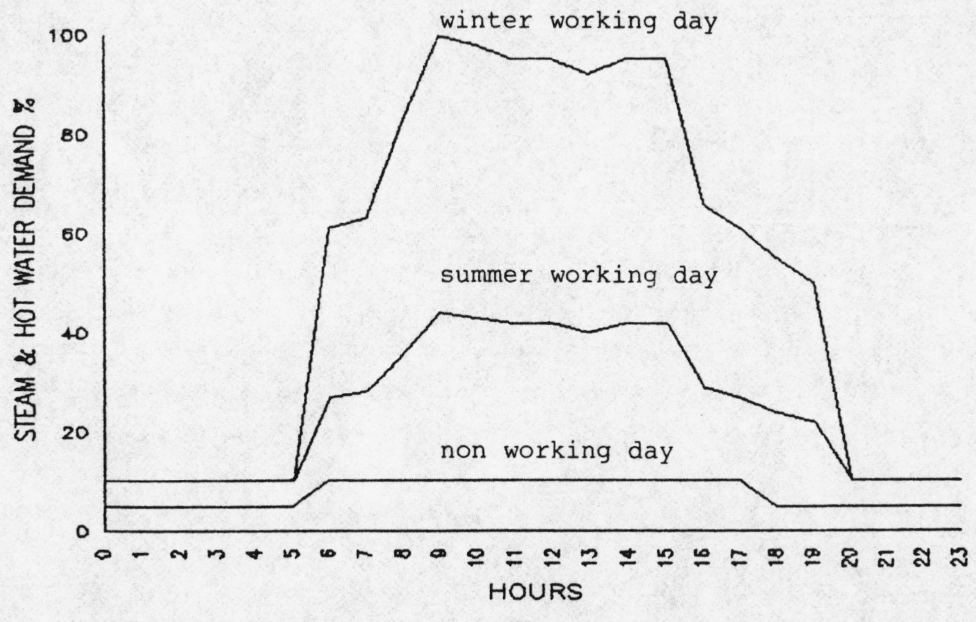

Liquid Milk Processing

Fig. 6

7. MILK AND MILK PRODUCTS - CREAMERIES [part of SIC 4130]

Profile of the Industry

Creameries convert liquid milk supplies which are in excess of immediate demand into forms suitable for medium and long term storage and resale, notably butter and skimmed milk powder. Some creameries are associated with milk pasteurising and bottling, and some with milk product manufacturing.

The milk received is refrigerated for storage purposes, and when required is passed through centrifuges which separate out the cream. Multi-effect plant is used to evaporate the skimmed milk and this is followed by drying to a powder in one or two stages. A significant amount of hot water is used for cleaning purposes. The cream may be made into butter.

Typical energy consumption and heat quality details are presented in Table 7.1 and typical profiles of electricity and heat demand are illustrated in Figure 7.1.

Energy Consumption and Heat Quality Details for Creameries

AVERAGE ANNUAL ENERGY CONSUMPTION PER SITE	Size Band 1			Size Band 2			Size Band 3		
	kTh	GWh	%	kTh	GWh	%	kTh	GWh	%
1) ELECTRICITY	71.0	2.1	19.47	238.4	7.0	15.01			
2) FUELS for STEAM & HOT WATER GENERATION	270.6	7.9	74.15	1243.2	36.4	78.25			
3) FUELS for OTHER USE	23.3	.7	6.38	107.0	3.1	6.74			
Totals	364.9	10.7	100	1588.7	46.6	100			

DETAILS OF STEAM AND HOT WATER GENERATION AND USE

Steam generation details: 100.0 psig saturated

Details of use: 67.5 % process heat
32.5 % space heat

Fraction of process heat demand which can be met by hot water at < or = 85 deg.C: 20.0 %

Boiler efficiency: 77.5 % Condensate return: 80.0 %

DETAILS OF APPROPRIATE DIRECT FIRED PLANT CONTAINED WITHIN 3)

Fraction of 3) used in the plant: 0 %

Operating temperature of plant: 0 deg.C

NUMBER OF SITES (all size bands): 49 NUMBER OF NON-WORKING DAYS PER YEAR: 64

Creameries

Table 7

Typical Daily Profiles of Heat and Power Demand for Creameries

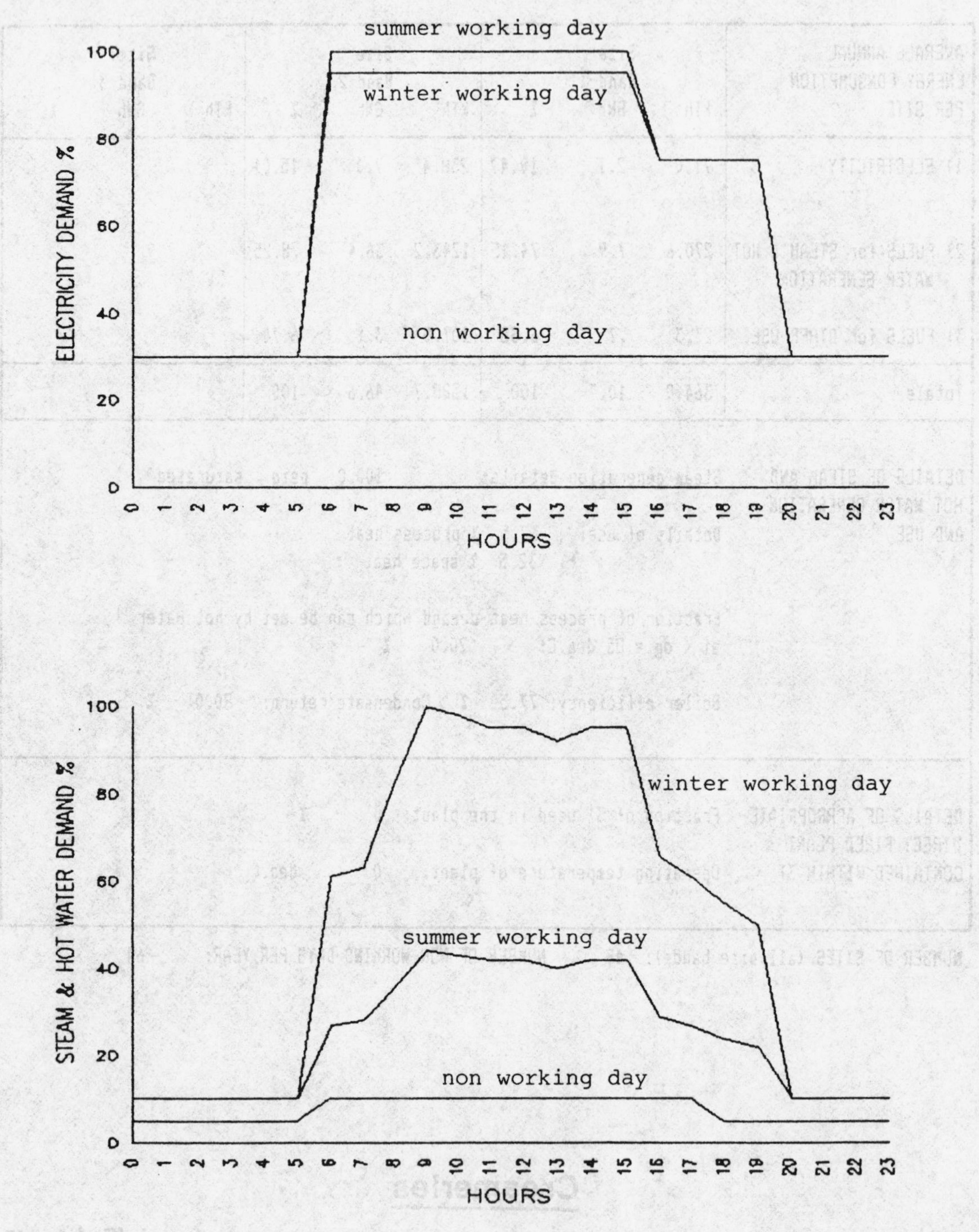

Creameries

Fig. 7

8. MILK AND MILK PRODUCTS - MILK PRODUCT MANUFACTURERS [part of SIC 4130]

Profile of the Industry

Over half of the UK milk production is used in manufacturing milk products which are principally cheese, condensed and evaporated milk, yoghurt and butter. Milk product manufacturing is about twice as energy intensive as liquid milk processing.

Most milk products involve separation of the milk solids into a more concentrated form leaving a less concentrated residue which is converted to manufactured products by means of evaporation and drying.

Typical energy consumption and heat quality details for milk product manufacturing are presented in Table 8.1 and Figure 8.1 presents typical daily profiles for electricity and heat demand.

Energy Consumption and Heat Quality Details for the Milk Products Industry

AVERAGE ANNUAL ENERGY CONSUMPTION PER SITE	Size Band 1			Size Band 2			Size Band 3		
	kTh	GWh	%	kTh	GWh	%	kTh	GWh	%
1) ELECTRICITY				238.4	7.0	12.02	1291.7	37.9	12.85
2) FUELS for STEAM & HOT WATER GENERATION				1539.0	45.1	77.60	7728.6	226.5	76.87
3) FUELS for OTHER USE				205.8	6.0	10.38	1033.7	30.3	10.28
Totals				1983.3	58.1	100	10054.0	294.7	100

DETAILS OF STEAM AND HOT WATER GENERATION AND USE

Steam generation details: 100.0 psig saturated

Details of use: 75.0 % process heat
25.0 % space heat

Fraction of process heat demand which can be met by hot water at < or = 85 deg.C: 20.0 %

Boiler efficiency: 77.5 % Condensate return: 80.0 %

DETAILS OF APPROPRIATE DIRECT FIRED PLANT CONTAINED WITHIN 3)

Fraction of 3) used in the plant: 0 %

Operating temperature of plant: 0 deg.C

NUMBER OF SITES (all size bands): 13 NUMBER OF NON-WORKING DAYS PER YEAR: 64

Milk Products

Table 8

Typical Daily Profiles of Heat and Power Demand for the Milk Products Industry

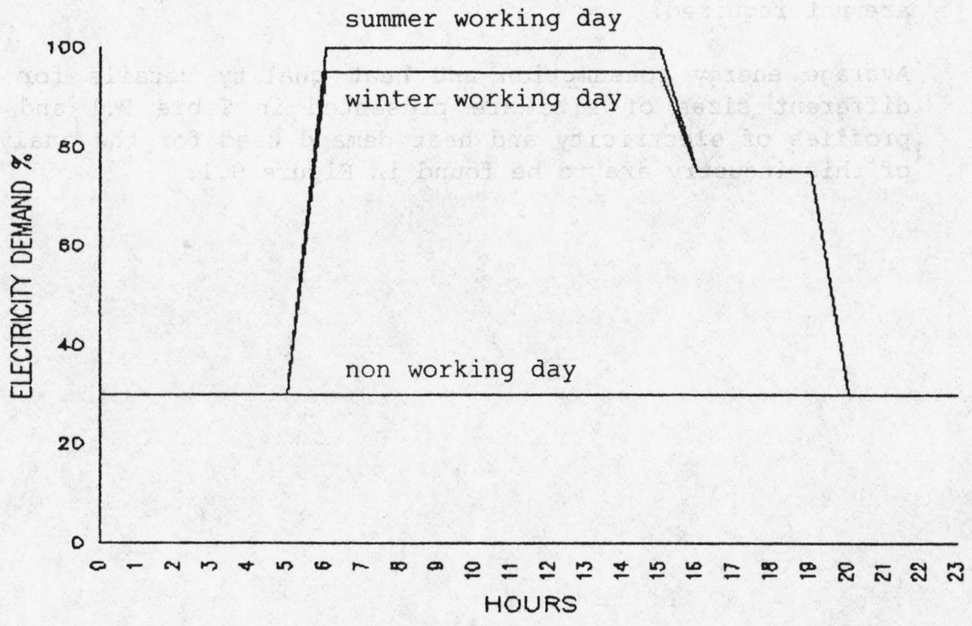

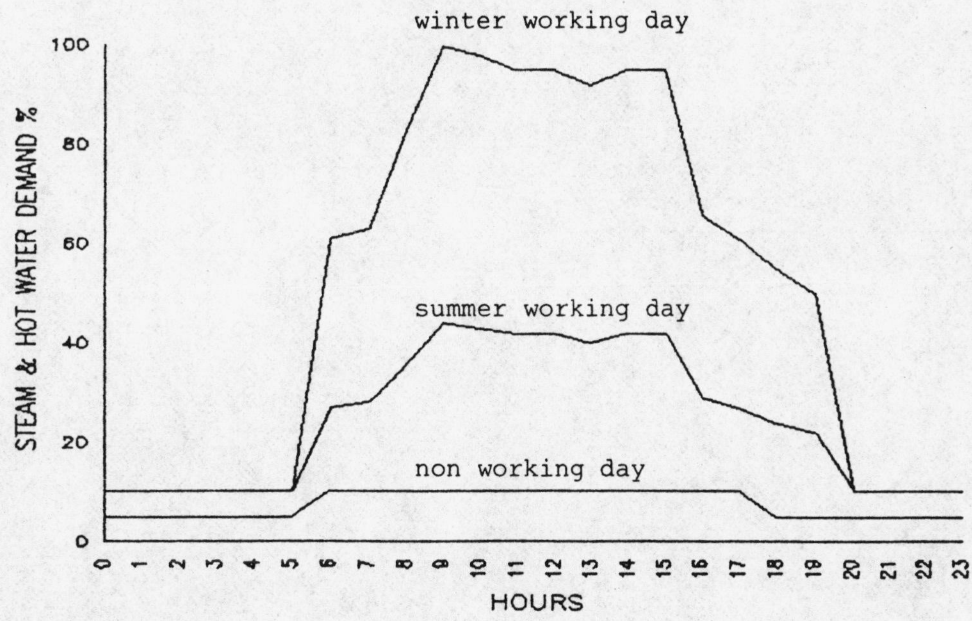

Milk Products

Fig. 8

9. FROZEN FRUITS AND VEGETABLES [part of SIC 4147]

Profile of the Industry

The frozen fruit and vegetable industry is seasonal in nature. Refrigeration is a major electricity user and heat is required by cooking operations, for the production of hot water for cleaning purposes, and for space heating. High temperatures are not required.

Average energy consumption and heat quality details for two different sizes of site are presented in Table 9.1 and the profiles of electricity and heat demand used for the analysis of this industry are to be found in Figure 9.1.

Energy Consumption and Heat Quality Details for the Frozen Fruit and Vegetable Industry

AVERAGE ANNUAL ENERGY CONSUMPTION PER SITE	Size Band 1 kTh	Size Band 1 GWh	Size Band 1 %	Size Band 2 kTh	Size Band 2 GWh	Size Band 2 %	Size Band 3 kTh	Size Band 3 GWh	Size Band 3 %
1) ELECTRICITY	31.0	.9	25.02	143.6	4.2	28.43			
2) FUELS for STEAM & HOT WATER GENERATION	81.9	2.4	66.10	318.7	9.3	63.10			
3) FUELS for OTHER USE	11.0	.3	8.88	42.8	1.3	8.47			
Totals	123.9	3.6	100	505.0	14.8	100			

DETAILS OF STEAM AND HOT WATER GENERATION AND USE

Steam generation details: 135.0 psig saturated

Details of use: 85.0 % process heat
15.0 % space heat

Fraction of process heat demand which can be met by hot water at < or = 85 deg.C: 13.0 %

Boiler efficiency: 77.5 % Condensate return: 80.0 %

DETAILS OF APPROPRIATE DIRECT FIRED PLANT CONTAINED WITHIN 3)

Fraction of 3) used in the plant: 0 %

Operating temperature of plant: 0 deg.C

NUMBER OF SITES (all size bands): 8 NUMBER OF NON-WORKING DAYS PER YEAR: 119

Frozen Fruit and Vegatables

Table 9

Typical Daily Profiles of Heat and Power Demand for the Frozen Fruit and Vegetable Industry

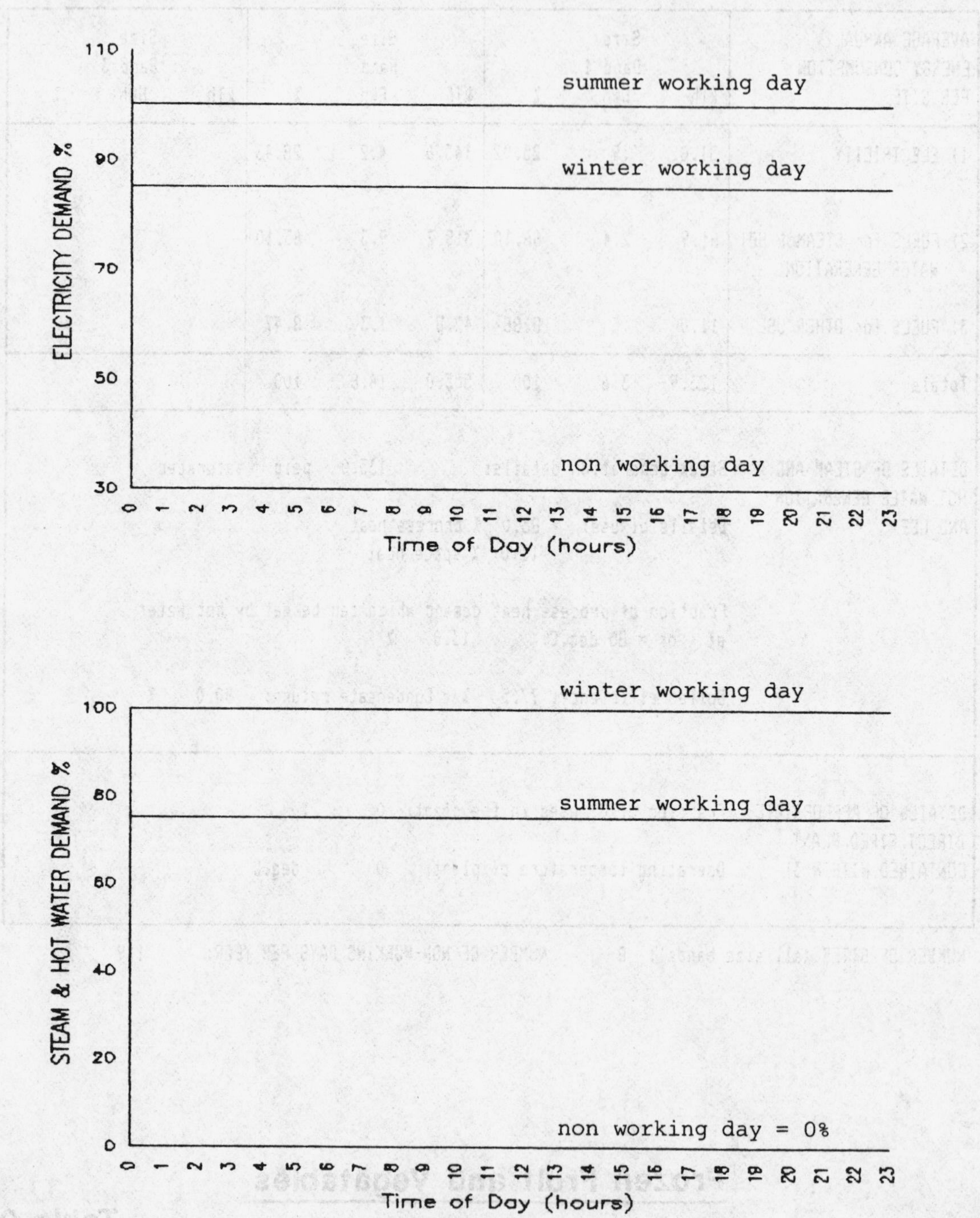

Frozen Fruit and Vegatables

Fig. 9

10. CANNED FRUIT AND VEGETABLES [part of SIC 4147]

Profile of the Industry

The main processes are washing, steam peeling, blanching, brine and syrup production, filling and closing of the cans, pasteurisation of fruits, and sterilisation of vegetables. Steam and hot water are required by most of these processes and a significant amount of hot water is required for cleaning purposes. High temperatures are not required.

Typical energy consumption and heat quality details for three different sizes of cannery are presented in Table 10.1 and representative profiles of electricity and heat demand are to be found in Figure 10.1.

Energy Consumption and Heat Quality Details for the Canned Fruit and Vegetable Industry

AVERAGE ANNUAL ENERGY CONSUMPTION PER SITE	Size Band 1 kTh	GWh	%	Size Band 2 kTh	GWh	%	Size Band 3 kTh	GWh	%
1) ELECTRICITY	31.0	.9	9.82	143.6	4.2	11.47	991.7	29.1	13.80
2) FUELS for STEAM & HOT WATER GENERATION	270.3	7.9	85.59	1052.0	30.8	84.02	5876.5	172.2	81.80
3) FUELS for OTHER USE	14.5	.4	4.60	56.5	1.7	4.51	315.7	9.3	4.39
Totals	315.8	9.3	100	1252.0	36.7	100	7183.8	210.5	100

DETAILS OF STEAM AND HOT WATER GENERATION AND USE

Steam generation details: 150.0 psig saturated

Details of use: 91.5 % process heat
8.5 % space heat

Fraction of process heat demand which can be met by hot water at < or = 85 deg.C: 10.0 %

Boiler efficiency: 81.0 % Condensate return: 45.0 %

DETAILS OF APPROPRIATE DIRECT FIRED PLANT CONTAINED WITHIN 3)

Fraction of 3) used in the plant: 0 %

Operating temperature of plant: 0 deg.C

NUMBER OF SITES (all size bands): 22 NUMBER OF NON-WORKING DAYS PER YEAR: 119

Canned Fruit and Vegatables

Table 10

Typical Daily Profiles of Heat and Power Demand for the Canned Fruit and Vegetable Industry

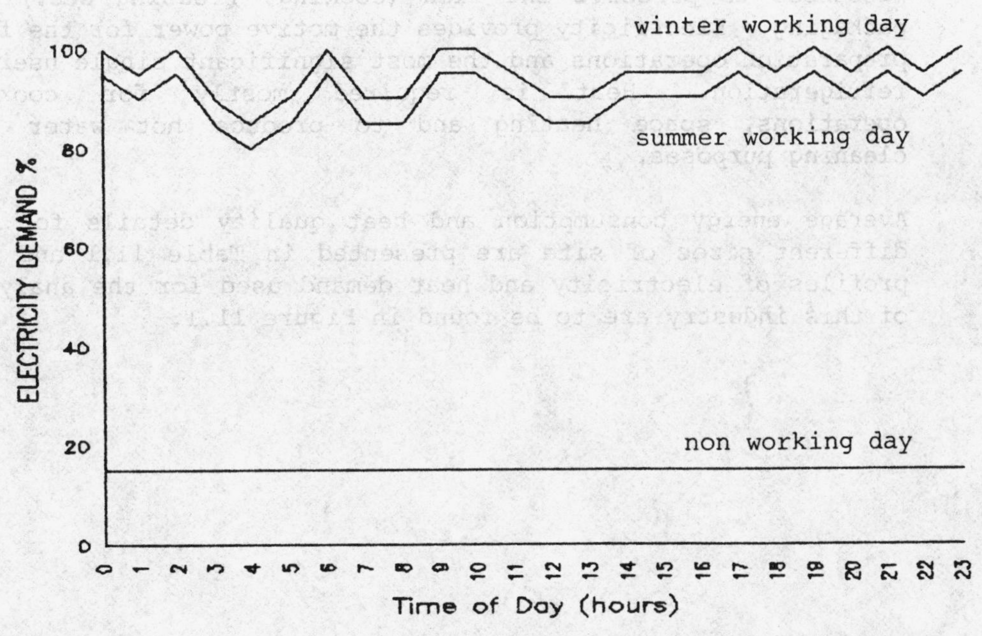

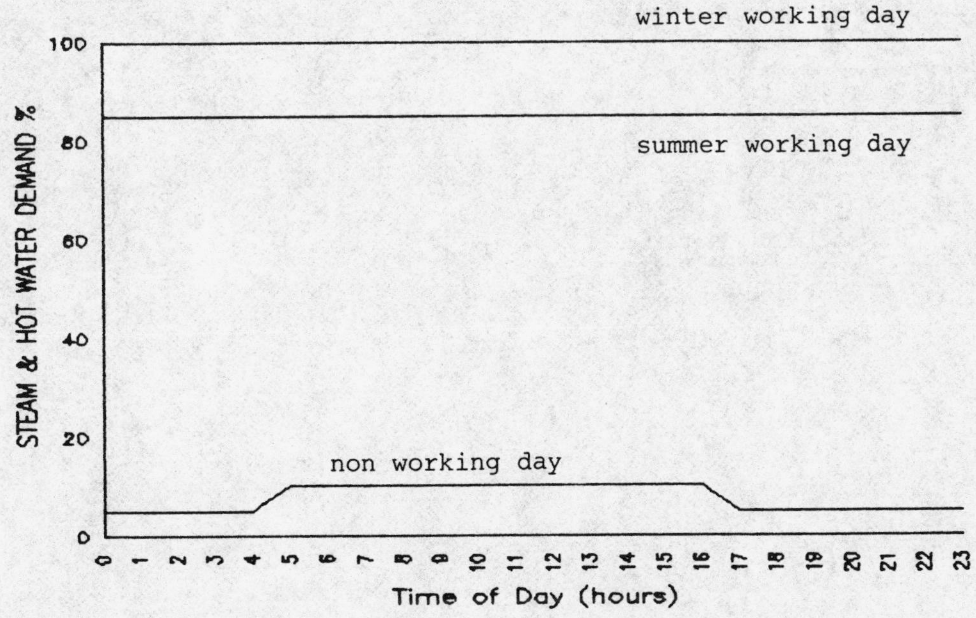

Canned Fruit and Vegatables **Fig. 10**

11. FISH AND FISH PRODUCTS [SIC 4150]

Profile of the Industry

The production of fish products consists principally of preparation (cutting, trimming, grinding, mixing etc.), processes to preserve the fish (cooking, freezing etc.) and packaging. Electricity provides the motive power for the fish preparation operations and the most significant single user is refrigeration. Heat is required mostly for cooking operations, space heating and to produce hot water for cleaning purposes.

Average energy consumption and heat quality details for two different sizes of site are presented in Table 11.1 and the profiles of electricity and heat demand used for the analysis of this industry are to be found in Figure 11.1.

Energy Consumption and Heat Quality Details for the Fish and Fish Products Industry

AVERAGE ANNUAL ENERGY CONSUMPTION PER SITE	Size Band 1 kTh	Size Band 1 GWh	Size Band 1 %	Size Band 2 kTh	Size Band 2 GWh	Size Band 2 %	Size Band 3 kTh	Size Band 3 GWh	Size Band 3 %
1) ELECTRICITY	80.0	2.3	20.56	2274.0	66.6	38.55			
2) FUELS for STEAM & HOT WATER GENERATION	230.6	6.8	59.28	2704.6	79.3	45.85			
3) FUELS for OTHER USE	78.4	2.3	20.16	919.5	27.0	15.59			
Totals	389.1	11.4	100	5898.1	172.9	100			

DETAILS OF STEAM AND HOT WATER GENERATION AND USE

Steam generation details: 100.0 psig saturated

Details of use: 90.0 % process heat
10.0 % space heat

Fraction of process heat demand which can be met by hot water at < or = 85 deg.C: 5.0 %

Boiler efficiency: 75.0 % Condensate return: 58.0 %

DETAILS OF APPROPRIATE DIRECT FIRED PLANT CONTAINED WITHIN 3)

Fraction of 3) used in the plant: 0 %

Operating temperature of plant: 0 deg.C

NUMBER OF SITES (all size bands): 8 NUMBER OF NON-WORKING DAYS PER YEAR: 119

Fish and Fish Products

Table 11

Typical Daily Profiles of Heat and Power Demand for the Fish and Fish Products Industry

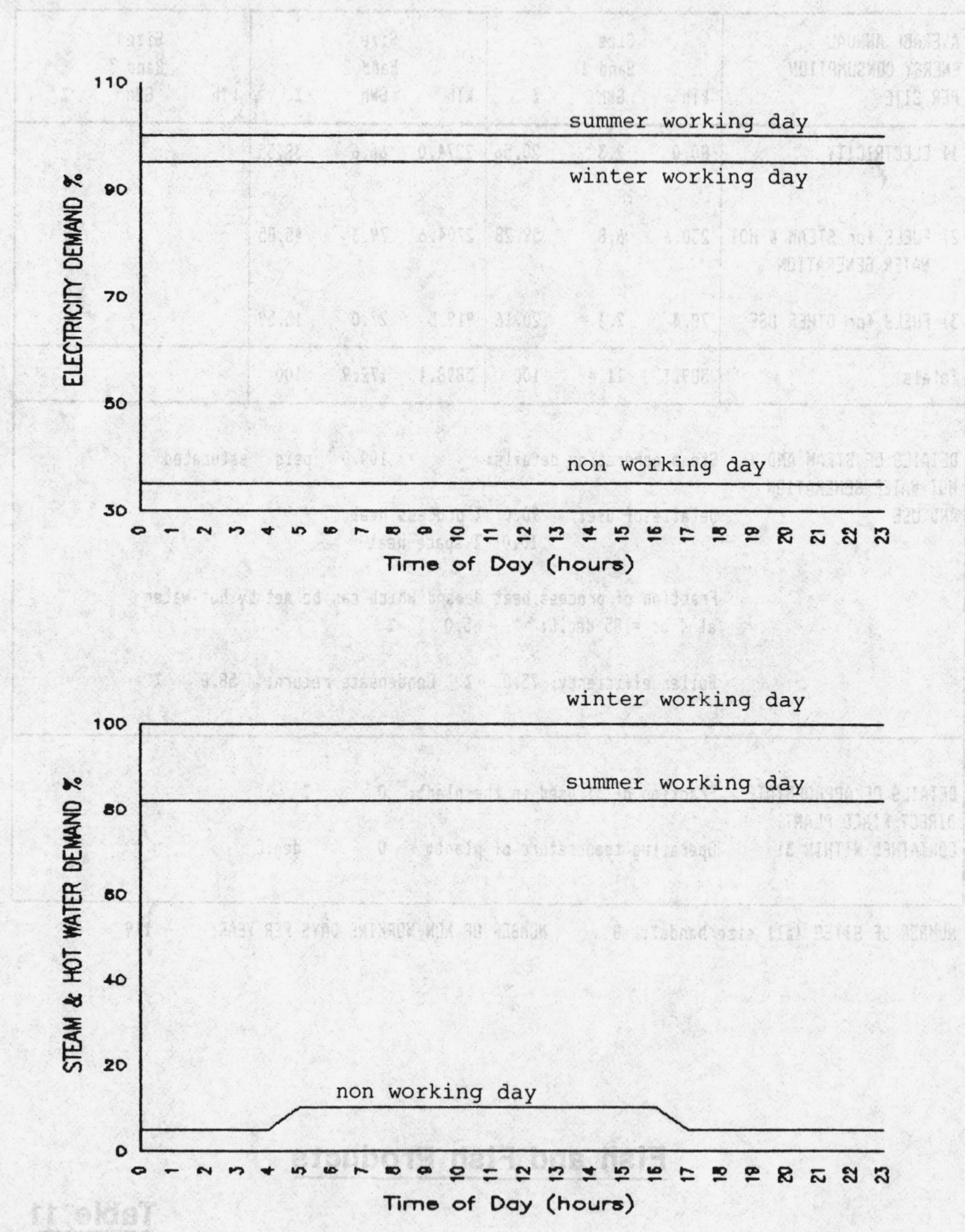

Fish and Fish Products **Fig. 11**

12. BREAD AND FLOUR CONFECTIONERY [SIC 4196]

Profile of the Industry

The production of bread and flour confectionery involves principally mixing, rolling, dividing and baking operations. Coatings may also be added after baking. Steam is used for proving, drying and to produce hot water for cleaning. Electricity provides power for the mixing, rolling and dividing operations and for refrigeration and conveyors. Both indirect and direct firing of ovens is to be found and natural gas is the most common fuel.

Typical energy consumption and heat quality details for three different sizes of non-CHP site are given in Table 12.1 and the electricity and heat demand profiles used for analysis of this subsector are presented in Figure 12.1.

Energy Consumption and Heat Quality Details for the Bread and Flour Confectionery Industry

AVERAGE ANNUAL ENERGY CONSUMPTION PER SITE	Size Band 1 kTh	GWh	%	Size Band 2 kTh	GWh	%	Size Band 3 kTh	GWh	%
1) ELECTRICITY	52.5	1.5	19.87	119.2	3.5	17.16	333.5	9.8	8.49
2) FUELS for STEAM & HOT WATER GENERATION	98.1	2.9	37.10	266.5	7.8	38.35	1665.0	48.8	42.37
3) FUELS for OTHER USE	113.8	3.3	43.03	309.1	9.1	44.49	1931.4	56.6	49.15
Totals	264.4	7.7	100	694.8	20.4	100	3929.9	115.2	100

DETAILS OF STEAM AND HOT WATER GENERATION AND USE	Steam generation details: 100.0 psig saturated Details of use: 57.0 % process heat; 43.0 % space heat Fraction of process heat demand which can be met by hot water at < or = 85 deg.C: 0 % Boiler efficiency: 75.0 % Condensate return: 55.0 %
DETAILS OF APPROPRIATE DIRECT FIRED PLANT CONTAINED WITHIN 3)	Fraction of 3) used in the plant: 75.0 % Operating temperature of plant: 300.0 deg.C

NUMBER OF SITES (all size bands): 114 NUMBER OF NON-WORKING DAYS PER YEAR: 62

Bread and Flour Confectionery

Table 12

Typical Daily Profiles of Heat and Power Demand for the Bread and Flour Confectionery Industry

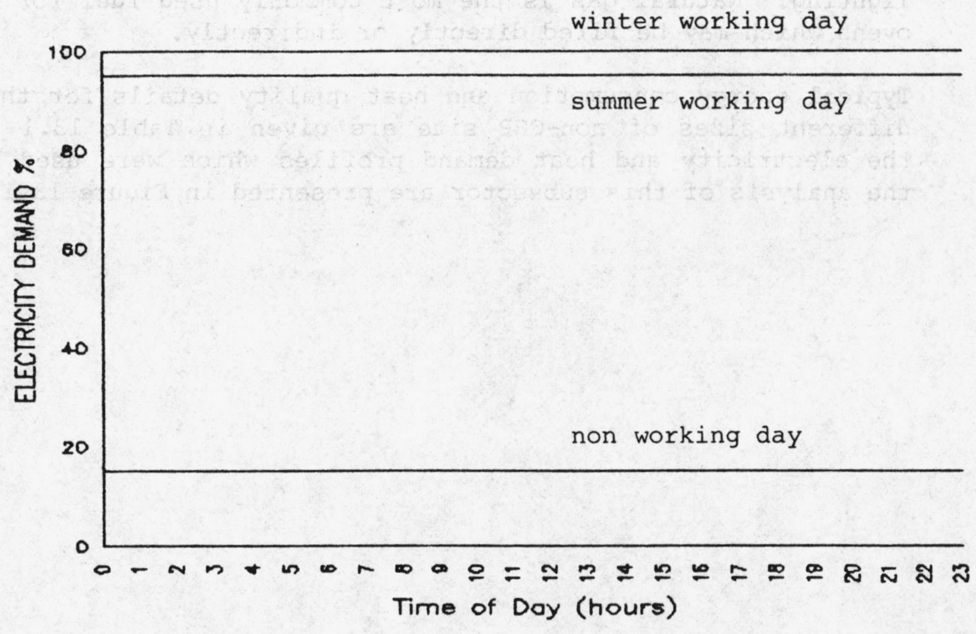

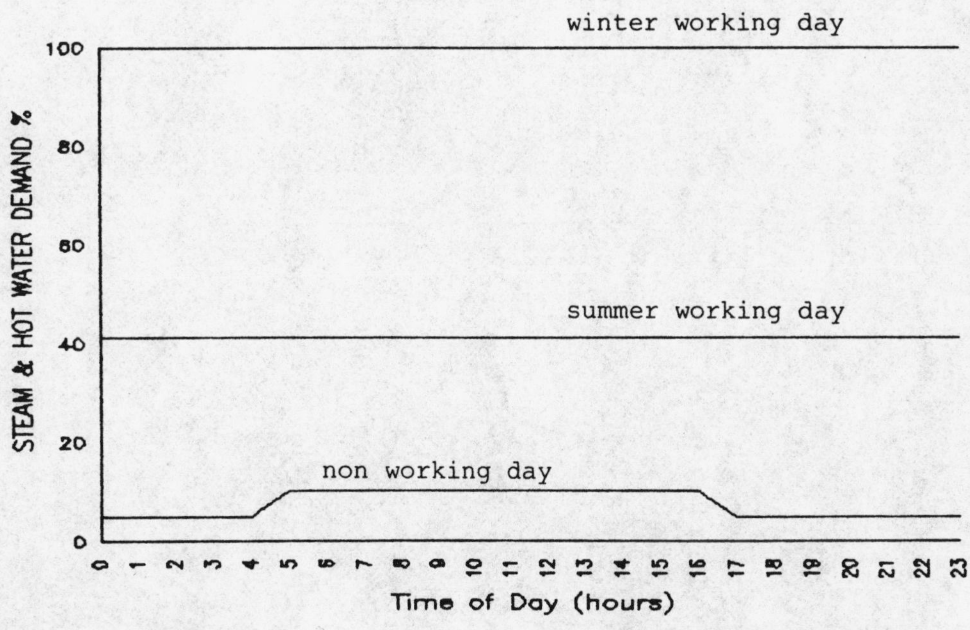

Bread and Flour Confectionery

Fig. 12

13. BISCUITS AND CRISPBREADS [SIC 4197]

Profile of the Industry

The operations involved in industrial biscuit and crispbread making are principally mixing of the ingredients, dividing and baking. Electricity is used to provide power for the mixing and dividing processes and for refrigeration, conveyors and lighting. Natural gas is the most commonly used fuel for the ovens which may be fired directly or indirectly.

Typical energy consumption and heat quality details for three different sizes of non-CHP site are given in Table 13.1 and the electricity and heat demand profiles which were used for the analysis of this subsector are presented in Figure 13.1.

Energy Consumption and Heat Quality Details for the Biscuit and Crispbread Industry

AVERAGE ANNUAL ENERGY CONSUMPTION PER SITE	Size Band 1			Size Band 2			Size Band 3		
	kTh	GWh	%	kTh	GWh	%	kTh	GWh	%
1) ELECTRICITY	52.3	1.5	21.30	175.6	5.1	14.36	1632.0	47.8	17.84
2) FUELS for STEAM & HOT WATER GENERATION	121.6	3.6	49.58	659.6	19.3	53.95	4735.2	138.8	51.76
3) FUELS for OTHER USE	71.4	2.1	29.12	387.4	11.4	31.69	2781.2	81.5	30.40
Totals	245.3	7.2	100	1222.6	35.8	100	9148.4	268.1	100

DETAILS OF STEAM AND HOT WATER GENERATION AND USE	Steam generation details: 150.0 psig saturated Details of use: 68.0 % process heat 32.0 % space heat Fraction of process heat demand which can be met by hot water at < or = 85 deg.C: 3.0 % Boiler efficiency: 77.5 % Condensate return: 86.0 %
DETAILS OF APPROPRIATE DIRECT FIRED PLANT CONTAINED WITHIN 3)	Fraction of 3) used in the plant: 72.0 % Operating temperature of plant: 300.0 deg.C

NUMBER OF SITES (all size bands): 18 NUMBER OF NON-WORKING DAYS PER YEAR: 119

Biscuits and Crispbreads

Table 13

Typical Daily Profiles of Heat and Power Demand for the Biscuit and Crispbread Industry

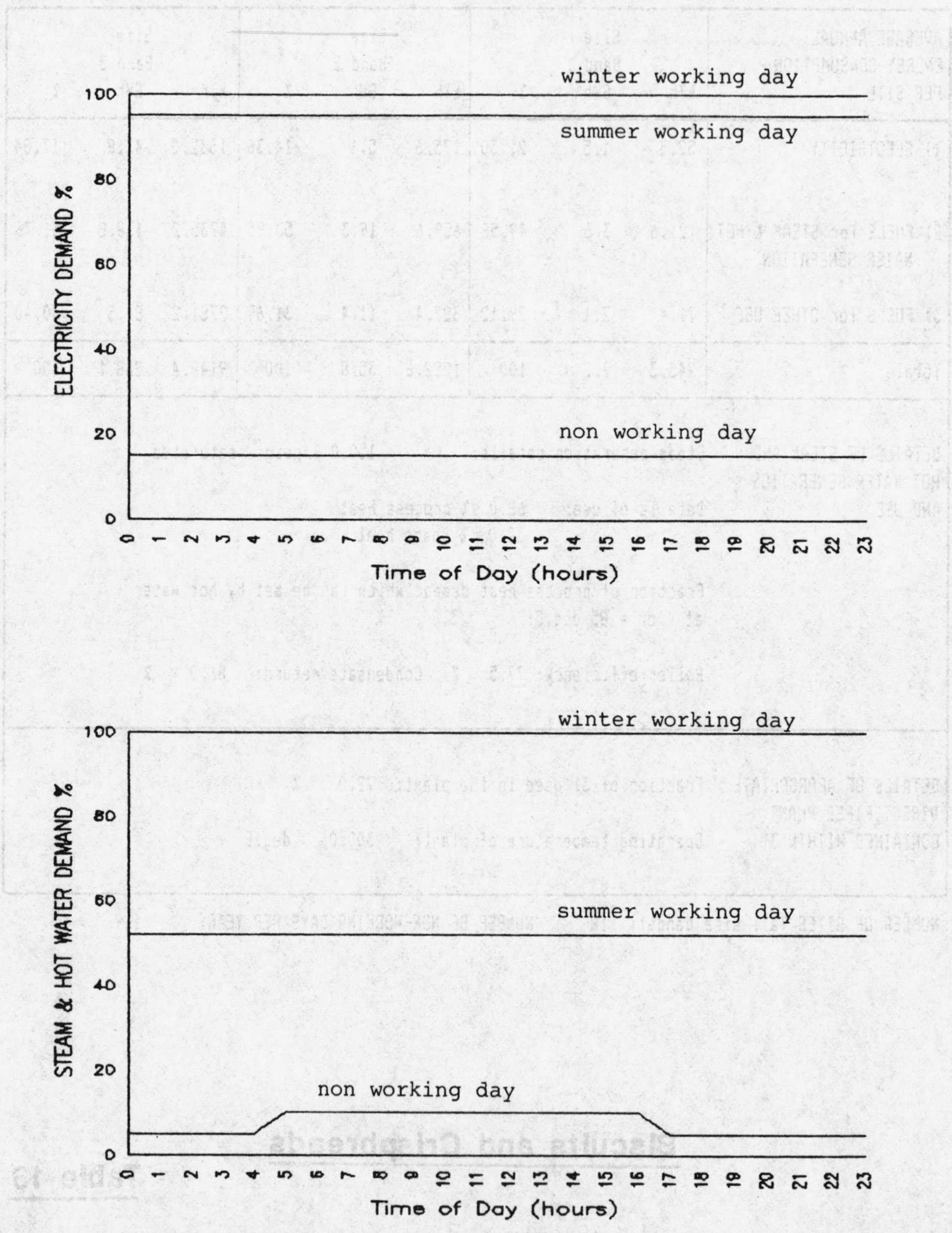

Biscuits and Crispbreads

Fig. 13

14. COCOA, CHOCOLATE AND SUGAR CONFECTIONERY INDUSTRY [SIC 4214]

Profile of the Industry

Chocolate manufacture principally involves mixing cocoa and sugar, refining and then conching for 24 hours or more. The resulting chocolate may be thinned or thickened depending on its use. Chocolate confectionery is usually chilled so that it sets quickly. Steam is used to heat the mixing and melting vessels and generally high temperatures are not required. The most important electricity uses are refrigeration, mixing, conveying and lighting.

Sugar confectionery covers a wide range of products including toffee, sweets, gums, jellies etc. Here, the main processes are: mixing, cooking, evaporation, shaping, cooling and packing. Again high temperatures are not required.

Average energy consumption and heat quality details for three different sizes of site in this subsector are given in Table 14.1 and representative electricity and heat profiles are presented in Figure 14.1.

Energy Consumption & Heat Quality Details for the Cocoa, Chocolate & Sugar Confectionery Industry

AVERAGE ANNUAL ENERGY CONSUMPTION PER SITE	Size Band 1 kTh	GWh	%	Size Band 2 kTh	GWh	%	Size Band 3 kTh	GWh	%
1) ELECTRICITY	109.2	3.2	23.13	264.5	7.8	20.96	2122.8	62.2	34.49
2) FUELS for STEAM & HOT WATER GENERATION	286.3	8.4	60.63	786.7	23.1	62.34	3180.7	93.2	51.67
3) FUELS for OTHER USE	76.7	2.2	16.24	210.8	6.2	16.70	852.2	25.0	13.84
Totals	472.2	13.8	100	1261.9	37.0	100	6155.7	180.4	100

DETAILS OF STEAM AND HOT WATER GENERATION AND USE

Steam generation details: 150.0 psig saturated

Details of use: 82.5 % process heat
17.5 % space heat

Fraction of process heat demand which can be met by hot water at < or = 85 deg.C: 0 %

Boiler efficiency: 72.0 % Condensate return: 55.0 %

DETAILS OF APPROPRIATE DIRECT FIRED PLANT CONTAINED WITHIN 3)

Fraction of 3) used in the plant: 0 %

Operating temperature of plant: 0 deg.C

NUMBER OF SITES (all size bands): 26 NUMBER OF NON-WORKING DAYS PER YEAR: 119

Cocoa, Chocolate, and Sugar Confectionery

Table 14

Typical Daily Profiles of Heat & Power Demand for the Cocoa, Chocolate & Sugar Confectionery Industry

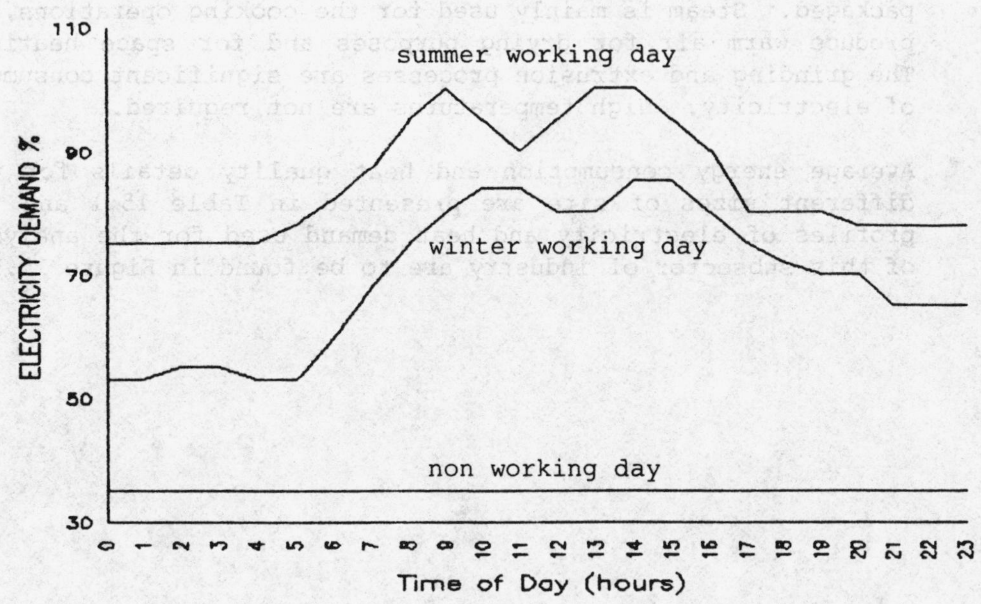

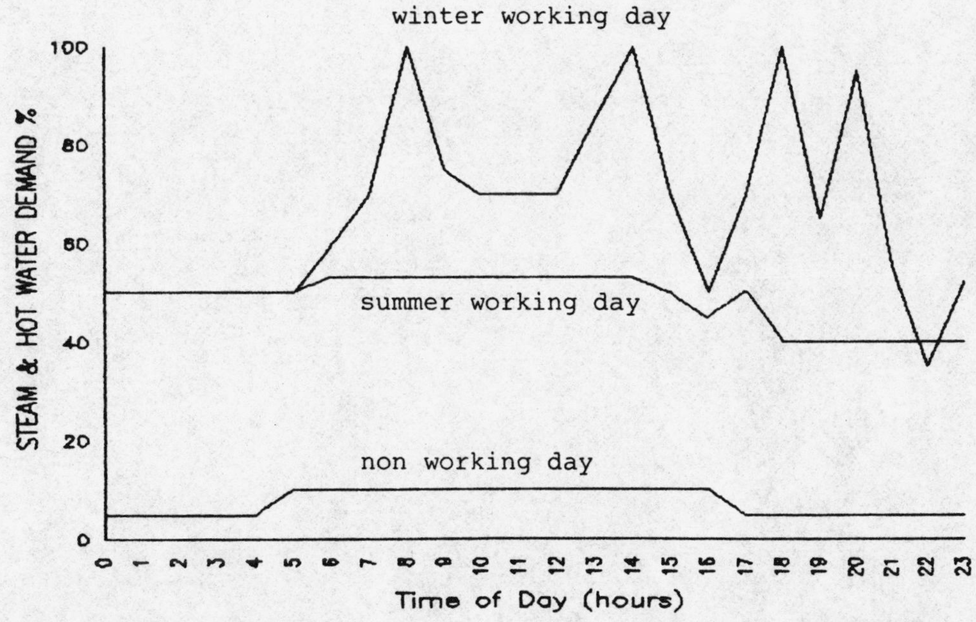

Cocoa, Chocolate, and Sugar Confectionery

Fig. 14

15. COMPOUND ANIMAL FEEDS [SIC 4221]

Profile of the Industry

Compound animal feeds are predominantly mixtures of fats and ground grains. The fats have to be heated to make them sufficiently liquid for the mixing process. The mixture is cooked in steam heated vessels, strained and dried using warm air. The dried mixture is finally extruded, pelletised and packaged. Steam is mainly used for the cooking operations, to produce warm air for drying purposes and for space heating. The grinding and extrusion processes are significant consumers of electricity. High temperatures are not required.

Average energy consumption and heat quality details for two different sizes of site are presented in Table 15.1 and the profiles of electricity and heat demand used for the analysis of this subsector of industry are to be found in Figure 15.1.

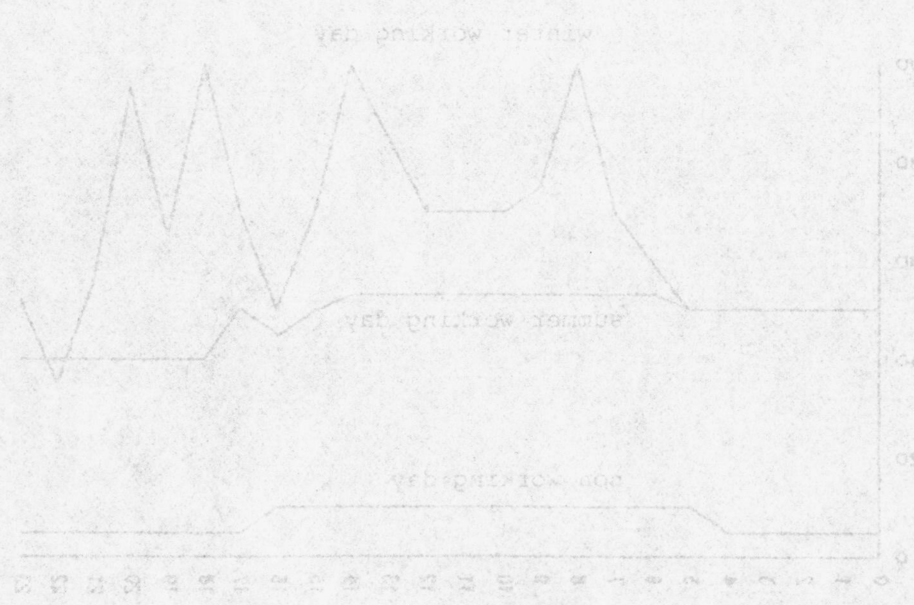

Energy Consumption and Heat Quality Details for the Compound Animal Feeds Industry

AVERAGE ANNUAL ENERGY CONSUMPTION PER SITE	Size Band 1 kTh	Size Band 1 GWh	Size Band 1 %	Size Band 2 kTh	Size Band 2 GWh	Size Band 2 %	Size Band 3 kTh	Size Band 3 GWh	Size Band 3 %
1) ELECTRICITY	111.9	3.3	51.45	388.3	11.4	49.46			
2) FUELS for STEAM & HOT WATER GENERATION	100.0	2.9	46.00	376.1	11.0	47.89			
3) FUELS for OTHER USE	5.5	.2	2.55	20.8	.6	2.65			
Totals	217.4	6.4	100	785.2	23.0	100			

DETAILS OF STEAM AND HOT WATER GENERATION AND USE

Steam generation details: 150.0 psig saturated

Details of use: 84.6 % process heat
15.4 % space heat

Fraction of process heat demand which can be met by hot water at < or = 85 deg.C: 0 %

Boiler efficiency: 77.5 % Condensate return: 80.0 %

DETAILS OF APPROPRIATE DIRECT FIRED PLANT CONTAINED WITHIN 3)

Fraction of 3) used in the plant: 0 %

Operating temperature of plant: 0 deg.C

NUMBER OF SITES (all size bands): 40 NUMBER OF NON-WORKING DAYS PER YEAR: 119

Compound Animal Feeds

Table 15

Typical Daily Profiles of Heat and Power Demand for the Compound Animal Feeds Industry

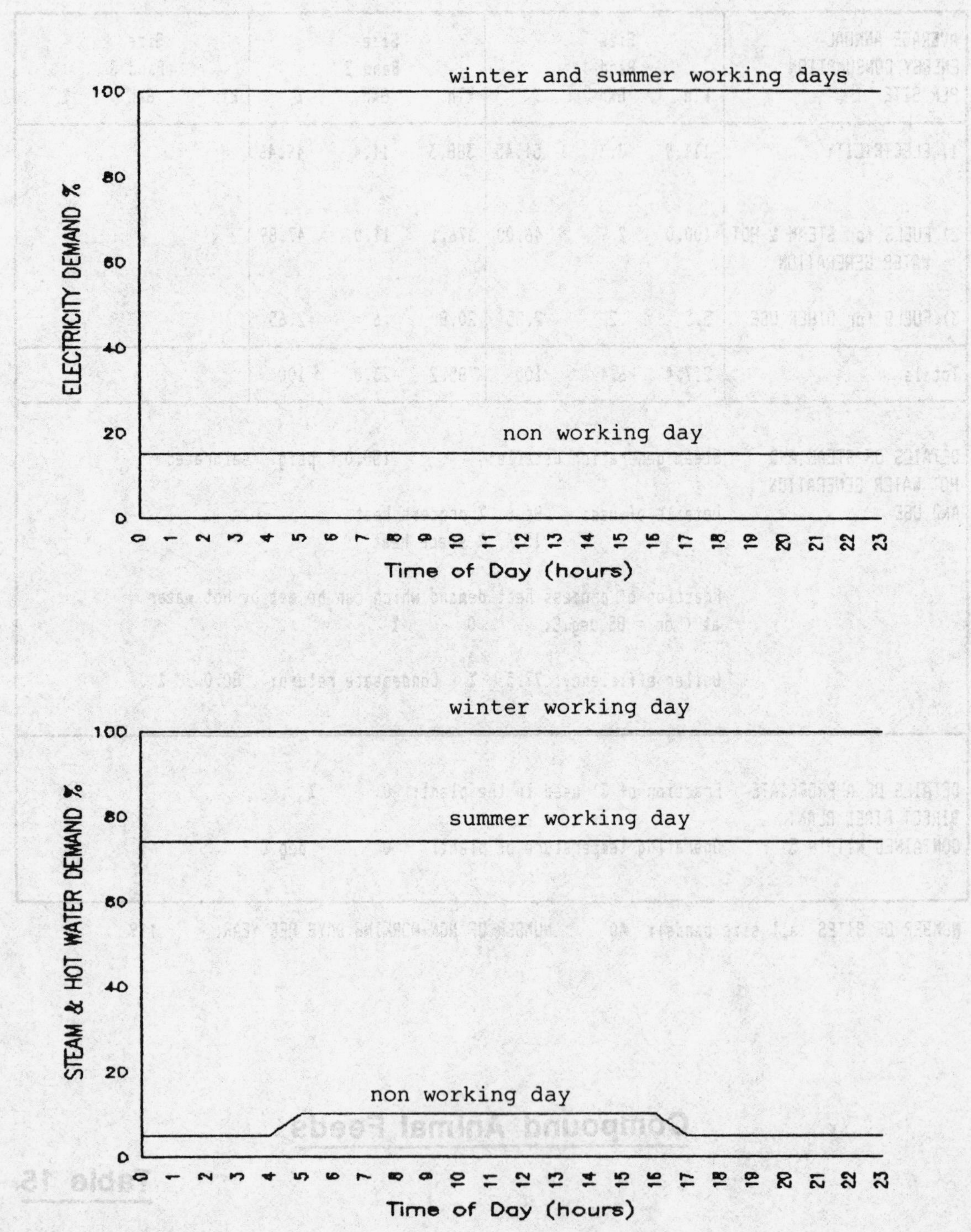

Compound Animal Feeds

Fig. 15

16. PETFOODS [SIC 4222]

Profile of the Industry

The ingredients, mainly meat and grains, are mixed, ground and extruded as lumps which are then either dried and packed or canned and sterilised. Petfood canneries tend to be large-scale continuous operations and product sterilisation accounts for a significant proportion of the requirement for steam. Sterilisation is carried out at about 130°C.

Typical energy consumption and heat quality details for two different sizes of non-CHP site are given in Table 1 and representative electricity and heat demand profiles are presented in Figure 16.1.

Energy Consumption and Heat Quality Details for the Petfoods Industry

AVERAGE ANNUAL ENERGY CONSUMPTION PER SITE	Size Band 1			Size Band 2			Size Band 3		
	kTh	GWh	%	kTh	GWh	%	kTh	GWh	%
1) ELECTRICITY	82.7	2.4	27.18				367.9	10.8	14.10
2) FUELS for STEAM & HOT WATER GENERATION	221.5	6.5	72.82				2240.7	65.7	85.90
3) FUELS for OTHER USE	0	0	0				0	0	0
Totals	304.1	8.9	100				2608.6	76.5	100

DETAILS OF STEAM AND HOT WATER GENERATION AND USE

Steam generation details: 100.0 psig saturated

Details of use: 80.0 % process heat
20.0 % space heat

Fraction of process heat demand which can be met by hot water at < or = 85 deg.C: 5.0 %

Boiler efficiency: 77.5 % Condensate return: 80.0 %

DETAILS OF APPROPRIATE DIRECT FIRED PLANT CONTAINED WITHIN 3)

Fraction of 3) used in the plant: 0 %

Operating temperature of plant: 0 deg.C

NUMBER OF SITES (all size bands): 17 NUMBER OF NON-WORKING DAYS PER YEAR: 21

Petfoods

Table 16

Typical Daily Profiles of Heat and Power Demand for the Petfoods Industry

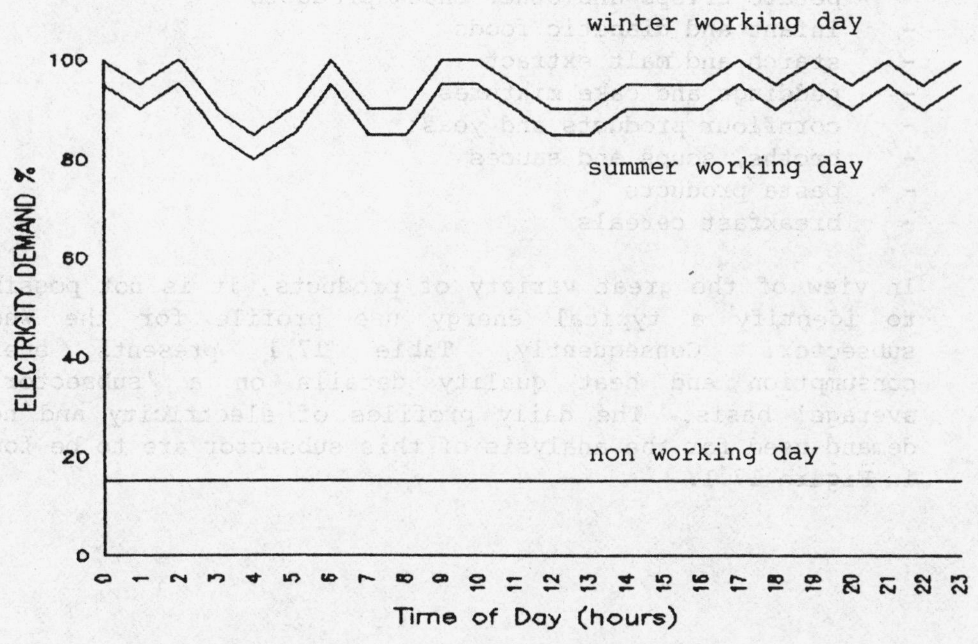

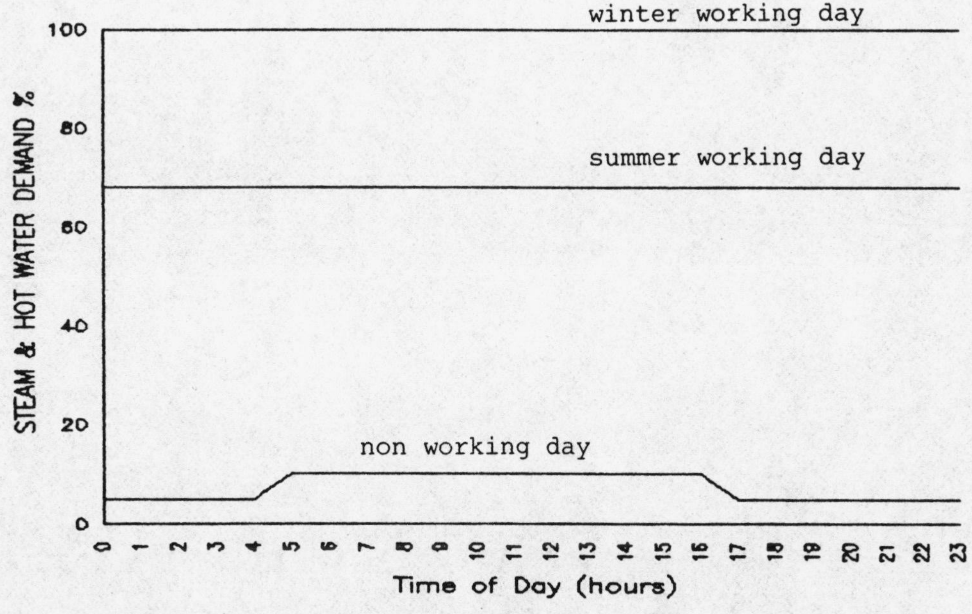

Petfoods

Fig. 16

17. MISCELLANEOUS FOODS [SIC 4239]

Profile of the Industry

This subsector covers a wide range of foods including:

- tea, coffee and coffee substitutes
- potato crisps and other snack products
- infant and diabetic foods
- starch and malt extract
- puddings and cake mixtures
- cornflour products and yeast
- broths, soups and sauces
- pasta products
- breakfast cereals

In view of the great variety of products, it is not possible to identify a typical energy use profile for the whole subsector. Consequently, Table 17.1 presents energy consumption and heat quality details on a 'subsector - average' basis. The daily profiles of electricity and heat demand used for the analysis of this subsector are to be found in Figure 17.1.

Energy Consumption and Heat Quality Details for the Miscellaneous Foods Industry

AVERAGE ANNUAL ENERGY CONSUMPTION PER SITE	Size Band 1 kTh	GWh	%	Size Band 2 kTh	GWh	%	Size Band 3 kTh	GWh	%
1) ELECTRICITY	99.9	2.9	26.54	347.6	10.2	15.89	1107.0	32.4	17.59
2) FUELS for STEAM & HOT WATER GENERATION	195.4	5.7	51.91	1300.4	38.1	59.44	3665.9	107.4	58.24
3) FUELS for OTHER USE	81.1	2.4	21.54	539.7	15.8	24.67	1521.3	44.6	24.17
Totals	376.3	11.0	100	2187.7	64.1	100	6294.2	184.5	100

DETAILS OF STEAM AND HOT WATER GENERATION AND USE

Steam generation details: 150.0 psig saturated

Details of use: 33.0 % process heat
67.0 % space heat

Fraction of process heat demand which can be met by hot water at < or = 85 deg.C: 0 %

Boiler efficiency: 80.0 % Condensate return: 67.0 %

DETAILS OF APPROPRIATE DIRECT FIRED PLANT CONTAINED WITHIN 3)

Fraction of 3) used in the plant: 0 %

Operating temperature of plant: 0 deg.C

NUMBER OF SITES (all size bands): 49 NUMBER OF NON-WORKING DAYS PER YEAR: 119

Miscellaneous Foods

Table 17

Typical Daily Profiles of Heat and Power Demand for the Miscellaneous Foods Industry

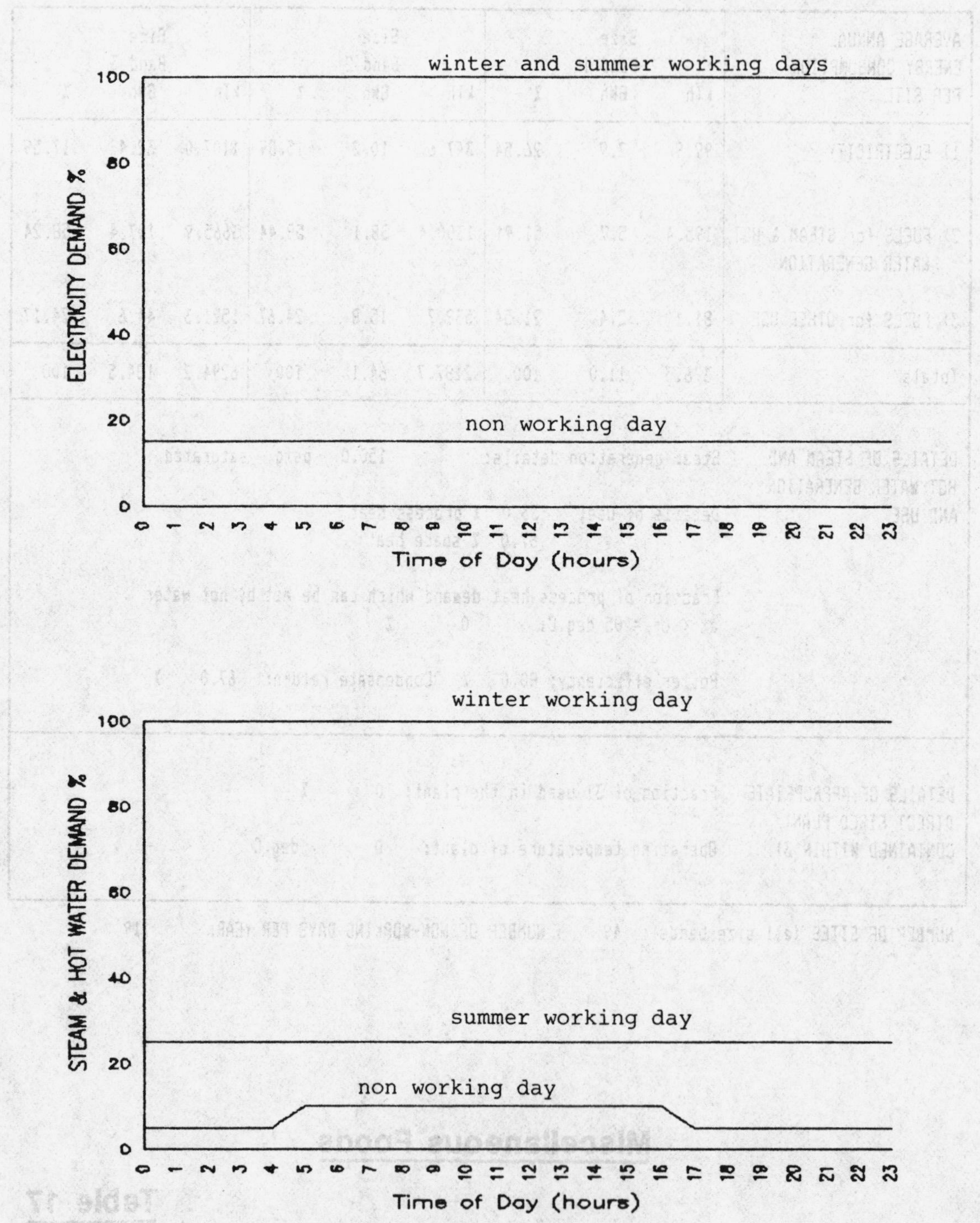

Miscellaneous Foods

Fig. 17

18. MALT WHISKY [SIC 4240]

Profile of the Industry

Malt whisky production is a batch process which varies little between sites and commences with the malt being ground and mixed with hot water in large vessels. The hot water causes the enzymes in the malt to convert the soluble starches to fermentable sugars. The product, known as wort, is decanted, cooled, dosed with yeast and allowed to ferment for about two days. The resulting liquid contains 4 to 7% alcohol and is distilled to achieve the required degree of concentration. Distillation is a major energy user in distilleries.

Typical energy consumption and heat quality details for two different sizes of non-CHP site are presented in Table 18.1 and typical profiles of electricity and heat demand are given in Figure 18.1.

Energy Consumption and Heat Quality Details for the Malt Whiskey Industry

AVERAGE ANNUAL ENERGY CONSUMPTION PER SITE	Size Band 1 kTh	GWh	%	Size Band 2 kTh	GWh	%	Size Band 3 kTh	GWh	%
1) ELECTRICITY	110.3	3.2	27.04	252.3	7.4	8.12			
2) FUELS for STEAM & HOT WATER GENERATION	267.5	7.8	65.58	2567.8	75.3	82.59			
3) FUELS for OTHER USE	30.1	.9	7.38	288.9	8.5	9.29			
Totals	407.9	12.0	100	3109.0	91.1	100			

DETAILS OF STEAM AND HOT WATER GENERATION AND USE	Steam generation details: 150.0 psig saturated Details of use: 87.5 % process heat 12.5 % space heat Fraction of process heat demand which can be met by hot water at < or = 85 deg.C: 10.0 % Boiler efficiency: 77.5 % Condensate return: 30.0 %
DETAILS OF APPROPRIATE DIRECT FIRED PLANT CONTAINED WITHIN 3)	Fraction of 3) used in the plant: 100.0 % Operating temperature of plant: 180.0 deg.C

NUMBER OF SITES (all size bands): 19 NUMBER OF NON-WORKING DAYS PER YEAR: 124

Malt Whiskey

Table 18

Typical Daily Profiles of Heat and Power Demand for the Malt Whiskey Industry

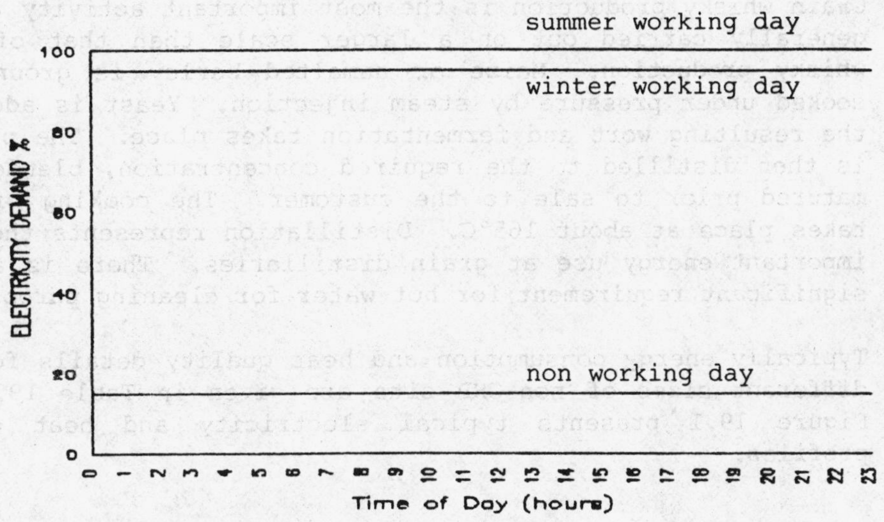

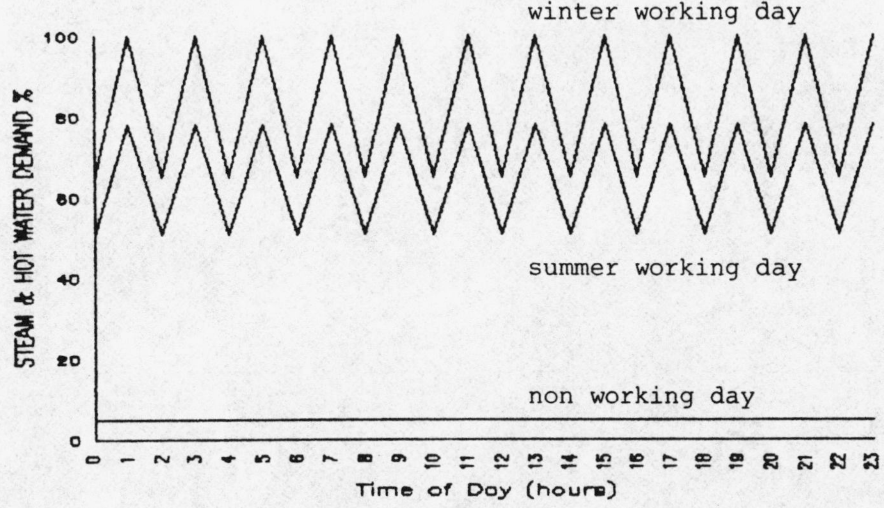

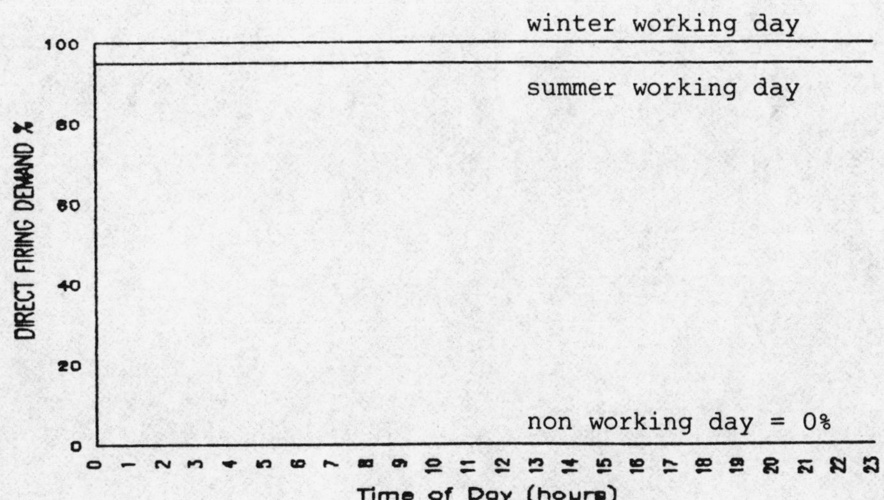

Malt Whiskey

Fig. 18

19. POTABLE SPIRITS (excluding malt whisky) [SIC 4240]

Profile of the Industry

Grain whisky production is the most important activity and is generally carried out on a larger scale than that of malt whisky production. Maize or unmalted barley is ground and cooked under pressure by steam injection. Yeast is added to the resulting wort and fermentation takes place. The product is then distilled to the required concentration, blended and matured prior to sale to the customer. The cooking of wort takes place at about 165°C. Distillation represents the most important energy use at grain distilleries. There is also a significant requirement for hot water for cleaning purposes.

Typically energy consumption and heat quality details for two different sizes of non-CHP site are given in Table 19.1 and Figure 19.1 presents typical electricity and heat demand profiles.

Energy Consumption and Heat Quality Details for the Potable Spirits Industry

AVERAGE ANNUAL ENERGY CONSUMPTION PER SITE	Size Band 1 kTh	GWh	%	Size Band 2 kTh	GWh	%	Size Band 3 kTh	GWh	%
1) ELECTRICITY				252.3	7.4	8.12	724.1	21.2	5.83
2) FUELS for STEAM & HOT WATER GENERATION				2567.8	75.3	82.59	10506.3	307.9	84.64
3) FUELS for OTHER USE				288.9	8.5	9.29	1182.0	34.6	9.52
Totals				3109.0	91.1	100	12412.4	363.8	100

DETAILS OF STEAM AND HOT WATER GENERATION AND USE	Steam generation details: 150.0 psig saturated Details of use: 87.5 % process heat 12.5 % space heat Fraction of process heat demand which can be met by hot water at < or = 85 deg.C: 10.0 % Boiler efficiency: 77.5 % Condensate return: 30.0 %
DETAILS OF APPROPRIATE DIRECT FIRED PLANT CONTAINED WITHIN 3)	Fraction of 3) used in the plant: 100.0 % Operating temperature of plant: 180.0 deg.C

NUMBER OF SITES (all size bands): 14 NUMBER OF NON-WORKING DAYS PER YEAR: 124

Potable Spirits

Table 19

Typical Daily Profiles of Heat and Power Demand for the Potable Spirits Industry

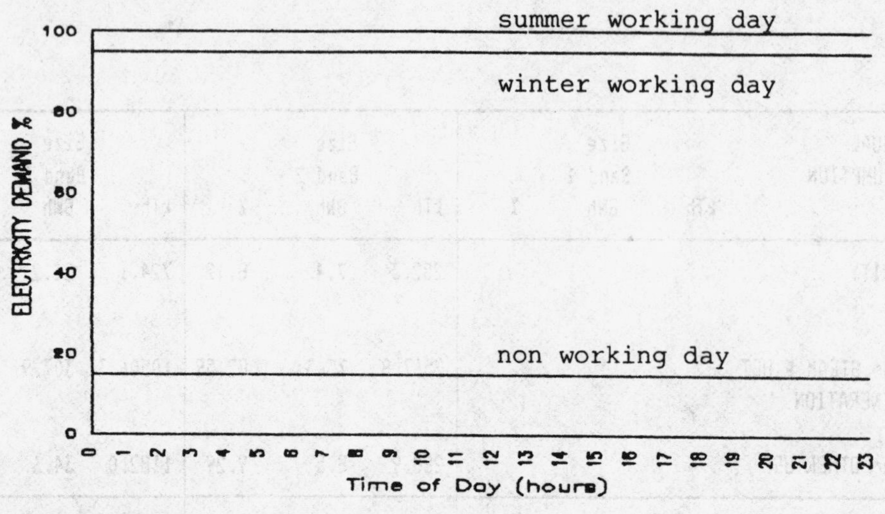

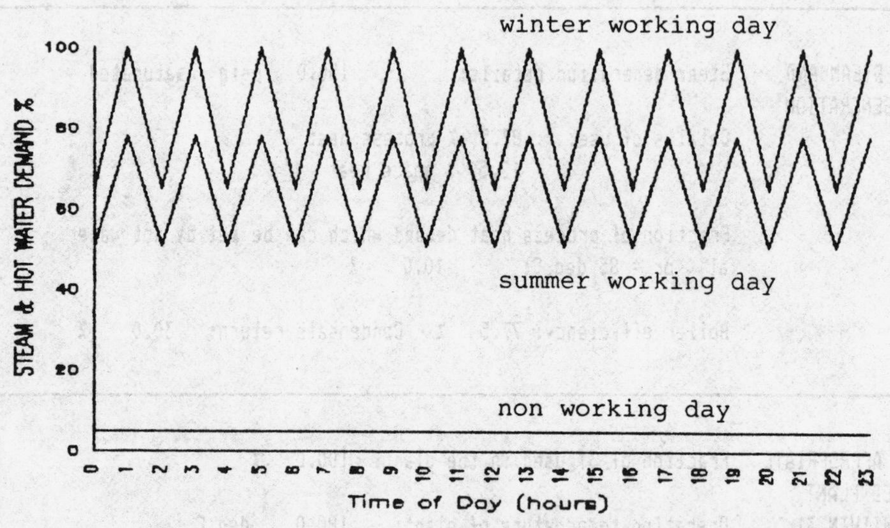

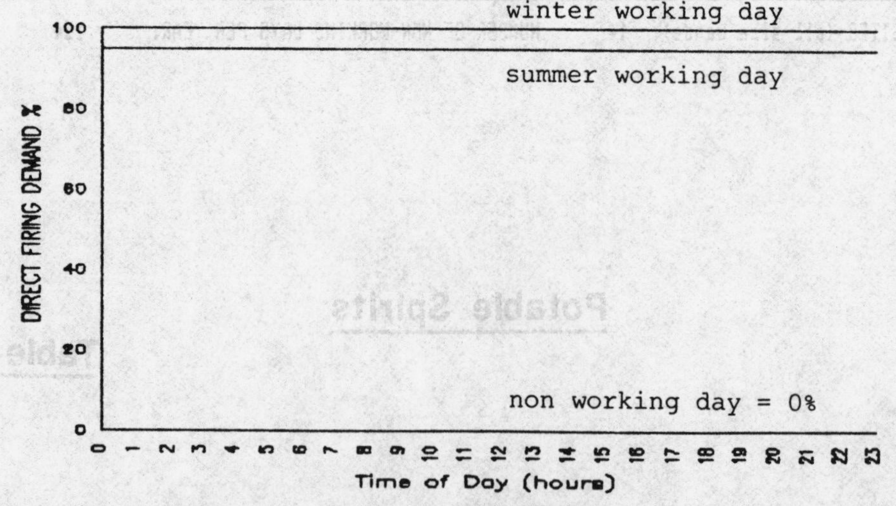

Potable Spirits

Fig. 19

20 BEER AND OTHER BREWING PRODUCTS
[Part of SIC 4270]

Profile of the Industry

Beer is made by alcoholic fermentation of carbohydrates which are derived mainly from cereals. No distillation is involved. Most beers in the UK are made from malted barley with the addition of hops and other grains to achieve the required taste. There are three main types of beer: ales, lagers and stouts. Stouts are brewed with roasted barley. The annual production of beer in the UK in 1980 was 6.6 x 10^9 litres and the industry is dominated by a small number of companies.

The brewing process can be summarised as follows: Malt is milled and mixed with hot liquor to produce a mash. Mashing then yields a sugar solution known as sweet wort. The wort is run off into a copper where the hops are added and the liquor is boiled for one or two hours. The wort is then cooled and transferred to another vessel where yeast is added and fermentation takes place. When the fermentation is complete, the beer is usually clarified, pasteurised and packaged.

Typical energy consumption and heat quality details of three different sizes of brewery are presented in Table 20.1. Brewing does not require high temperatures and approximately 75% of the energy consumed (on a heat supplied basis) is used to raise saturated steam. A significant amount of hot water is required throughout for cleaning purposes. Electricity is used primarily for motive power (60%) and refrigeration (30%). Lager production tends to use more electricity as a percentage of the total energy consumption than other beers.

Typical profiles for electricity and heat demand are presented in Figure 20.1.

Energy Consumption & Heat Quality Details for the Beer & Beer Products Industry

AVERAGE ANNUAL ENERGY CONSUMPTION PER SITE	Size Band 1 kTh	GWh	%	Size Band 2 kTh	GWh	%	Size Band 3 kTh	GWh	%
1) ELECTRICITY	90.3	2.6	20.17	352.1	10.3	15.87	2648.5	77.6	15.67
2) FUELS for STEAM & HOT WATER GENERATION	321.2	9.4	71.76	1678.6	49.2	75.63	12808.1	375.4	75.80
3) FUELS for OTHER USE	36.1	1.1	8.07	188.8	5.5	8.51	1440.9	42.2	8.53
Totals	447.6	13.1	100	2219.6	65.1	100	16897.6	495.2	100

DETAILS OF STEAM AND HOT WATER GENERATION AND USE

Steam generation details: 100.0 psig saturated

Details of use: 87.5 % process heat
12.5 % space heat

Fraction of process heat demand which can be met by hot water at < or = 85 deg.C: 10.0 %

Boiler efficiency: 77.5 % Condensate return: 30.0 %

DETAILS OF APPROPRIATE DIRECT FIRED PLANT CONTAINED WITHIN 3)

Fraction of 3) used in the plant: 0 %

Operating temperature of plant: 0 deg.C

NUMBER OF SITES (all size bands): 45 NUMBER OF NON-WORKING DAYS PER YEAR: 62

Beer and Beer Products

Table 20

Typical Daily Profiles of Heat & Power Demand for the Beer & Beer Products Industry

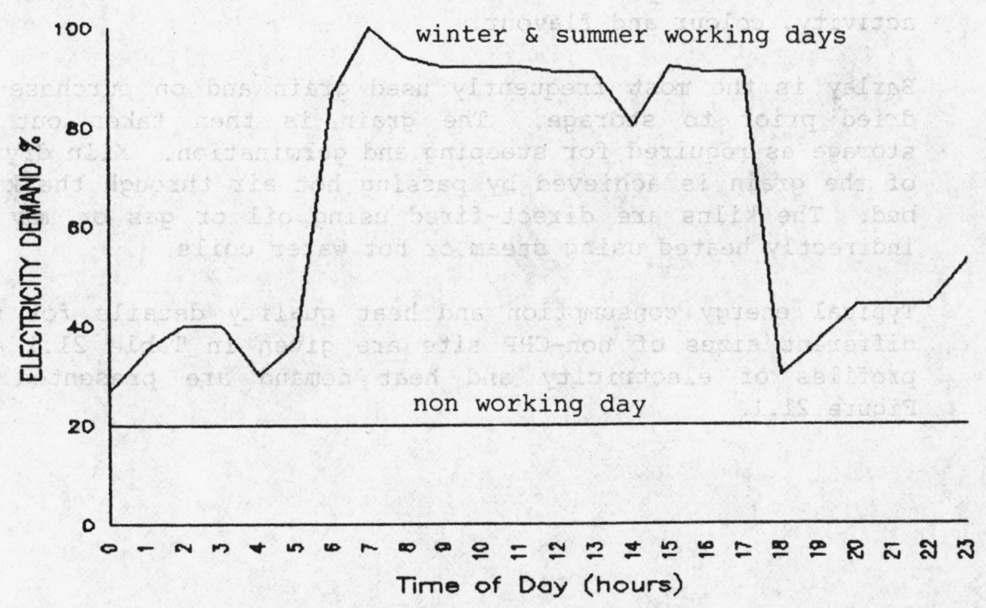

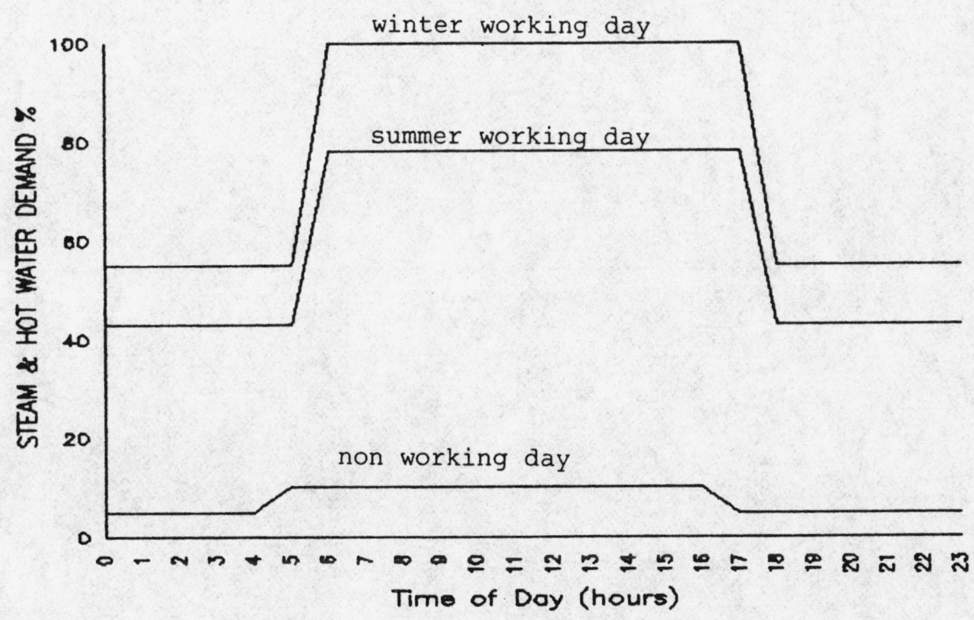

Beer and Beer Products

Fig. 20

21 MALT AND MALT PRODUCTS [Brewing and Malting SIC 4270]

Profile of the Industry

Malt is a cereal grain which has been steeped in water and allowed to germinate to a certain state, when it is stabilized by kilning. During germination, enzymes are produced which are required by brewers and distillers. The malted grain is required to have particular characteristics in terms of enzyme activity, colour and flavour.

Barley is the most frequently used grain and on purchase is dried prior to storage. The grain is then taken out of storage as required for steeping and germination. Kiln drying of the grain is achieved by passing hot air through the kiln bed. The kilns are direct-fired using oil or gas or may be indirectly heated using steam or hot water coils.

Typical energy consumption and heat quality details for two different sizes of non-CHP site are given in Table 21.1 and profiles of electricity and heat demand are presented in Figure 21.1.

Energy Consumption & Heat Quality Details for the Malt & Malt Products Industry

AVERAGE ANNUAL ENERGY CONSUMPTION PER SITE	Size Band 1			Size Band 2			Size Band 3		
	kTh	GWh	%	kTh	GWh	%	kTh	GWh	%
1) ELECTRICITY	90.3	2.6	20.50	352.1	10.3	16.14			
2) FUELS for STEAM & HOT WATER GENERATION	138.0	4.0	31.34	721.3	21.1	33.05			
3) FUELS for OTHER USE	212.1	6.2	48.17	1108.7	32.5	50.81			
Totals	440.4	12.9	100	2182.1	64.0	100			

DETAILS OF STEAM AND HOT WATER GENERATION AND USE

Steam generation details: 100.0 psig saturated

Details of use: 85.7 % process heat
14.3 % space heat

Fraction of process heat demand which can be met by hot water at < or = 85 deg.C: 5.0 %

Boiler efficiency: 77.5 % Condensate return: 80.0 %

DETAILS OF APPROPRIATE DIRECT FIRED PLANT CONTAINED WITHIN 3)

Fraction of 3) used in the plant: 100.0 %

Operating temperature of plant: 90.0 deg.C

NUMBER OF SITES (all size bands): 22 NUMBER OF NON-WORKING DAYS PER YEAR: 62

Malt and Malt Products

Table 21

Typical Daily Profiles of Heat & Power Demand for the Malt & Malt Products Industry

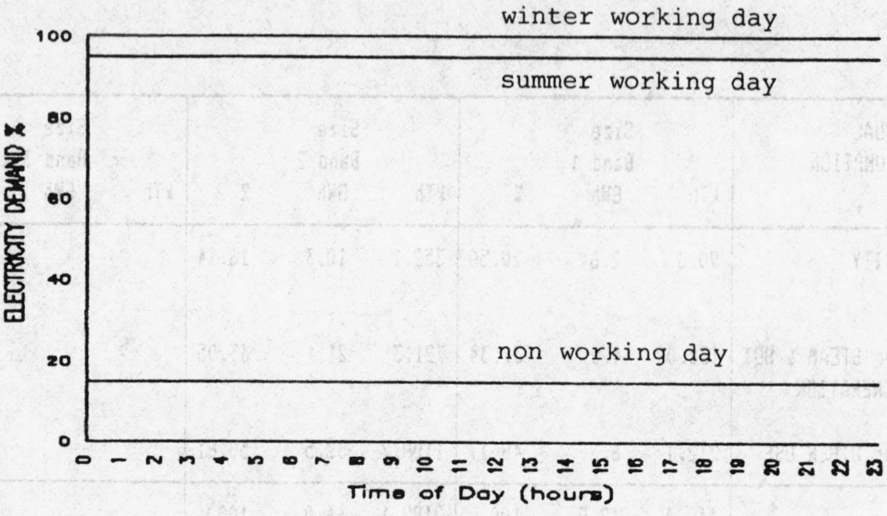

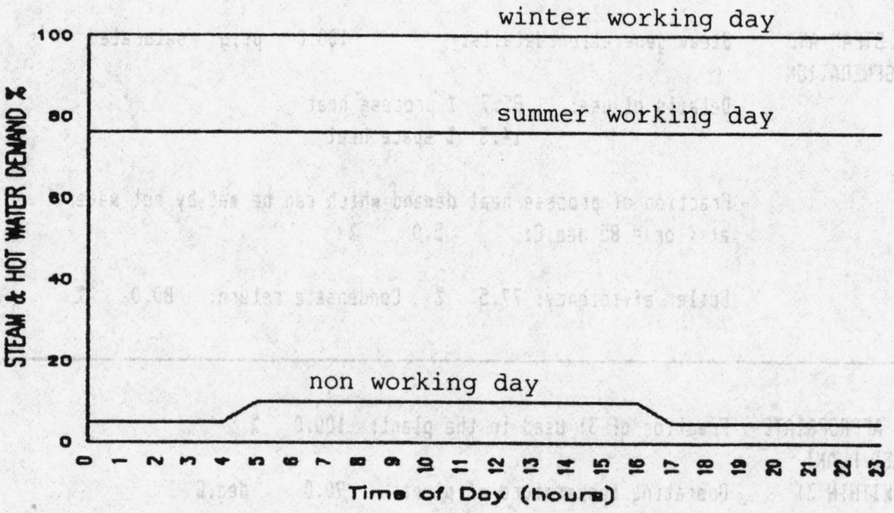

winter working day
summer working day
non working day
ELECTRICITY DEMAND %
0
20
40
60
80
100
0 1 2 3 4 5 6 7 8 9 10 11 12 13 14 15 16 17 18 19 20 21 22 23
Time of Day (hours)

winter working day
summer working day
non working day
STEAM & HOT WATER DEMAND %
0
20
40
60
80
100
0 1 2 3 4 5 6 7 8 9 10 11 12 13 14 15 16 17 18 19 20 21 22 23
Time of Day (hours)

winter working day
summer working day
non working day = 0%
DIRECT FIRING DEMAND %
0
20
40
60
80
100
0 1 2 3 4 5 6 7 8 9 10 11 12 13 14 15 16 17 18 19 20 21 22 23
Time of Day (hours)

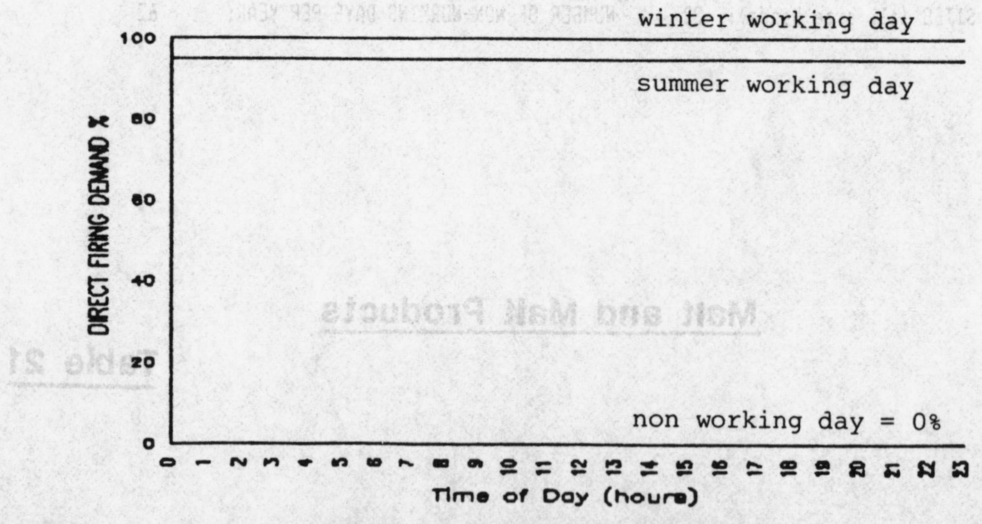

Malt and Malt Products

Fig. 21

22. TOBACCO [SIC 4290]

Profile of the Industry

The main activities of the tobacco industry are to cure and prepare the imported tobacco leaf which is then cut and made into pipe tobacco, cigars or cigarettes and finally packed. The imported tobacco arrives in the form of dry and highly compressed bales and has to be moistened by water sprays and steamed to render it pliable for cutting. The cut tobacco is then dried to a controlled moisture content. Cigarette making and packaging is generally carried out on high-speed machines.

Typical energy consumption and heat quality details for two different sizes of site are provided in Table 22.1 and representative profiles for electricity and heat demand are presented in Figure 22.1. Steam is typically used for process heating (paticularly the humidification of raw leaf) and space heating.

Energy Consumption & Heat Quality Details for the Tobacco Industry

AVERAGE ANNUAL ENERGY CONSUMPTION PER SITE	Size Band 1 kTh	Size Band 1 GWh	Size Band 1 %	Size Band 2 kTh	Size Band 2 GWh	Size Band 2 %	Size Band 3 kTh	Size Band 3 GWh	Size Band 3 %
1) ELECTRICITY	257.3	7.5	28.43	1050.3	30.8	17.76			
2) FUELS for STEAM & HOT WATER GENERATION	432.3	12.7	47.77	3245.8	95.1	54.89			
3) FUELS for OTHER USE	215.4	6.3	23.80	1617.0	47.4	27.35			
Totals	905.0	26.5	100	5913.1	173.3	100			

DETAILS OF STEAM AND HOT WATER GENERATION AND USE

Steam generation details: 150.0 psig saturated

Details of use: 18.0 % process heat
82.0 % space heat

Fraction of process heat demand which can be met by hot water at < or = 85 deg.C: 0 %

Boiler efficiency: 77.5 % Condensate return: 80.0 %

DETAILS OF APPROPRIATE DIRECT FIRED PLANT CONTAINED WITHIN 3)

Fraction of 3) used in the plant: 75.0 %

Operating temperature of plant: 120.0 deg.C

NUMBER OF SITES (all size bands): 15 NUMBER OF NON-WORKING DAYS PER YEAR: 119

Tobacco

Table 22

Typical Daily Profiles of Heat & Power Demand for the Tobacco Industry

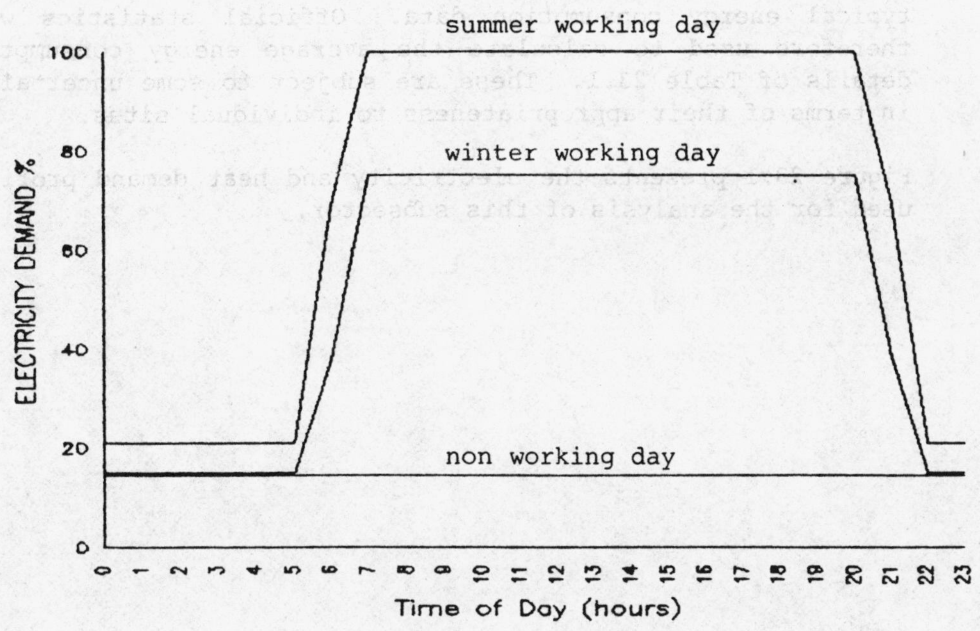

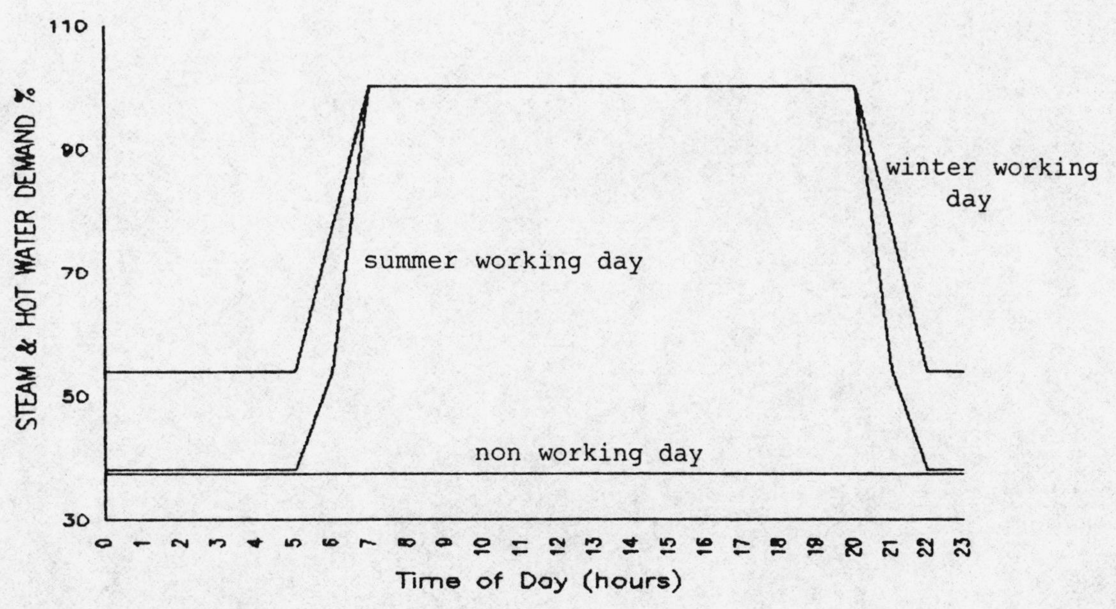

Tobacco

Fig. 22

23. INORGANIC CHEMICALS [SIC 2511]

Profile of the Industry

The industry is largely engaged in the production of intermediate chemicals. The larger sites tend to be integrated and can employ a great variety of different processes, some net energy producers and others net energy consumers. Consequently, it is extremely difficult to derive typical energy consumption data. Official statistics were therefore used to calculate the average energy consumption details of Table 23.1. These are subject to some uncertainty in terms of their appropriateness to individual sites.

Figure 23.1 presents the electricity and heat demand profiles used for the analysis of this subsector.

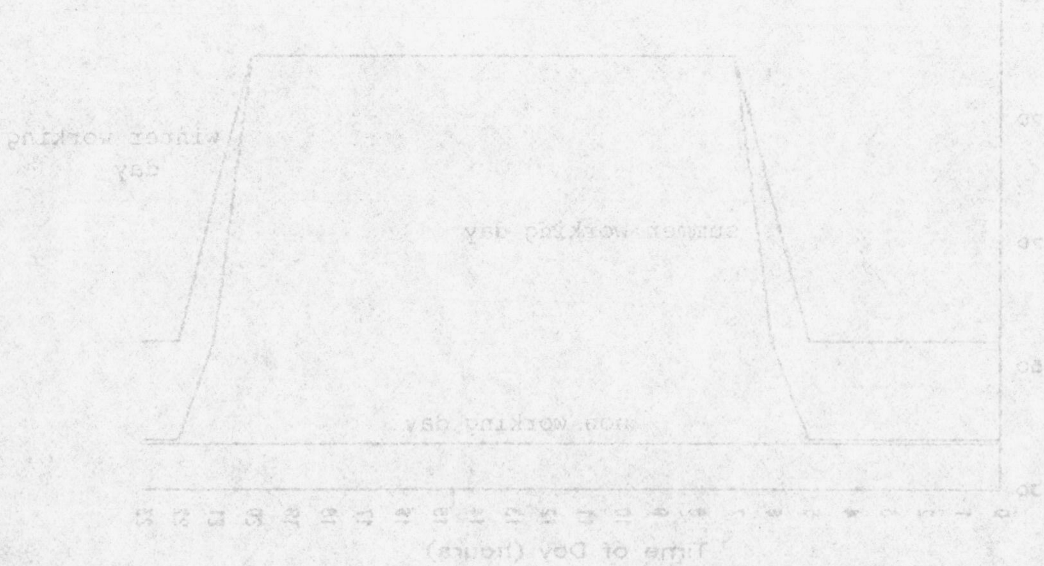

Energy Consumption and Heat Quality Details for the Inorganic Chemicals Industry

AVERAGE ANNUAL ENERGY CONSUMPTION PER SITE	Size Band 1			Size Band 2			Size Band 3		
	kTh	GWh	%	kTh	GWh	%	kTh	GWh	%
1) ELECTRICITY	203.9	6.0	17.47	377.8	11.1	12.59			
2) FUELS for STEAM & HOT WATER GENERATION	913.2	26.8	78.23	2485.6	72.8	82.85			
3) FUELS for OTHER USE	50.2	1.5	4.30	136.7	4.0	4.56			
Totals	1167.2	34.2	100	3000.1	87.9	100			

DETAILS OF STEAM AND HOT WATER GENERATION AND USE	Steam generation details: 250.0 psig saturated Details of use: 96.2 % process heat 3.8 % space heat Fraction of process heat demand which can be met by hot water at < or = 85 deg.C: 0 % Boiler efficiency: 77.5 % Condensate return: 80.0 %
DETAILS OF APPROPRIATE DIRECT FIRED PLANT CONTAINED WITHIN 3)	Fraction of 3) used in the plant: 0 % Operating temperature of plant: 0 deg.C

NUMBER OF SITES (all size bands): 20 NUMBER OF NON-WORKING DAYS PER YEAR: 23

Inorganic Chemicals

Table 23

Typical Daily Profiles of Heat and Power Demand for the Inorganic Chemicals Industry

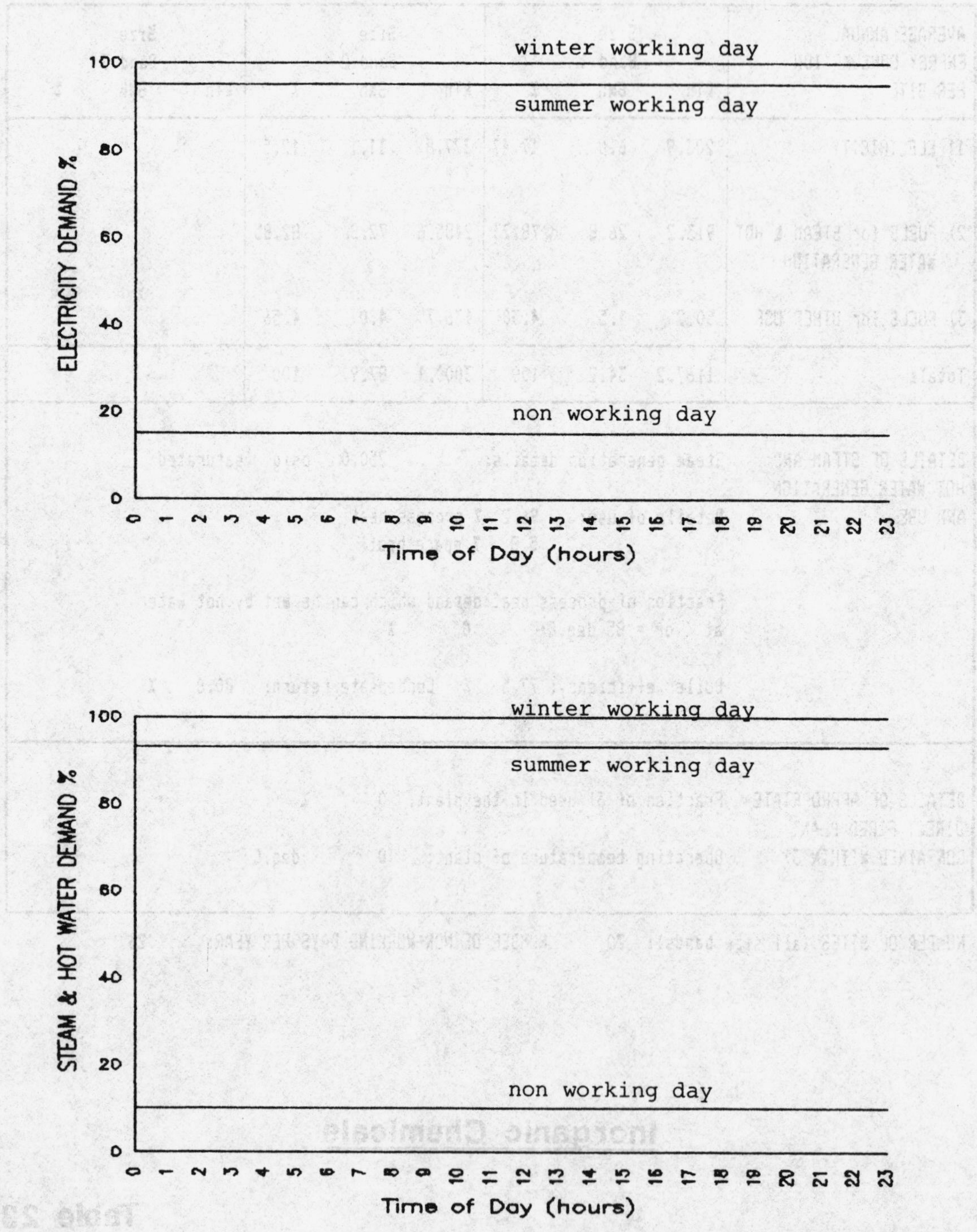

Inorganic Chemicals

Fig. 23

24. ORGANIC CHEMICALS [SIC 2512]

Profile of the Industry

The organics sector of the chemical industry accounts for a large proportion of the conventional fossil fuel consumption of the chemical industry as a whole. It uses petroleum products both as feedstocks and as energy sources. Plants tend to be highly integrated.

Energy consumption and heat quality details for an average non-CHP site are given in Table 24.1. The daily profiles of electricity and heat demand used for the analysis of this subsector of industry are given in Figure 24.1.

Energy Consumption and Heat Quality Details for the Organic Chemicals Industry

AVERAGE ANNUAL ENERGY CONSUMPTION PER SITE	Size Band 1			Size Band 2			Size Band 3		
	kTh	GWh	%	kTh	GWh	%	kTh	GWh	%
1) ELECTRICITY	350.5	10.3	26.93	531.0	15.6	17.70			
2) FUELS for STEAM & HOT WATER GENERATION	950.7	27.9	73.07	2468.9	72.4	82.30			
3) FUELS for OTHER USE	0	0	0	0	0	0			
Totals	1301.1	38.1	100	2999.9	87.9	100			

DETAILS OF STEAM AND HOT WATER GENERATION AND USE	Steam generation details: 250.0 psig saturated Details of use: 94.8 % process heat 5.2 % space heat Fraction of process heat demand which can be met by hot water at < or = 85 deg.C: 0 % Boiler efficiency: 77.5 % Condensate return: 80.0 %
DETAILS OF APPROPRIATE DIRECT FIRED PLANT CONTAINED WITHIN 3)	Fraction of 3) used in the plant: 0 % Operating temperature of plant: 0 deg.C

NUMBER OF SITES (all size bands): 15 NUMBER OF NON-WORKING DAYS PER YEAR: 23

Organic Chemicals

Table 24

Typical Daily Profiles of Heat and Power Demand for the Organic Chemicals Industry

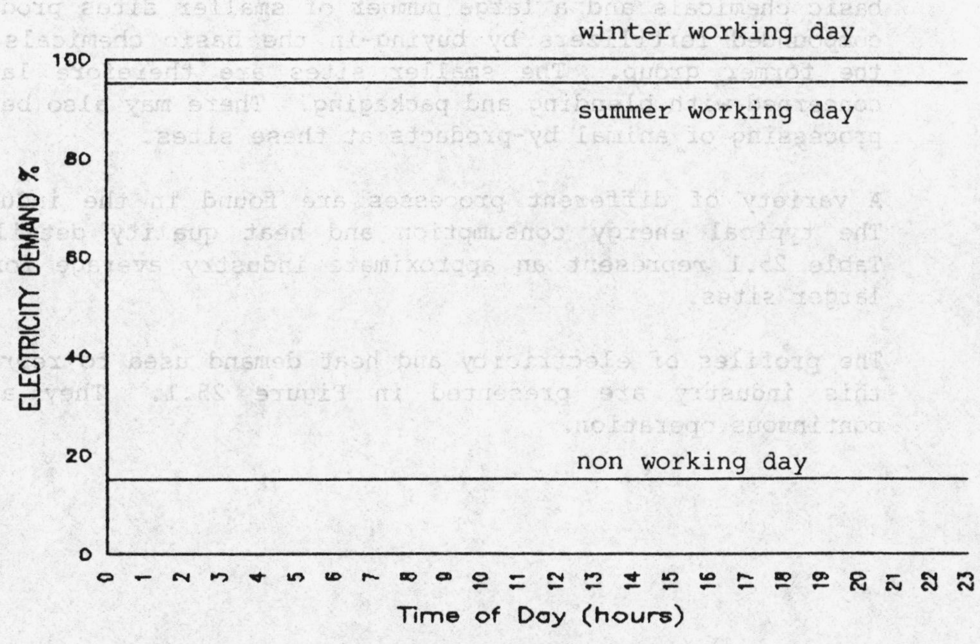

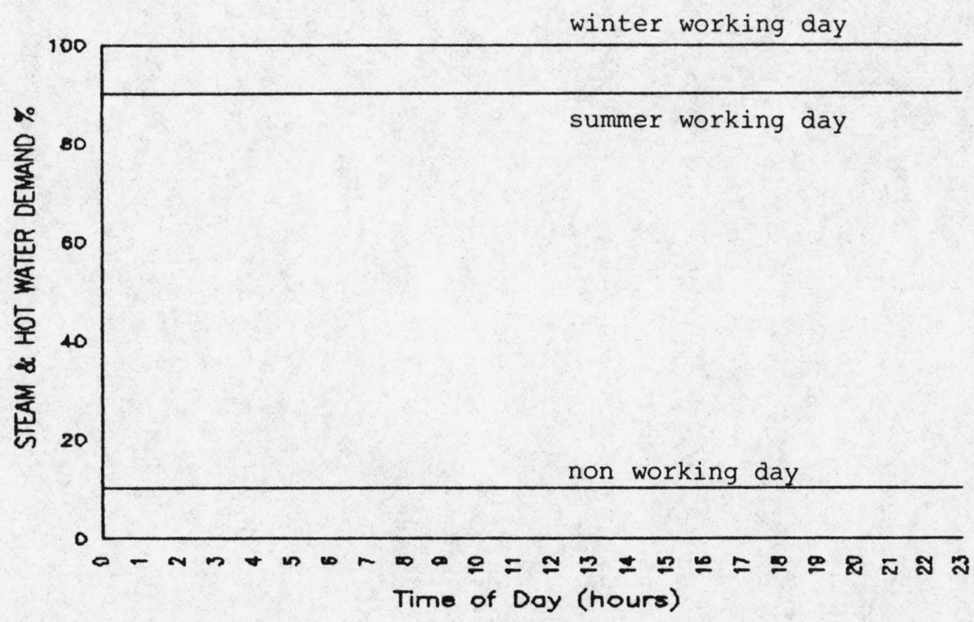

Organic Chemicals

Fig. 24

25 FERTILIZERS [SIC 2513]

Profile of the Industry

The principal chemicals produced are ammonium salts, urea, calcium superphosphate and potassium chloride. The industry comprises a small number of large sites which manufacture the basic chemicals and a large number of smaller sites producing compounded fertilizers by buying-in the basic chemicals from the former group. The smaller sites are therefore largely concerned with blending and packaging. There may also be some processing of animal by-products at these sites.

A variety of different processes are found in the industry. The typical energy consumption and heat quality details of Table 25.1 represent an approximate industry average for the larger sites.

The profiles of electricity and heat demand used to represent this industry are presented in Figure 25.1. They assume continuous operation.

Energy Consumption and Heat Quality Details for the Fertilizers Industry

AVERAGE ANNUAL ENERGY CONSUMPTION PER SITE	Size Band 1 kTh	Size Band 1 GWh	Size Band 1 %	Size Band 2 kTh	Size Band 2 GWh	Size Band 2 %	Size Band 3 kTh	Size Band 3 GWh	Size Band 3 %
1) ELECTRICITY	169.7	5.0	15.81	7233.2	212.0	26.61			
2) FUELS for STEAM & HOT WATER GENERATION	827.4	24.2	77.11	18270.8	535.5	67.22			
3) FUELS for OTHER USE	75.9	2.2	7.07	1675.7	49.1	6.17			
Totals	1072.9	31.4	100	27179.6	796.6	100			

DETAILS OF STEAM AND HOT WATER GENERATION AND USE

Steam generation details: 250.0 psig saturated

Details of use: 85.2 % process heat
14.8 % space heat

Fraction of process heat demand which can be met by hot water at < or = 85 deg.C: 0 %

Boiler efficiency: 77.5 % Condensate return: 50.0 %

DETAILS OF APPROPRIATE DIRECT FIRED PLANT CONTAINED WITHIN 3)

Fraction of 3) used in the plant: 0 %

Operating temperature of plant: 0 deg.C

NUMBER OF SITES (all size bands): 12 NUMBER OF NON-WORKING DAYS PER YEAR: 14

Fertilisers

Table 25

Typical Daily Profiles of Heat and Power Demand for the Fertilizers Industry

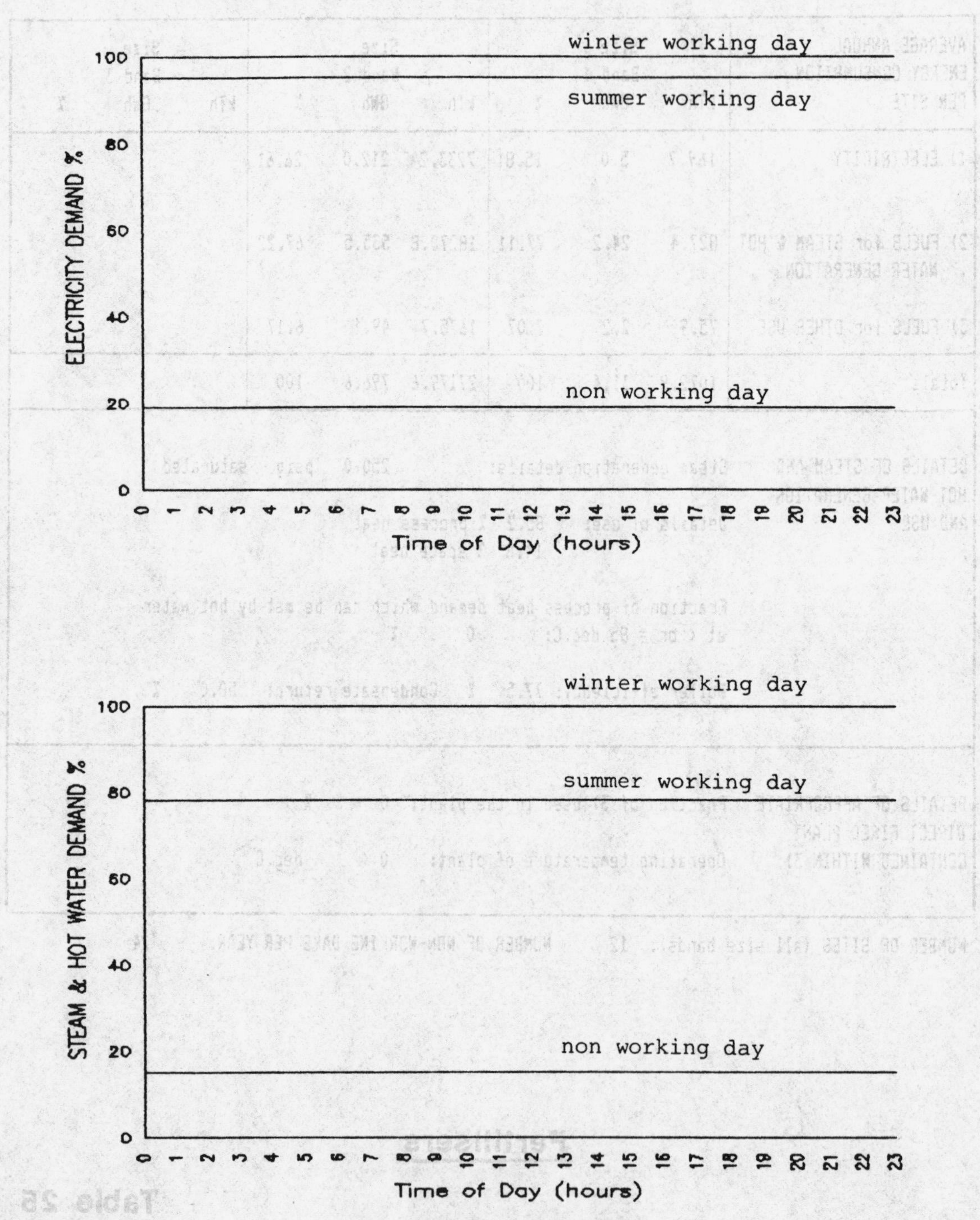

Fertilisers

Fig. 25

26. SYNTHETIC RESINS AND PLASTICS [SIC 2514]

Profile of the Industry

The manufacture of synthetic resins and plastics is achieved by means of polymerisation reactions using a variety of monomers and production methods. Heat is required principally for process heating, evaporation and distillation operations. Electricity is used mostly for motive power.

Average energy consumption and heat quality details for two sizes of site in this subsector of industry are presented in Table 26.1 and representative profiles of electricity and heat demand are given in Figure 26.1.

Energy Consumption and Heat Quality Details for theSynthetic Resins and Plastics Industry

AVERAGE ANNUAL ENERGY CONSUMPTION PER SITE	Size Band 1 kTh	GWh	%	Size Band 2 kTh	GWh	%	Size Band 3 kTh	GWh	%
1) ELECTRICITY	155.5	4.6	20.75	2313.0	67.8	32.83			
2) FUELS for STEAM & HOT WATER GENERATION	559.1	16.4	74.59	4453.3	130.5	63.22			
3) FUELS for OTHER USE	34.9	1.0	4.66	278.3	8.2	3.95			
Totals	749.5	22.0	100	7044.6	206.5	100			

DETAILS OF STEAM AND HOT WATER GENERATION AND USE	Steam generation details: 200.0 psig saturated Details of use: 75.0 % process heat 25.0 % space heat Fraction of process heat demand which can be met by hot water at < or = 85 deg.C: 0 % Boiler efficiency: 77.5 % Condensate return: 60.0 %
DETAILS OF APPROPRIATE DIRECT FIRED PLANT CONTAINED WITHIN 3)	Fraction of 3) used in the plant: 0 % Operating temperature of plant: 0 deg.C

NUMBER OF SITES (all size bands): 30 NUMBER OF NON-WORKING DAYS PER YEAR: 14

Synthetic Resins and Plastics

Table 26

Typical Daily Profiles of Heat and Power Demand for the Synthetic Resins and Plastics Industry

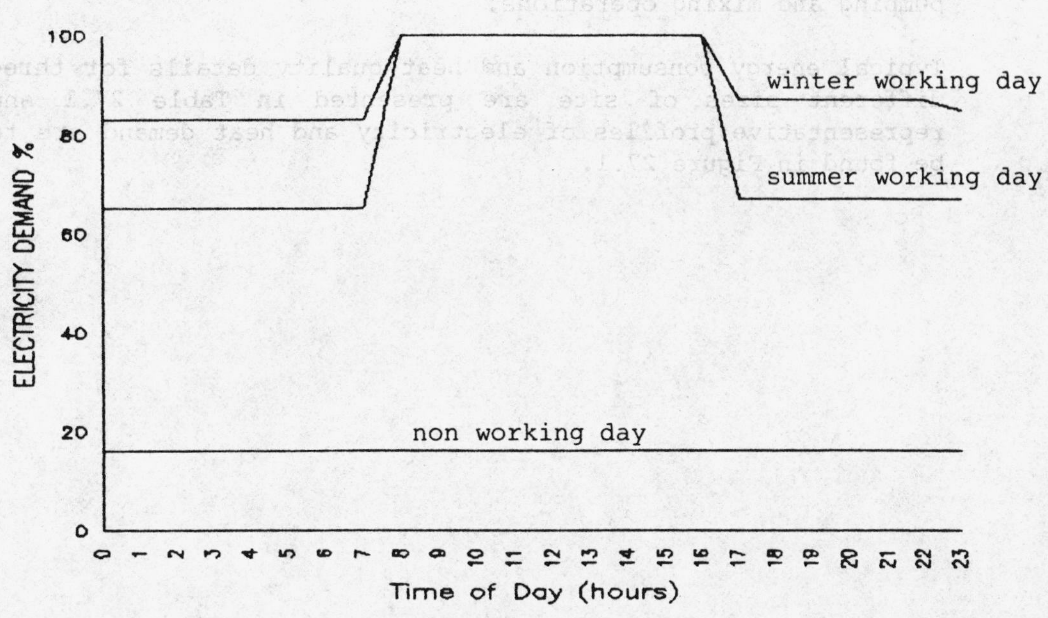

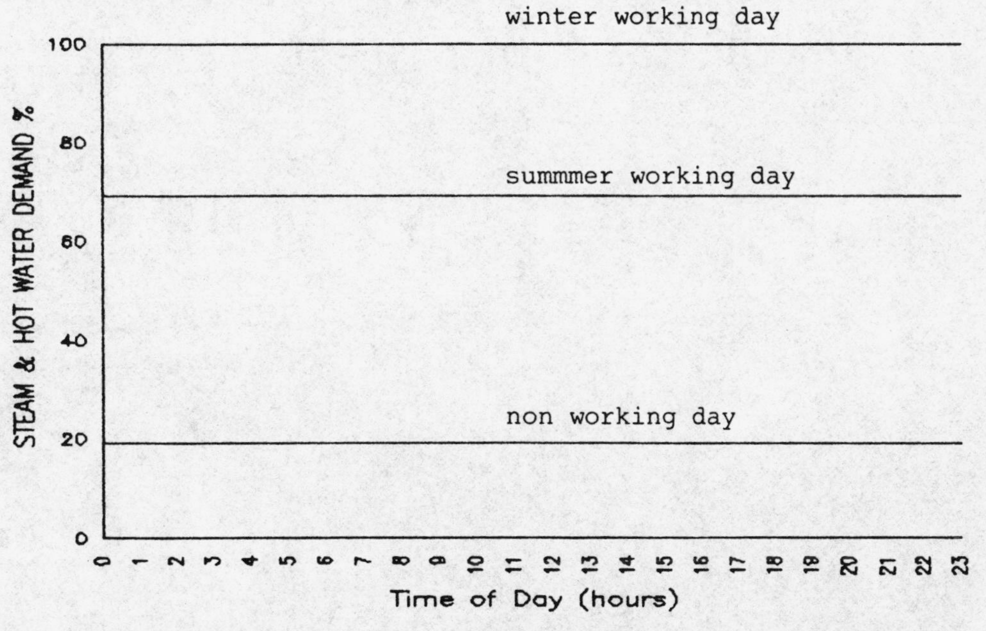

Synthetic Resins and Plastics

Fig. 26

27 SYNTHETIC RUBBER [SIC 2515]

Profile of the Industry

The manufacture of synthetic rubber is achieved by the polymerization of monomomers and heat is required mostly for process heating, evaporation and distillation operations. Electricity is used primarily to provide motive power for pumping and mixing operations.

Typical energy consumption and heat quality details for three different sizes of site are presented in Table 27.1 and representative profiles of electricity and heat demand are to be found in Figure 27.1.

Energy Consumption and Heat Quality Details for the Synthetic Rubber Industry

AVERAGE ANNUAL ENERGY CONSUMPTION PER SITE	Size Band 1 kTh	Size Band 1 GWh	Size Band 1 %	Size Band 2 kTh	Size Band 2 GWh	Size Band 2 %	Size Band 3 kTh	Size Band 3 GWh	Size Band 3 %
1) ELECTRICITY	83.0	2.4	10.23	2010.0	58.9	55.53	1573.3	46.1	10.29
2) FUELS for STEAM & HOT WATER GENERATION	673.3	19.7	83.02	1488.9	43.6	41.13	12689.2	371.9	82.97
3) FUELS for OTHER USE	54.7	1.6	6.75	121.0	3.5	3.34	1031.0	30.2	6.74
Totals	811.1	23.8	100	3619.9	106.1	100	15293.6	448.2	100

DETAILS OF STEAM AND HOT WATER GENERATION AND USE

Steam generation details: 200.0 psig saturated

Details of use: 75.0 % process heat
25.0 % space heat

Fraction of process heat demand which can be met by hot water at < or = 85 deg.C: 0 %

Boiler efficiency: 77.5 % Condensate return: 60.0 %

DETAILS OF APPROPRIATE DIRECT FIRED PLANT CONTAINED WITHIN 3)

Fraction of 3) used in the plant: 0 %

Operating temperature of plant: 0 deg.C

NUMBER OF SITES (all size bands): 7 NUMBER OF NON-WORKING DAYS PER YEAR: 14

Synthetic Rubber

Table 27

Typical Daily Profiles of Heat and Power Demand for the Synthetic Rubber Industry

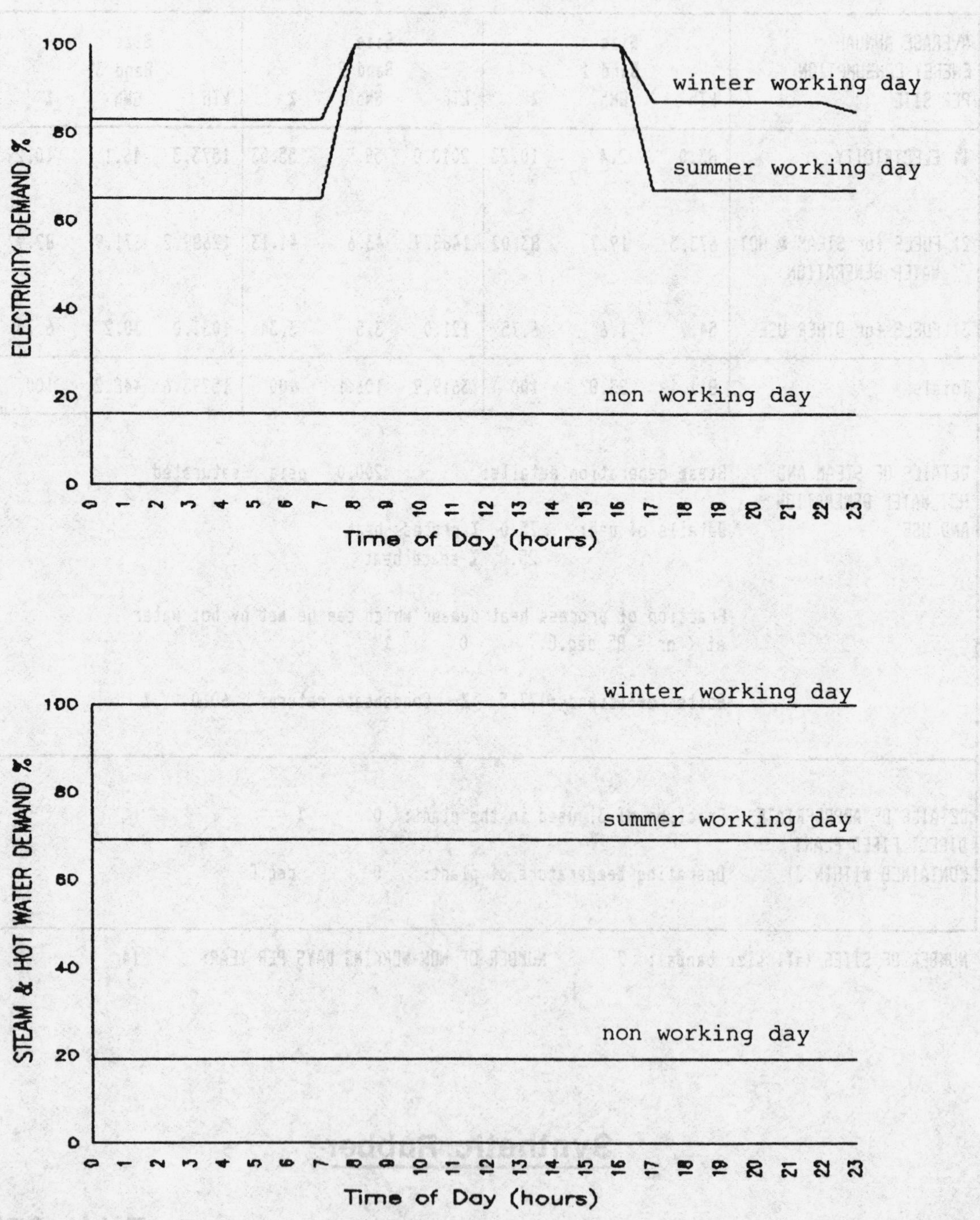

Synthetic Rubber

Fig. 27

28. PAINTS [SIC 2551]

Profile of the Industry

Paint manufacture principally concerns the dispersion of pigment particles in a "vehicle" or "binder". Space heating tends to be the single most important heat demand. Electricity is required to provide motive power for mixing and grinding operations.

Table 28.1 presents average energy consumption and heat quality details for two sizes of paint making site and the profiles of electricity and heat demand used for the analysis of this industry are to be found in Figure 28.1.

Energy Consumption and Heat Quality Details for the Paints Industry

AVERAGE ANNUAL ENERGY CONSUMPTION PER SITE	Size Band 1			Size Band 2			Size Band 3		
	kTh	GWh	%	kTh	GWh	%	kTh	GWh	%
1) ELECTRICITY	61.1	1.8	14.72	341.4	10.0	14.39			
2) FUELS for STEAM & HOT WATER GENERATION	283.1	8.3	68.23	1624.7	47.6	68.49			
3) FUELS for OTHER USE	70.8	2.1	17.06	406.2	11.9	17.12			
Totals	415.0	12.2	100	2372.3	69.5	100			

DETAILS OF STEAM AND HOT WATER GENERATION AND USE

Steam generation details: 140.0 psig saturated

Details of use: 33.4 % process heat
66.6 % space heat

Fraction of process heat demand which can be met by hot water at < or = 85 deg.C: 0 %

Boiler efficiency: 77.5 % Condensate return: 80.0 %

DETAILS OF APPROPRIATE DIRECT FIRED PLANT CONTAINED WITHIN 3)

Fraction of 3) used in the plant: 0 %

Operating temperature of plant: 0 deg.C

NUMBER OF SITES (all size bands): 24 NUMBER OF NON-WORKING DAYS PER YEAR: 119

Paints

Table 28

Typical Daily Profiles of Heat and Power Demand for the Paints Industry

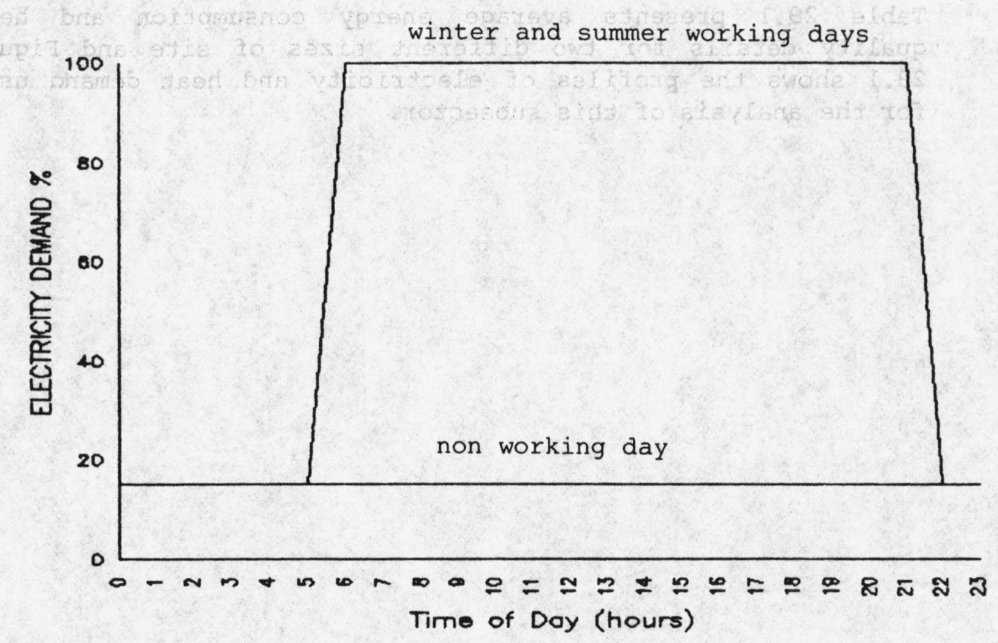

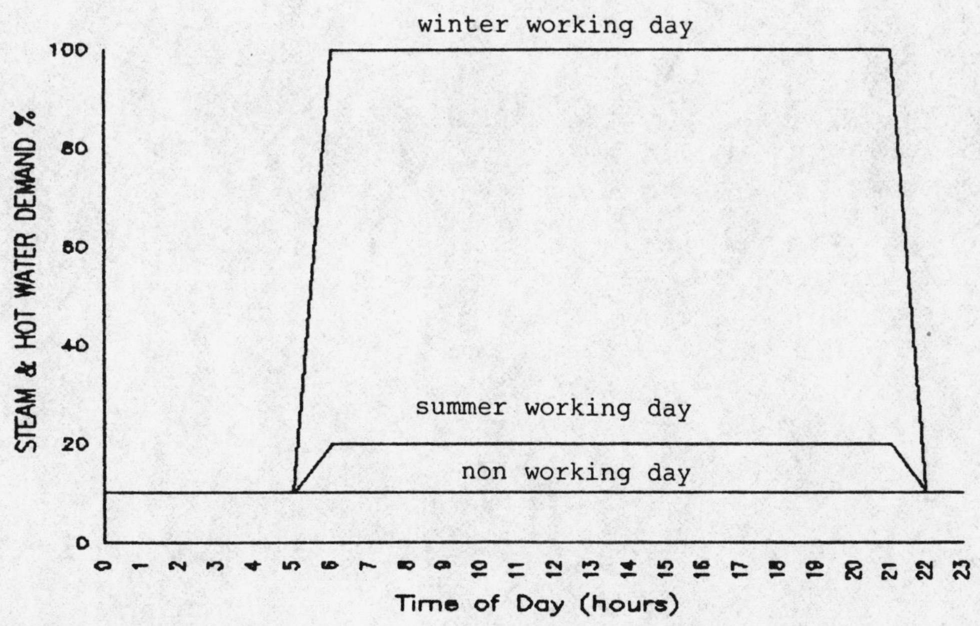

Paints

Fig. 28

29 FORMULATED ADHESIVES AND SEALANTS [SIC 2562]

Profile of the Industry

This subsector encompasses the manufacture of a wide variety of adhesives and sealants, and it is difficult to generalize on the patterns of energy use at the sites in this industry.

Table 29.1 presents average energy consumption and heat quality details for two different sizes of site and Figure 29.1 shows the profiles of electricity and heat demand used for the analysis of this subsector.

Energy Consumption and Heat Quality Details for the Formulated Adhesives and Sealants Industry

AVERAGE ANNUAL ENERGY CONSUMPTION PER SITE	Size Band 1 kTh	Size Band 1 GWh	Size Band 1 %	Size Band 2 kTh	Size Band 2 GWh	Size Band 2 %	Size Band 3 kTh	Size Band 3 GWh	Size Band 3 %
1) ELECTRICITY	109.7	3.2	17.07	239.6	7.0	10.85			
2) FUELS for STEAM & HOT WATER GENERATION	486.0	14.2	75.65	1794.7	52.6	81.32			
3) FUELS for OTHER USE	46.8	1.4	7.28	172.7	5.1	7.83			
Totals	642.5	18.8	100	2207.0	64.7	100			

DETAILS OF STEAM AND HOT WATER GENERATION AND USE

Steam generation details: 150.0 psig saturated

Details of use: 79.0 % process heat
21.0 % space heat

Fraction of process heat demand which can be met by hot water at < or = 85 deg.C: 0 %

Boiler efficiency: 77.5 % Condensate return: 80.0 %

DETAILS OF APPROPRIATE DIRECT FIRED PLANT CONTAINED WITHIN 3)

Fraction of 3) used in the plant: 0 %

Operating temperature of plant: 0 deg.C

NUMBER OF SITES (all size bands): 13 NUMBER OF NON-WORKING DAYS PER YEAR: 28

Formulated Adhesives and Sealants

Table 29

Typical Daily Profiles of Heat and Power Demand for the Formulated Adhesives and Sealants Industry

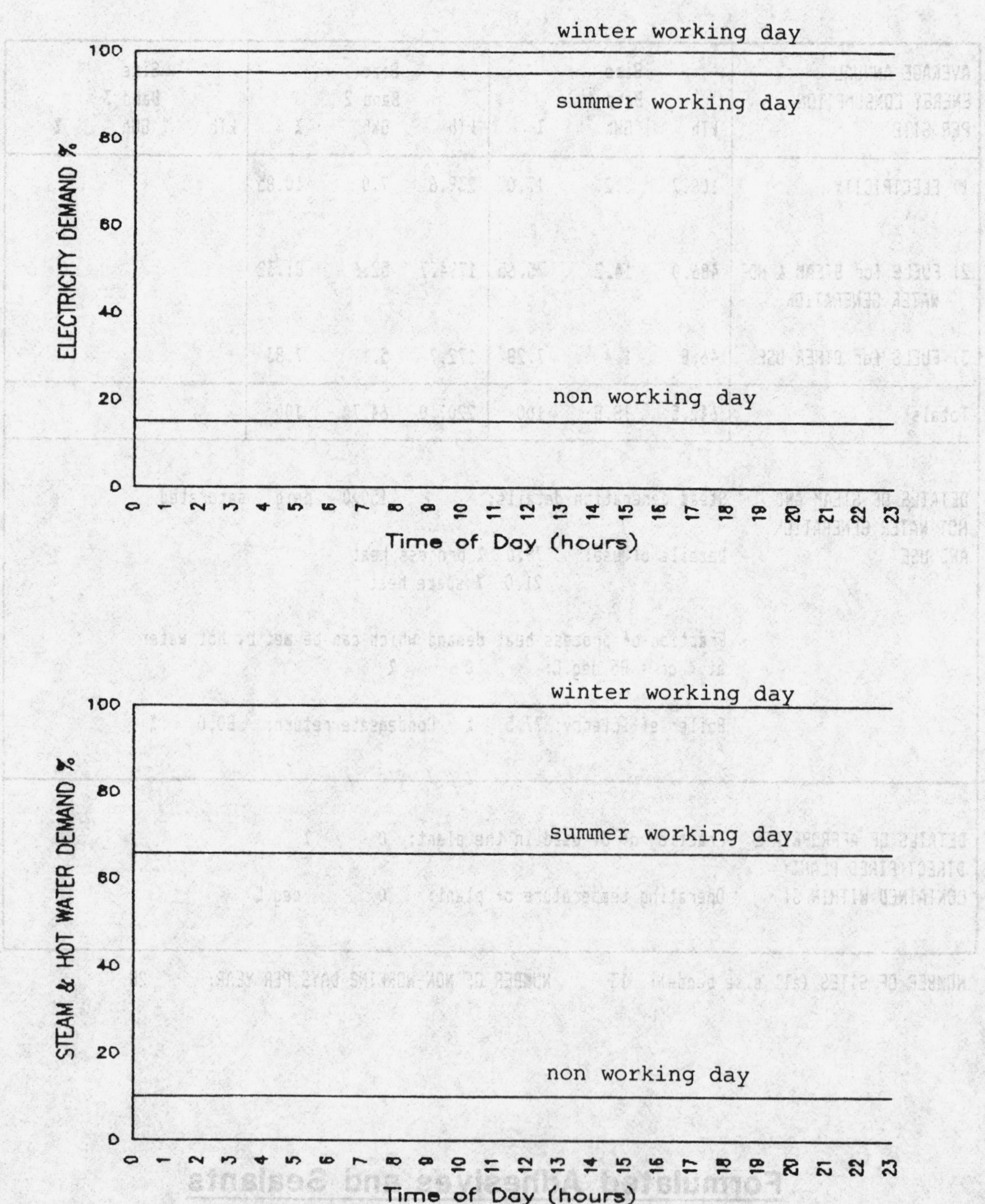

Formulated Adhesives and Sealants

Fig. 29

30. CHEMICAL TREATMENT OF OILS AND FATS [SIC 2563]

Profile of the Industry

Average energy consumption and heat quality details are given in Table 30.1 and the profiles of electricity and heat demand used for the analysis of this subsector are presented in Figure 30.1.

Energy Consumption and Heat Quality Details for the Chemical Treatment of Oils and Fats Industry

AVERAGE ANNUAL ENERGY CONSUMPTION PER SITE	Size Band 1 kTh	GWh	%	Size Band 2 kTh	GWh	%	Size Band 3 kTh	GWh	%
1) ELECTRICITY	376.0	11.0	10.30						
2) FUELS for STEAM & HOT WATER GENERATION	3029.9	88.8	83.00						
3) FUELS for OTHER USE	244.6	7.2	6.70						
Totals	3650.5	107.0	100						

DETAILS OF STEAM AND HOT WATER GENERATION AND USE	Steam generation details: 250.0 psig 290.0 deg.C Details of use: 88.0 % process heat 12.0 % space heat Fraction of process heat demand which can be met by hot water at < or = 85 deg.C: 0 % Boiler efficiency: 77.0 % Condensate return: 53.0 %
DETAILS OF APPROPRIATE DIRECT FIRED PLANT CONTAINED WITHIN 3)	Fraction of 3) used in the plant: 0 % Operating temperature of plant: 0 deg.C

NUMBER OF SITES (all size bands): 3 NUMBER OF NON-WORKING DAYS PER YEAR: 28

Chemical Treatment of Oils and Fats

Table 30

Typical Daily Profiles of Heat and Power Demand for the Chemical Treatment of Oils and Fats Industry

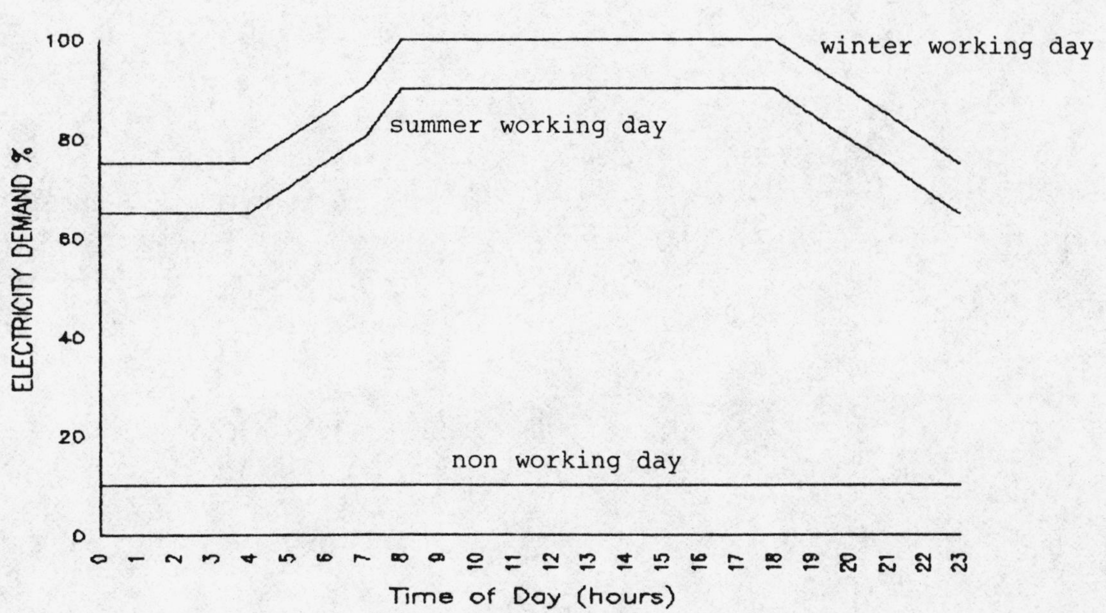

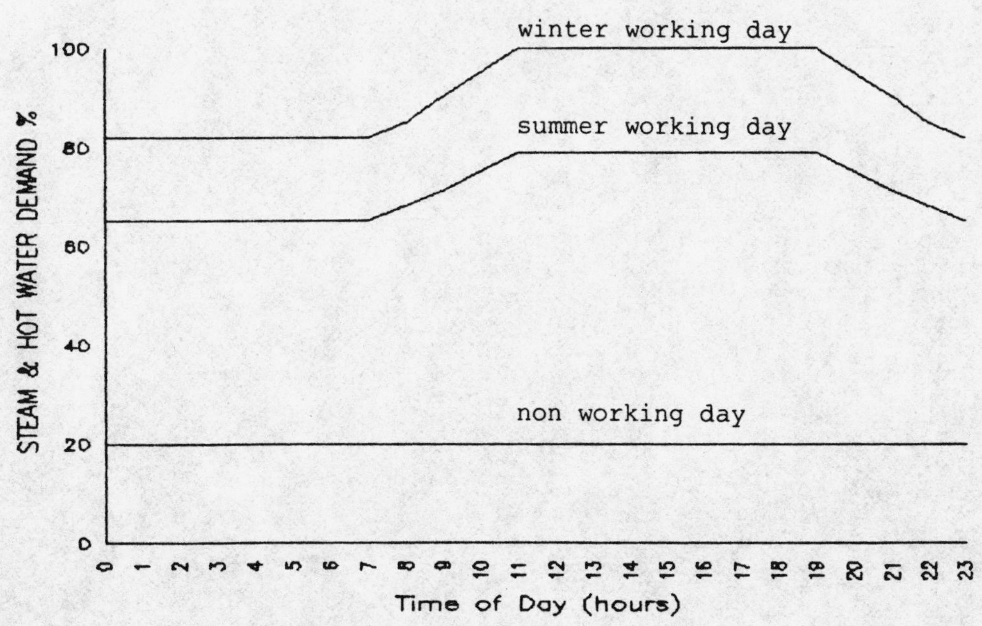

Chemical Treatment of Oils and Fats

Fig. 30

31. ESSENTIAL OILS AND FLAVOURINGS [SIC 2564]

Profile of the Industry

Essential oils and flavourings are produced by complex solvent extraction, distillation and chemical transformation steps.

Average energy consumption and heat quality details are given in Table 31.1 and the profiles of the electricity and heat demand used for the analysis of this subsector are presented in Figure 31.1.

Energy Consumption and Heat Quality Details for the Essential Oils and Flavourings Industry

AVERAGE ANNUAL ENERGY CONSUMPTION PER SITE	Size Band 1 kTh	Size Band 1 GWh	Size Band 1 %	Size Band 2 kTh	Size Band 2 GWh	Size Band 2 %	Size Band 3 kTh	Size Band 3 GWh	Size Band 3 %
1) ELECTRICITY	36.5	1.1	6.80	268.7	7.9	9.36			
2) FUELS for STEAM & HOT WATER GENERATION	440.5	12.9	82.03	2290.1	67.1	79.77			
3) FUELS for OTHER USE	60.0	1.8	11.18	312.0	9.1	10.87			
Totals	537.0	15.7	100	2870.8	84.1	100			

DETAILS OF STEAM AND HOT WATER GENERATION AND USE

Steam generation details: 150.0 psig saturated

Details of use: 50.0 % process heat
50.0 % space heat

Fraction of process heat demand which can be met by hot water at < or = 85 deg.C: 0 %

Boiler efficiency: 77.5 % Condensate return: 80.0 %

DETAILS OF APPROPRIATE DIRECT FIRED PLANT CONTAINED WITHIN 3)

Fraction of 3) used in the plant: 0 %

Operating temperature of plant: 0 deg.C

NUMBER OF SITES (all size bands): 5 NUMBER OF NON-WORKING DAYS PER YEAR: 119

Essential Oils and Flavourings

Table 31

Typical Daily Profiles of Heat and Power Demand for the Essential Oils and Flavourings Industry

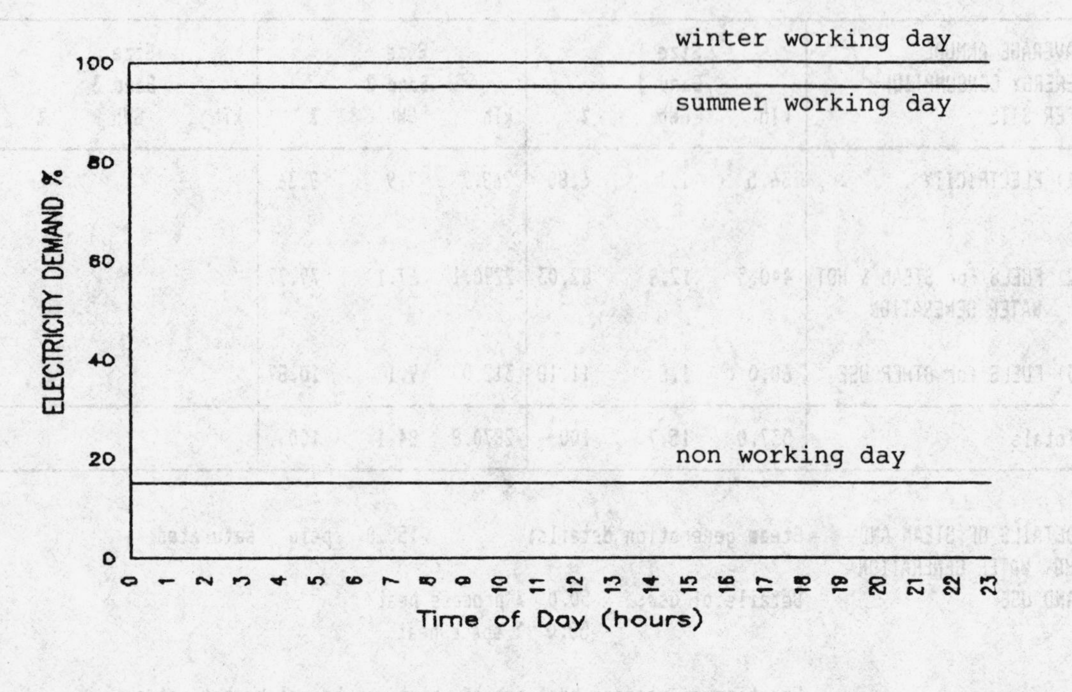

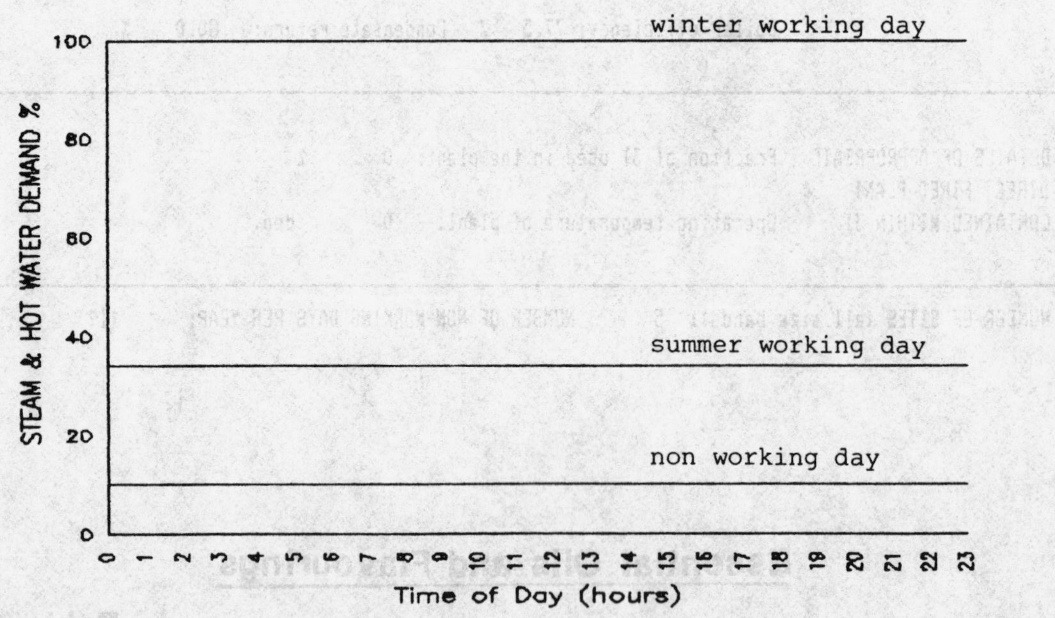

Essential Oils and Flavourings

Fig. 31

32. MISCELLANEOUS CHEMICALS [SIC 2567]

Profile of the Industry

The subsector encompasses the manufacture of a wide variety of industrial chemicals. Department of Energy statistics have been used to derive the average energy consumption details which are presented in Table 32.1. The profiles of electricity and heat demand which were used for the analysis of this subsector are to be found in Figure 32.1.

Energy Consumption and Heat Quality Details for the Miscellaneous Chemicals Industry

AVERAGE ANNUAL ENERGY CONSUMPTION PER SITE	Size Band 1			Size Band 2			Size Band 3		
	kTh	GWh	%	kTh	GWh	%	kTh	GWh	%
1) ELECTRICITY	134.0	3.9	17.30						
2) FUELS for STEAM & HOT WATER GENERATION	512.3	15.0	66.12						
3) FUELS for OTHER USE	128.5	3.8	16.58						
Totals	774.8	22.7	100						

DETAILS OF STEAM AND HOT WATER GENERATION AND USE	Steam generation details: 200.0 psig saturated Details of use: 78.0 % process heat 22.0 % space heat Fraction of process heat demand which can be met by hot water at < or = 85 deg.C: 0 % Boiler efficiency: 77.5 % Condensate return: 80.0 %
DETAILS OF APPROPRIATE DIRECT FIRED PLANT CONTAINED WITHIN 3)	Fraction of 3) used in the plant: 0 % Operating temperature of plant: 0 deg.C

NUMBER OF SITES (all size bands): 21 NUMBER OF NON-WORKING DAYS PER YEAR: 21

Miscellaneous Chemicals

Table 32

Typical Daily Profiles of Heat and Power Demand for the Miscellaneous Chemicals Industry

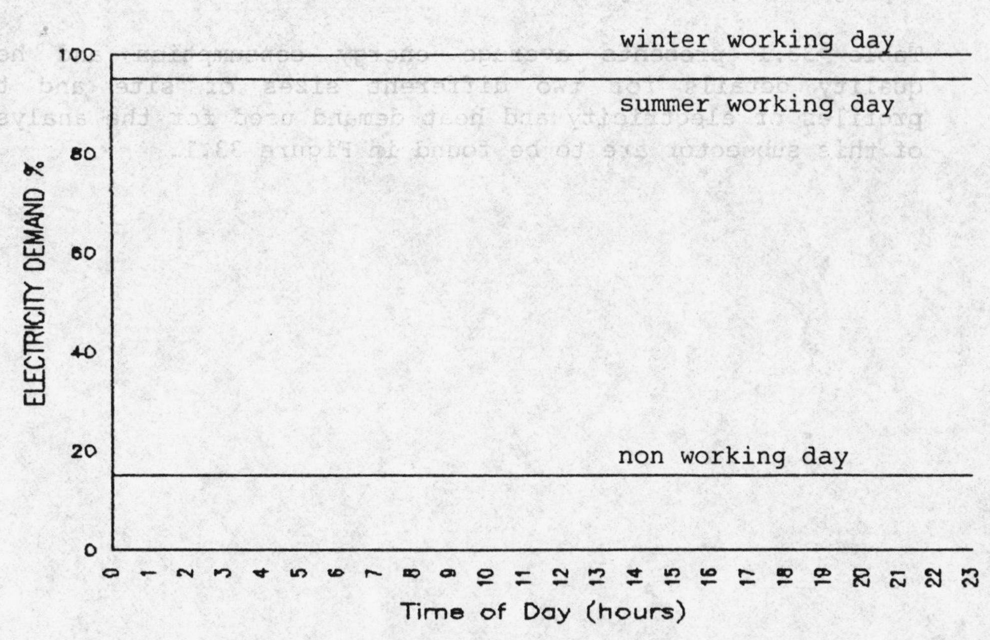

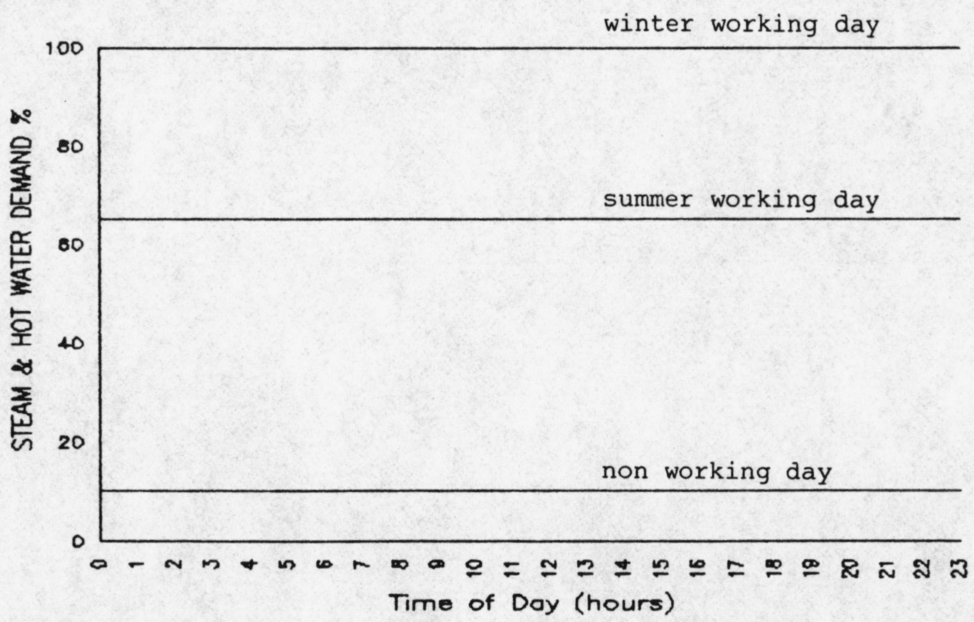

Miscellaneous Chemicals

Fig. 32

33. FORMULATED PESTICIDES [SIC 2568]

Profile of the Industry

The preparation of formulated pesticides involves the use of agents such as dust carriers, solvents, emulsifiers, wetting and dispersing agents, sticking agents, and deodorants. The toxicants, diluants and agents are generally combined by a variety of operations, including: milling, solvent impregnation, fusing etc.

Table 33.1 presents average energy consumption and heat quality details for two different sizes of site and the profiles of electricity and heat demand used for the analysis of this subsector are to be found in Figure 33.1.

Energy Consumption and Heat Quality Details for the Formulated Pesticide Industry

AVERAGE ANNUAL ENERGY CONSUMPTION PER SITE	Size Band 1			Size Band 2			Size Band 3		
	kTh	GWh	%	kTh	GWh	%	kTh	GWh	%
1) ELECTRICITY	119.7	3.5	15.36	254.0	7.4	9.26			
2) FUELS for STEAM & HOT WATER GENERATION	601.6	17.6	77.21	2269.6	66.5	82.77			
3) FUELS for OTHER USE	57.9	1.7	7.43	218.4	6.4	7.97			
Totals	779.3	22.8	100	2742.0	80.4	100			

DETAILS OF STEAM AND HOT WATER GENERATION AND USE	Steam generation details: 150.0 psig saturated Details of use: 50.0 % process heat 50.0 % space heat Fraction of process heat demand which can be met by hot water at < or = 85 deg.C: 0 % Boiler efficiency: 77.5 % Condensate return: 80.0 %
DETAILS OF APPROPRIATE DIRECT FIRED PLANT CONTAINED WITHIN 3)	Fraction of 3) used in the plant: 0 % Operating temperature of plant: 0 deg.C

NUMBER OF SITES (all size bands): 9 NUMBER OF NON-WORKING DAYS PER YEAR: 119

Formulated Pesticides

Table 33

Typical Daily Profiles of Heat and Power Demand for the Formulated Pesticide Industry

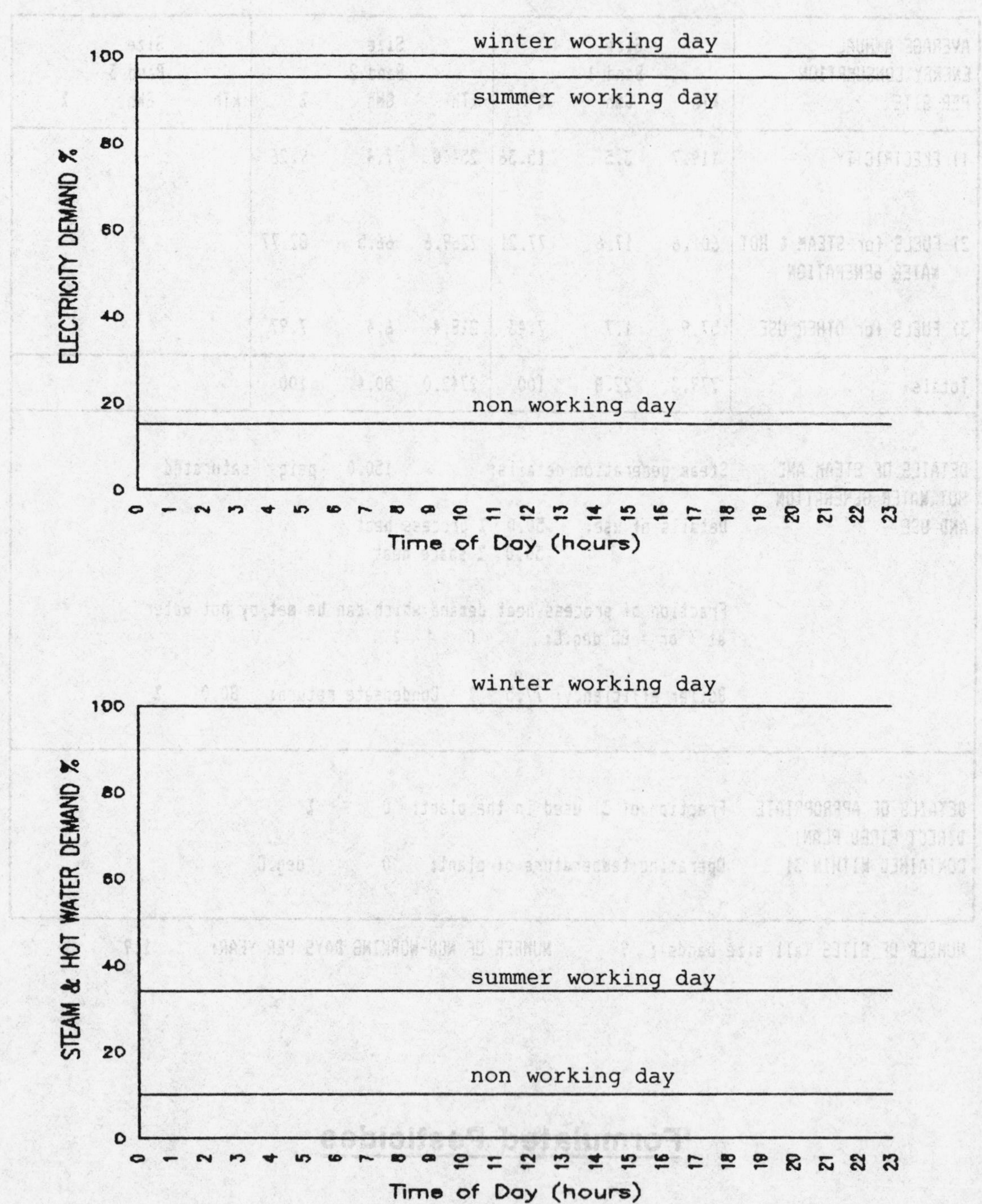

Formulated Pesticides

Fig. 33

34 ANTIBIOTICS [Pharmaceutical Industry SIC 2570]

Profile of the Industry

Antibiotics production begins with the production of an active culture which is then 'grown' and made up to around 10,000 litres of inoculum. This is added to large fermentation vessels whose temperatures are accurately controlled. Typically, continuous air agitation is employed and fermentation lasts for 100 to 200 hours. Following filtration the broth is cooled and the product is separated using specialised extraction processes, washed and dried. Additional energy-consuming operations include steam sterilisation of the fermentation plant and refrigeration to cool the broth.

Typical energy consumption and heat quality details for three sizes of non-CHP site are given in Table 34.1 and typical profiles of electricity and heat demand are presented in Figure 34.1

Energy Consumption and Heat Quality Details for the Antibiotics Industry

AVERAGE ANNUAL ENERGY CONSUMPTION PER SITE	Size Band 1 kTh	GWh	%	Size Band 2 kTh	GWh	%	Size Band 3 kTh	GWh	%
1) ELECTRICITY	209.7	6.1	38.91	1063.5	31.2	24.42	4943.7	144.9	22.47
2) FUELS for STEAM & HOT WATER GENERATION	329.2	9.6	61.09	3291.9	96.5	75.58	17057.3	499.9	77.53
3) FUELS for OTHER USE	0	0	0	0	0	0	0	0	0
Totals	538.8	15.8	100	4355.4	127.6	100	22001.0	644.8	100

DETAILS OF STEAM AND HOT WATER GENERATION AND USE

Steam generation details: 150.0 psig saturated

Details of use: 89.1 % process heat
10.9 % space heat

Fraction of process heat demand which can be met by hot water at < or = 85 deg.C: 0 %

Boiler efficiency: 77.5 % Condensate return: 80.0 %

DETAILS OF APPROPRIATE DIRECT FIRED PLANT CONTAINED WITHIN 3)

Fraction of 3) used in the plant: 0 %

Operating temperature of plant: 0 deg.C

NUMBER OF SITES (all size bands): 8 NUMBER OF NON-WORKING DAYS PER YEAR: 14

Antibiotics

Table 34

Typical Daily Profiles of Heat and Power Demand for the Antibiotics Industry

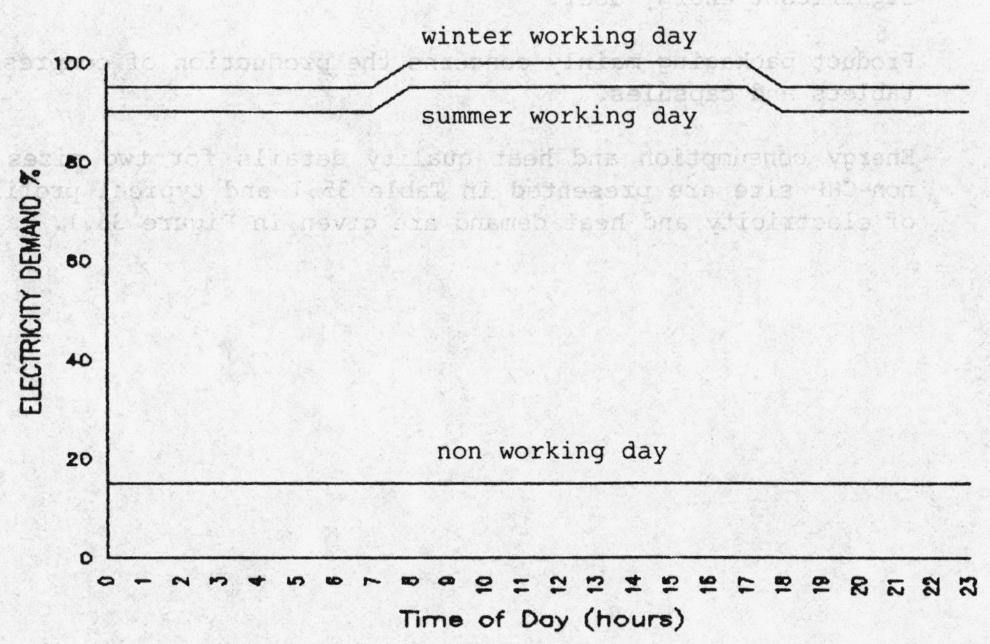

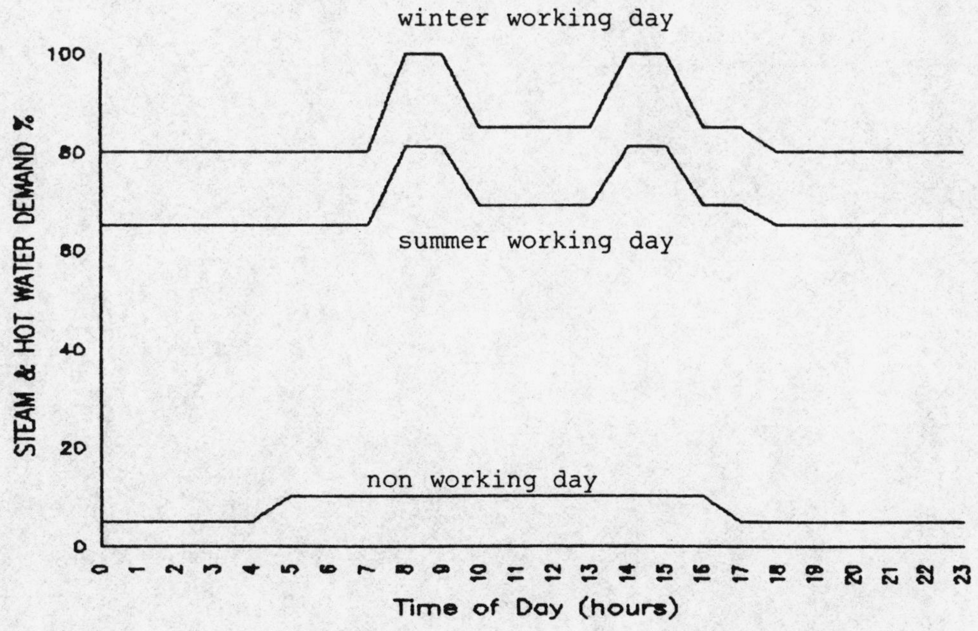

Antibiotics

Fig. 34

35 PHARMACEUTICAL INDUSTRY (excluding antibiotics) [SIC 2570]

Profile of the Industry

The range of fine chemicals produced is very large. However, there are many similar component operations involved. Typically, steam-heated reaction vessels are employed and product separation is usually by filtration, solvent extraction or centrifugation. Product drying is also a significant energy user.

Product packaging mainly concerns the production of compressed tablets and capsules.

Energy consumption and heat quality details for two sizes of non-CHP site are presented in Table 35.1 and typical profiles of electricity and heat demand are given in Figure 35.1.

Energy Consumption and Heat Quality Details for the Pharmaceutical Industry (excluding antibiotics)

AVERAGE ANNUAL ENERGY CONSUMPTION PER SITE	Size Band 1 kTh	GWh	%	Size Band 2 kTh	GWh	%	Size Band 3 kTh	GWh	%
1) ELECTRICITY	209.7	6.1	22.34	1063.5	31.2	12.73			
2) FUELS for STEAM & HOT WATER GENERATION	712.1	20.9	75.86	7121.2	208.7	85.24			
3) FUELS for OTHER USE	17.0	.5	1.81	169.6	5.0	2.03			
Totals	938.7	27.5	100	8354.2	244.8	100			

DETAILS OF STEAM AND HOT WATER GENERATION AND USE

Steam generation details: 150.0 psig saturated

Details of use: 56.0 % process heat
44.0 % space heat

Fraction of process heat demand which can be met by hot water at < or = 85 deg.C: 0 %

Boiler efficiency: 80.0 % Condensate return: 70.0 %

DETAILS OF APPROPRIATE DIRECT FIRED PLANT CONTAINED WITHIN 3)

Fraction of 3) used in the plant: 0 %

Operating temperature of plant: 0 deg.C

NUMBER OF SITES (all size bands): 29 NUMBER OF NON-WORKING DAYS PER YEAR: 14

Other Pharmaceuticals

Table 35

Typical Daily Profiles of Heat and Power Demand for the Pharmaceutical Industry (excluding antibiotics)

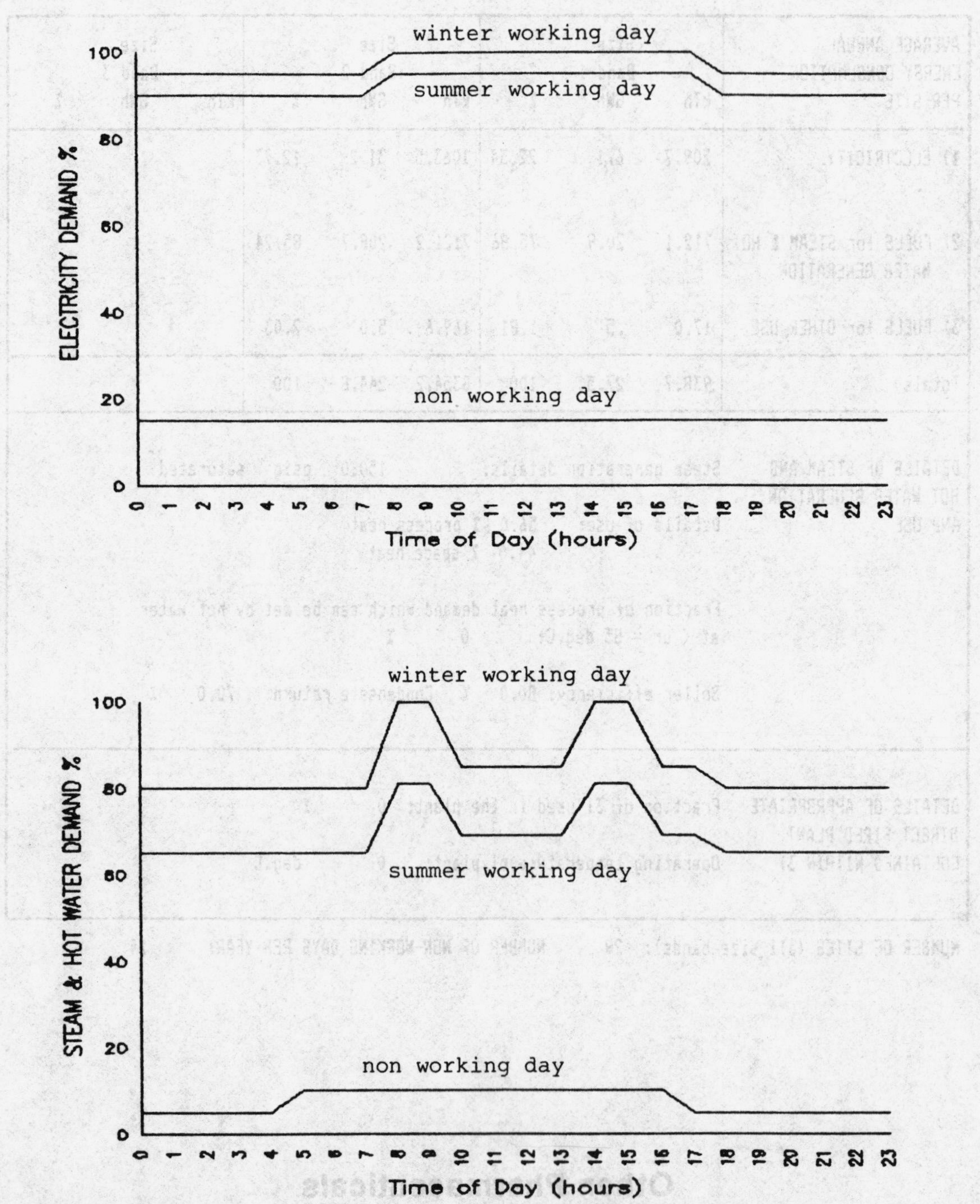

Other Pharmaceuticals

Fig. 35

SOAP AND SYNTHETIC DETERGENTS [SIC 2581]

Profile of the Industry

Soap manufacture involves two main chemical stages. The first stage consists of fat splitting which is the separation of fatty acids from the glycerol with which they are chemically combined. The second stage is saponification in which the fatty acids are reacted with alkali (usually caustic soda) to produce fatty acid salts which we recognise as soap. Various finishing operations may be employed. The smaller sites in this section of industry are often only involved in soap finishing and packaging.

A small number of companies make up the largest part of the industry.

Typical energy consumption and heat quality details for two different sizes of non-CHP site are presented in Table 36.1 and profiles of electricity and heat demand are given in Figure 36.1.

Energy Consumption and Heat Quality Details for the Soap and Detergents Industry

AVERAGE ANNUAL ENERGY CONSUMPTION PER SITE	Size Band 1 kTh	GWh	%	Size Band 2 kTh	GWh	%	Size Band 3 kTh	GWh	%
1) ELECTRICITY	212.0	6.2	25.21	1088.5	31.9	13.48			
2) FUELS for STEAM & HOT WATER GENERATION	480.9	14.1	57.19	5344.2	156.6	66.17			
3) FUELS for OTHER USE	148.0	4.3	17.60	1644.4	48.2	20.36			
Totals	840.9	24.6	100	8077.0	236.7	100			

DETAILS OF STEAM AND HOT WATER GENERATION AND USE	Steam generation details: 400.0 psig Details of use: 81.5 % process heat 18.5 % space heat Fraction of process heat demand which can be met by hot water at < or = 85 deg.C: 0 % Boiler efficiency: 77.5 % Condensate return: 80.0 %
DETAILS OF APPROPRIATE DIRECT FIRED PLANT CONTAINED WITHIN 3)	Fraction of 3) used in the plant: 100.0 % Operating temperature of plant: 650.0 deg.C

NUMBER OF SITES (all size bands): 5 NUMBER OF NON-WORKING DAYS PER YEAR: 14

Soaps and Detergents

Table 36

Typical Daily Profiles of Heat and Power Demand for the Soap and Detergents Industry

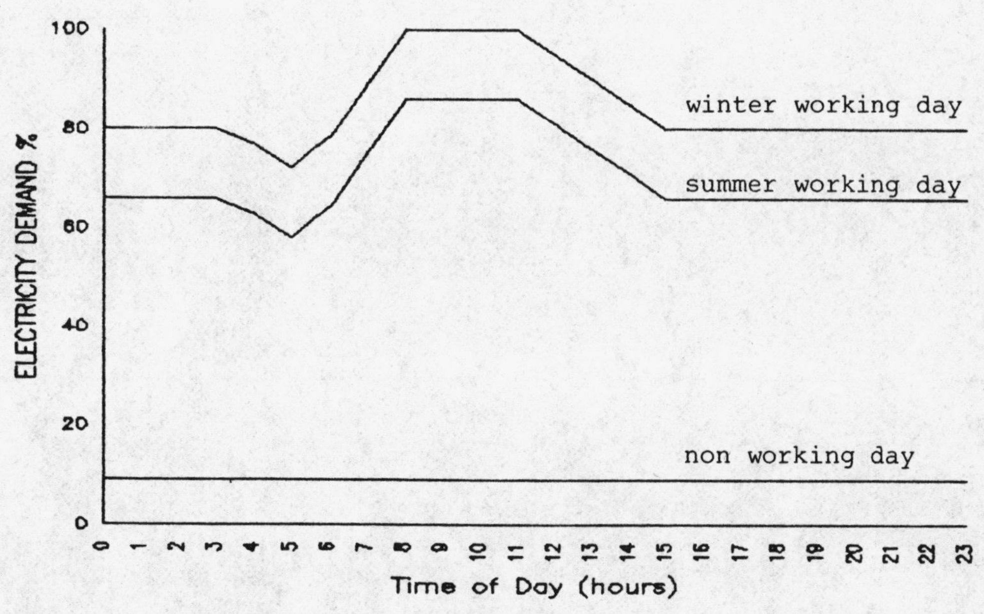

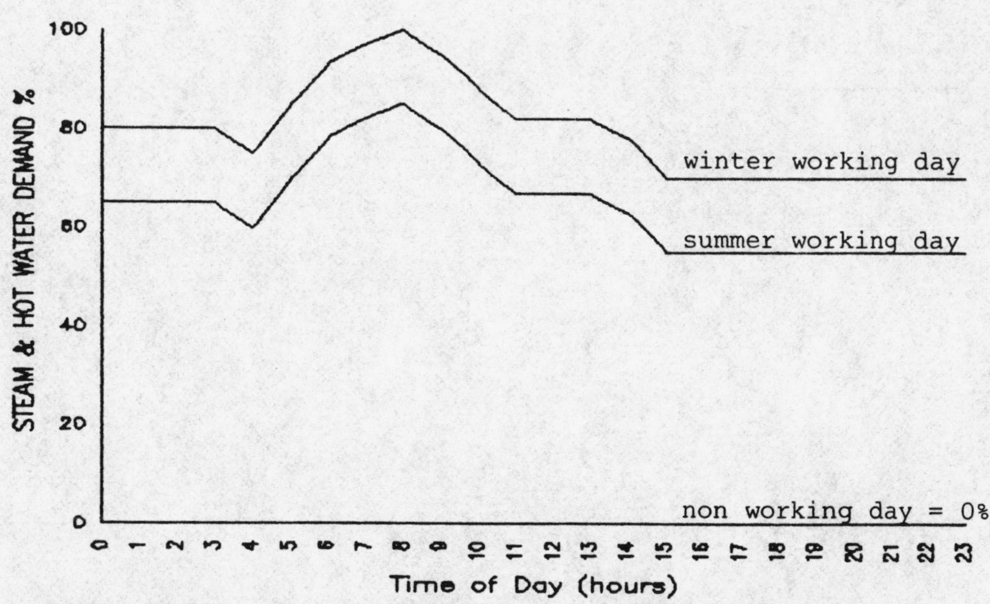

Soaps and Detergents

Fig. 36

37. PERFUMES AND COSMETICS [SIC 2582]

Profile of the Industry

Energy consumption and heat quality details for two sizes of non-CHP site are presented in Table 37.1 and the daily profiles of electricity and heat demand used for the analysis of this subsector are given in Figure 37.1.

Energy Consumption and Heat Quality Details for the Perfumes and Cosmetics Industry

AVERAGE ANNUAL ENERGY CONSUMPTION PER SITE	Size Band 1			Size Band 2			Size Band 3		
	kTh	GWh	%	kTh	GWh	%	kTh	GWh	%
1) ELECTRICITY	130.1	3.8	25.38	255.7	7.5	18.32			
2) FUELS for STEAM & HOT WATER GENERATION	382.5	11.2	74.62	1140.0	33.4	81.68			
3) FUELS for OTHER USE	0	0	0	0	0	0			
Totals	512.6	15.0	100	1395.7	40.9	100			

DETAILS OF STEAM AND HOT WATER GENERATION AND USE

Steam generation details: 150.0 psig saturated

Details of use: 50.0 % process heat
50.0 % space heat

Fraction of process heat demand which can be met by hot water at < or = 85 deg.C: 0 %

Boiler efficiency: 77.5 % Condensate return: 80.0 %

DETAILS OF APPROPRIATE DIRECT FIRED PLANT CONTAINED WITHIN 3)

Fraction of 3) used in the plant: 0 %

Operating temperature of plant: 0 deg.C

NUMBER OF SITES (all size bands): 14 NUMBER OF NON-WORKING DAYS PER YEAR: 119

Perfumes and Cosmetics

Table 37

Typical Daily Profiles of Heat and Power Demand for the Perfumes and Cosmetics Industry

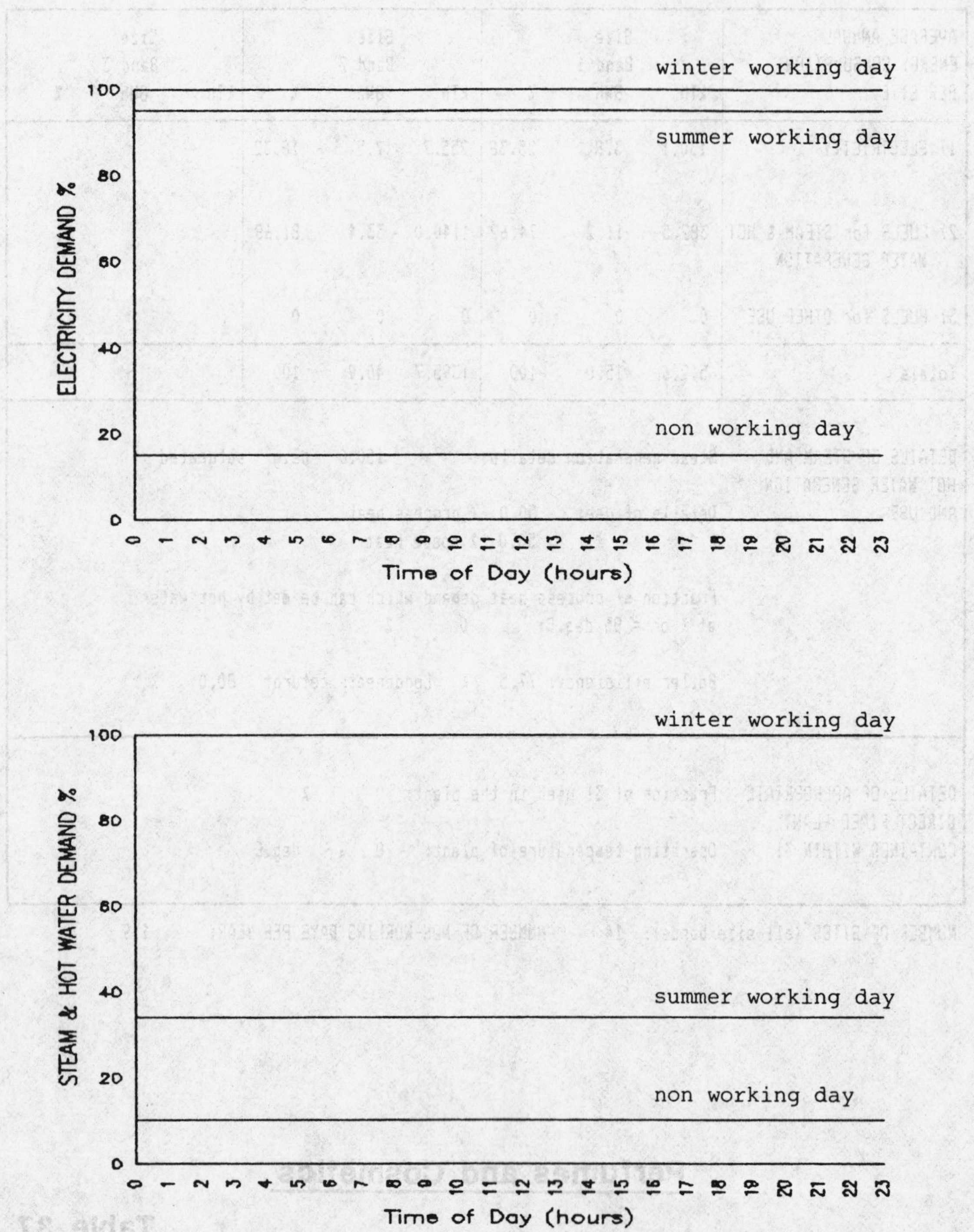

Perfumes and Cosmetics

Fig. 37

Profile of the Industry

The preparation of photographic films, paper and plates involves the growth of halide micro-crystals in large reaction vessels. The resulting suspension of crystals, referred to as an emulsion, is sensitised and coated onto film, paper or plates. The emulsion is usually chilled for long term storage.

Typical energy consumption and heat quality details for two sizes of site are presented in Table 38.1 and representative profiles of electricity and heat demand are to be found in Figure 38.1.

Energy Consumption and Heat Quality Details for the Photographic Materials Industry

AVERAGE ANNUAL ENERGY CONSUMPTION PER SITE	Size Band 1 kTh	GWh	%	Size Band 2 kTh	GWh	%	Size Band 3 kTh	GWh	%
1) ELECTRICITY	269.7	7.9	30.85	1140.5	33.4	10.34			
2) FUELS for STEAM & HOT WATER GENERATION	544.3	16.0	62.27	8903.0	260.9	80.74			
3) FUELS for OTHER USE	60.1	1.8	6.88	982.9	28.8	8.91			
Totals	874.1	25.6	100	11026.5	323.2	100			

DETAILS OF STEAM AND HOT WATER GENERATION AND USE	Steam generation details: 150.0 psig saturated Details of use: 77.8 % process heat 22.2 % space heat Fraction of process heat demand which can be met by hot water at < or = 85 deg.C: 0 % Boiler efficiency: 77.5 % Condensate return: 80.0 %
DETAILS OF APPROPRIATE DIRECT FIRED PLANT CONTAINED WITHIN 3)	Fraction of 3) used in the plant: 0 % Operating temperature of plant: 0 deg.C

NUMBER OF SITES (all size bands): 5 NUMBER OF NON-WORKING DAYS PER YEAR: 119

Photographic Materials

Table 38

Typical Daily Profiles of Heat and Power Demand for the Photographic Materials Industry

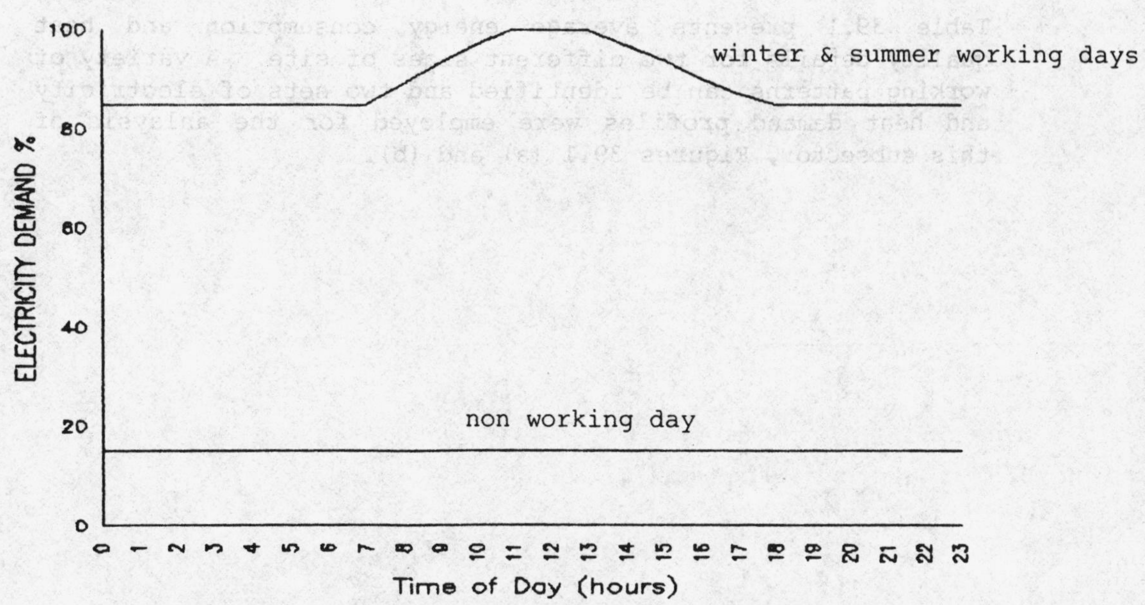

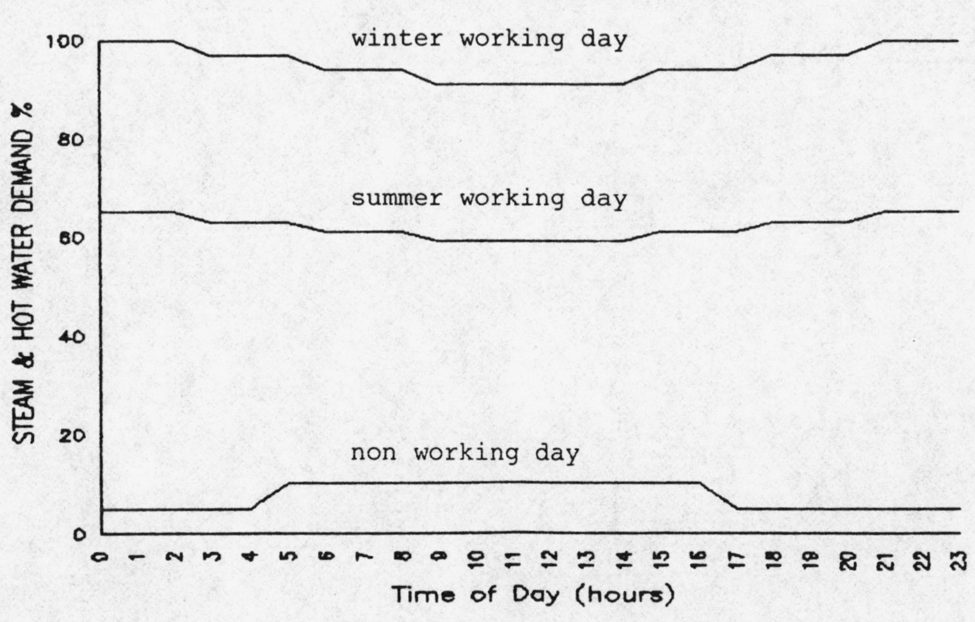

Photographic Materials

Fig. 38

39. MECHANICAL ENGINEERING [SIC 32]

Profile of the Industry

Although the products are diverse, the pattern of energy use across the mechanical engineering industry has certain characteristics which are representative of the whole sector. Most importantly, space heating is the major heat user at sites. Furthermore, electricity is used mostly for motive power and a significant part of this is for air compression.

Table 39.1 presents average energy consumption and heat quality details for two different sizes of site. A variety of working patterns can be identified and two sets of electricity and heat demand profiles were employed for the anlaysis of this subsector, Figures 39.1 (a) and (b).

Energy Consumption and Heat Quality Details for the Mechanical Engineering Industry

AVERAGE ANNUAL ENERGY CONSUMPTION PER SITE	Size Band 1 kTh	GWh	%	Size Band 2 kTh	GWh	%	Size Band 3 kTh	GWh	%
1) ELECTRICITY	139.3	4.1	24.06	868.3	25.4	21.05			
2) FUELS for STEAM & HOT WATER GENERATION	375.2	11.0	64.82	2780.3	81.5	67.40			
3) FUELS for OTHER USE	64.3	1.9	11.11	476.6	14.0	11.55			
Totals	578.8	17.0	100	4125.2	120.9	100			

DETAILS OF STEAM AND HOT WATER GENERATION AND USE

Steam generation details: 150.0 psig saturated

Details of use: 21.4 % process heat
78.6 % space heat

Fraction of process heat demand which can be met by hot water at < or = 85 deg.C: 0 %

Boiler efficiency: 77.5 % Condensate return: 80.0 %

DETAILS OF APPROPRIATE DIRECT FIRED PLANT CONTAINED WITHIN 3)

Fraction of 3) used in the plant: 0 %

Operating temperature of plant: 0 deg.C

NUMBER OF SITES (all size bands): 224 NUMBER OF NON-WORKING DAYS PER YEAR: 119

Mechanical Engineering

Table 39

Typical Daily Profile of Heat and Power Demand for Single Shift Operation

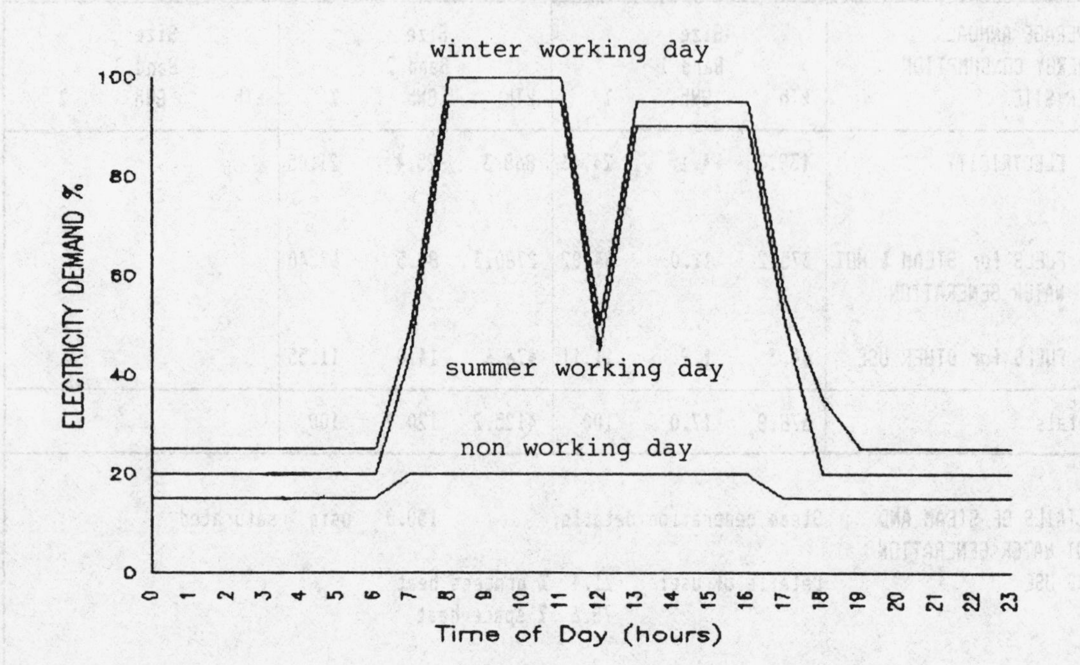

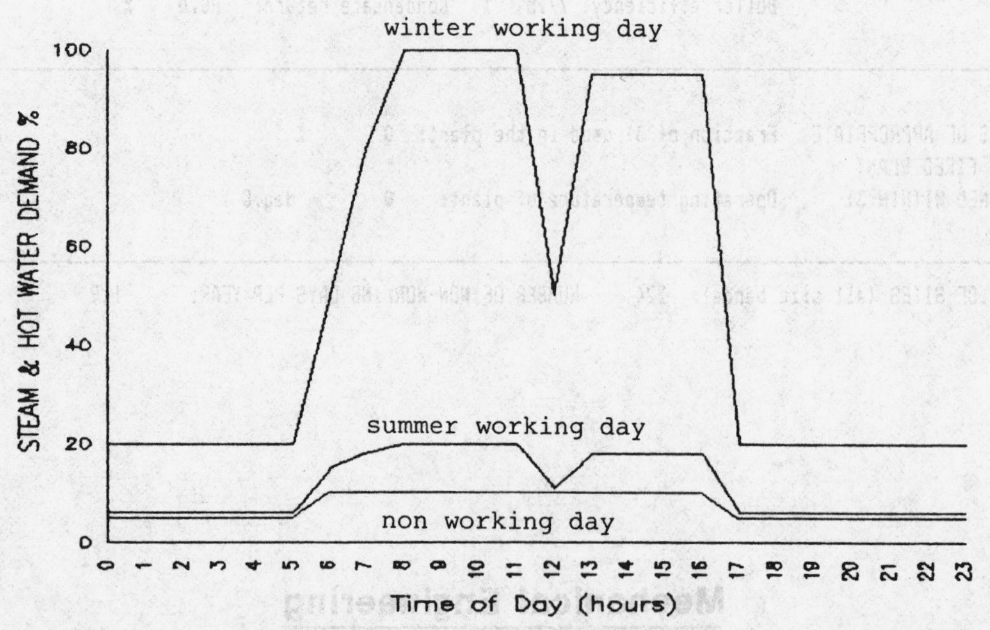

Mechanical Engineering

Fig. 39a

Typical Daily Profiles of Heat and Power Demand for Double Shift Operation

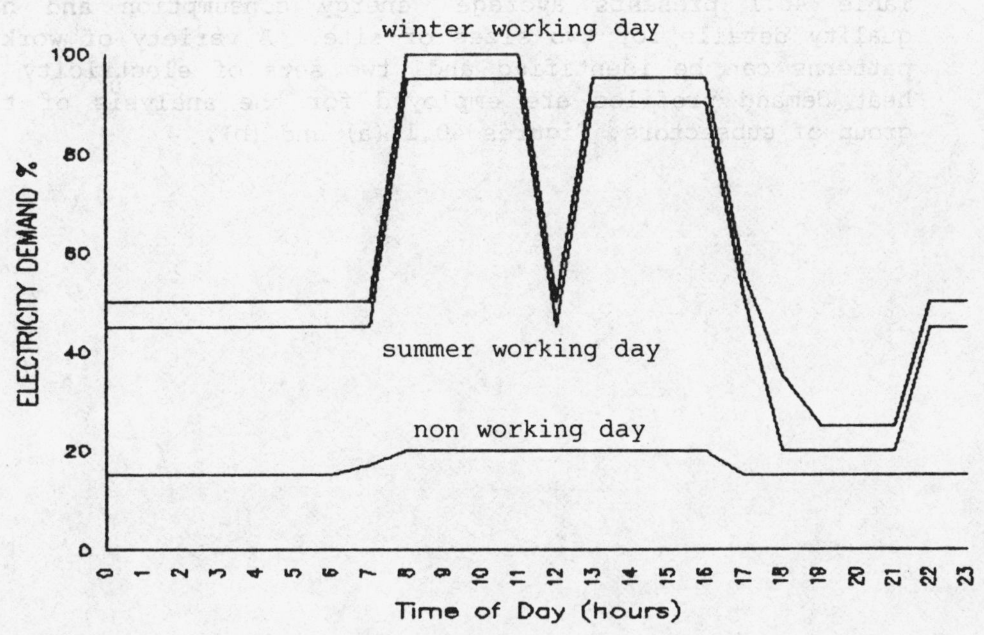

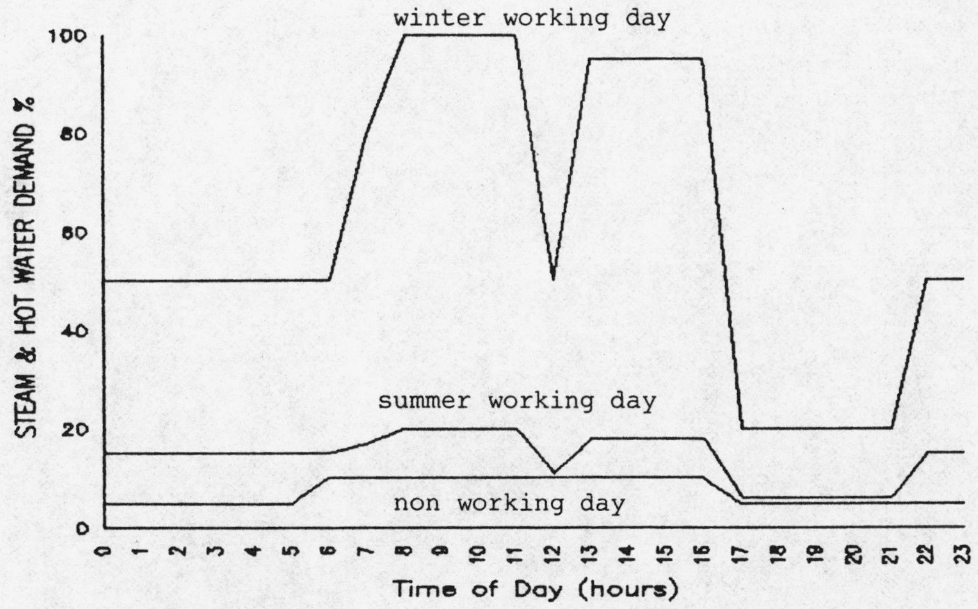

Mechanical Engineering

Fig. 39b

40. ELECTRICAL, ELECTRONIC AND INSTRUMENT ENGINEERING
[SIC 33, 34 and 37]

Profile of the Industry

The electrical, electronic and instrument engineering industries require heat primarily for space heating and electricity is used mostly for motive power and lighting. Table 40.1 presents average energy consumption and heat quality details for two sizes of site. A variety of working patterns can be identified and two sets of electricity and heat demand profiles are employed for the analysis of this group of subsectors, Figures 40.1 (a) and (b).

Energy Consumption and Heat Quality Details for the Electrical, Electronic and Instrument Engineering Industry

AVERAGE ANNUAL ENERGY CONSUMPTION PER SITE	Size Band 1 kTh	Size Band 1 GWh	Size Band 1 %	Size Band 2 kTh	Size Band 2 GWh	Size Band 2 %	Size Band 3 kTh	Size Band 3 GWh	Size Band 3 %
1) ELECTRICITY	209.7	6.1	37.55	1060.7	31.1	30.86			
2) FUELS for STEAM & HOT WATER GENERATION	348.7	10.2	62.45	2375.9	69.6	69.14			
3) FUELS for OTHER USE	0	0	0	0	0	0			
Totals	558.4	16.4	100	3436.5	100.7	100			

DETAILS OF STEAM AND HOT WATER GENERATION AND USE

Steam generation details: 150.0 psig saturated

Details of use: 7.0 % process heat
93.0 % space heat

Fraction of process heat demand which can be met by hot water at < or = 85 deg.C: 0 %

Boiler efficiency: 77.5 % Condensate return: 80.0 %

DETAILS OF APPROPRIATE DIRECT FIRED PLANT CONTAINED WITHIN 3)

Fraction of 3) used in the plant: 0 %

Operating temperature of plant: 0 deg.C

NUMBER OF SITES (all size bands): 24 NUMBER OF NON-WORKING DAYS PER YEAR: 119

Electrical, Electronic, and Instrument Engineering

Table 40

Typical Daily Profiles of Heat and Power Demand for Single Shift Operation

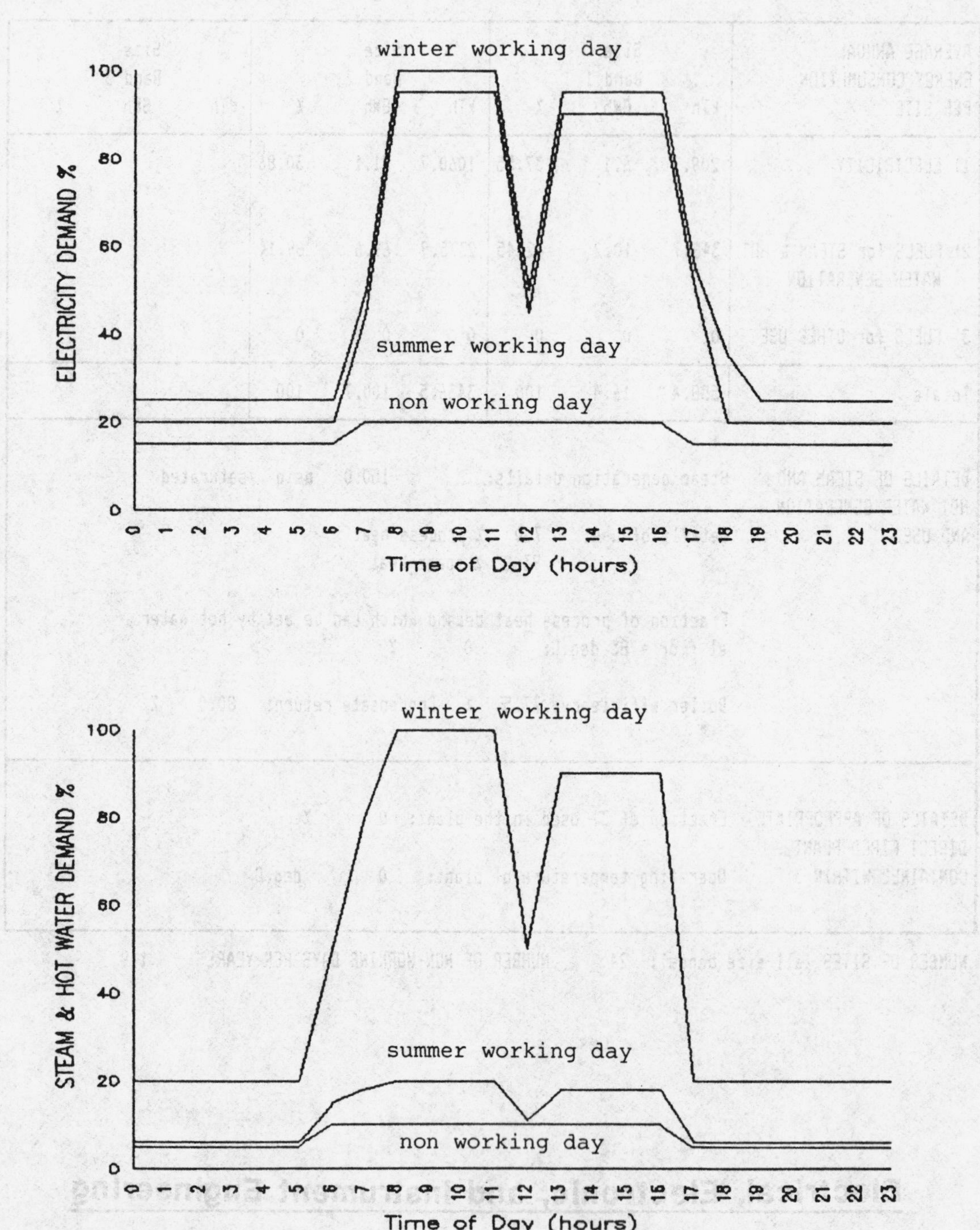

Electrical, Electronic, and Instrument Engineering

Fig. 40a

Typical Daily Profiles of Heat and Power Demand for Double Shift Operation

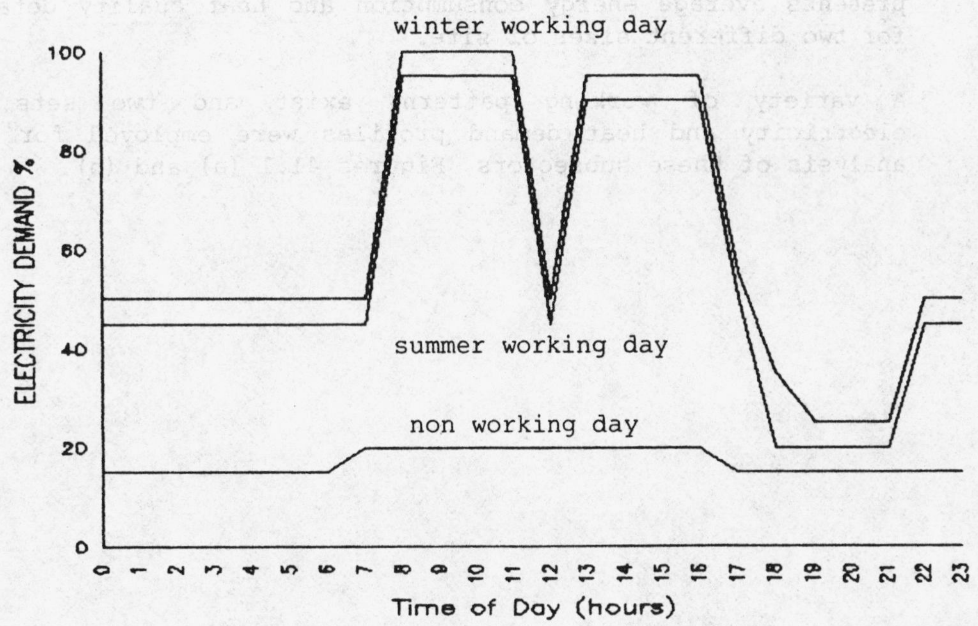

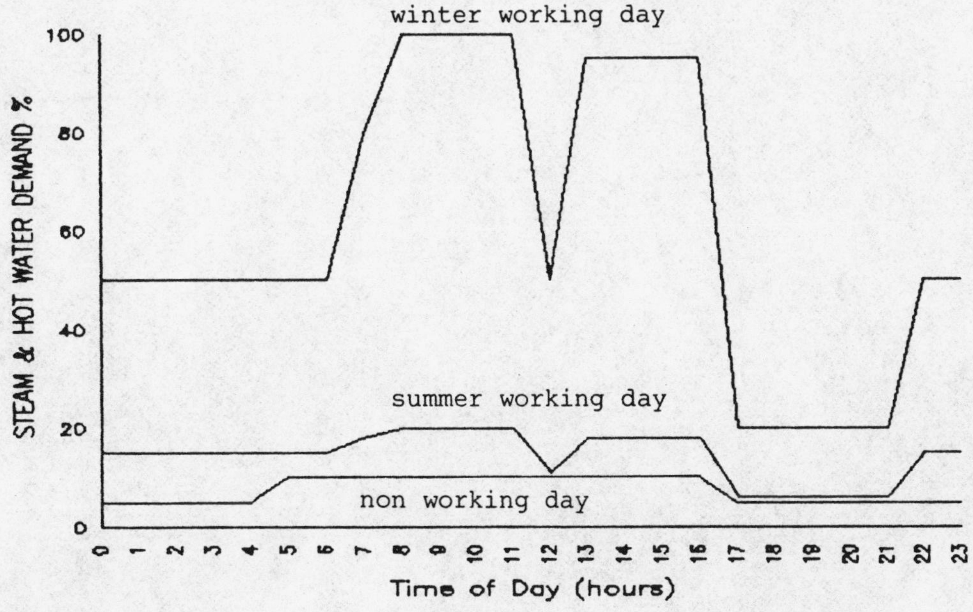

Electrical, Electronic, and Instrument Engineering

Fig. 40b

41 VEHICLES, AIRCRAFT, SHIPS AND METAL GOODS [SIC 35, 36 and 3120 - 3169]

Profile of the Industry

Space heating is a major heat user in these subsectors and this gives rise to large seasonal variations in heat demand. Electricity is used mostly for motive power and a significant proportion of this is for air compression. Table 41.1 presents average energy consumption and heat quality details for two different sizes of site.

A variety of working patterns exist and two sets of electricity and heat demand profiles were employed for the analysis of these subsectors, Figures 41.1 (a) and (b).

Energy Consumption and Heat Quality Details for the Vehicles, Aircraft, Ships and Metal Goods Industry

AVERAGE ANNUAL ENERGY CONSUMPTION PER SITE	Size Band 1 kTh	Size Band 1 GWh	Size Band 1 %	Size Band 2 kTh	Size Band 2 GWh	Size Band 2 %	Size Band 3 kTh	Size Band 3 GWh	Size Band 3 %
1) ELECTRICITY	138.4	4.1	22.79	1137.8	33.3	21.59	13204.3	387.0	16.79
2) FUELS for STEAM & HOT WATER GENERATION	400.2	11.7	65.91	3526.8	103.4	66.93	55847.6	1636.8	71.03
3) FUELS for OTHER USE	68.6	2.0	11.30	604.6	17.7	11.47	9573.9	280.6	12.18
Totals	607.2	17.8	100	5269.2	154.4	100	78625.8	2304.4	100

DETAILS OF STEAM AND HOT WATER GENERATION AND USE

Steam generation details: 150.0 psig saturated

Details of use: 21.4 % process heat
78.6 % space heat

Fraction of process heat demand which can be met by hot water at < or = 85 deg.C: 0 %

Boiler efficiency: 77.5 % Condensate return: 80.0 %

DETAILS OF APPROPRIATE DIRECT FIRED PLANT CONTAINED WITHIN 3)

Fraction of 3) used in the plant: 0 %

Operating temperature of plant: 0 deg.C

NUMBER OF SITES (all size bands): 227 NUMBER OF NON-WORKING DAYS PER YEAR: 119

Vehicles, Aircraft, Ships, and Metal Goods

Table 41

Typical Daily Profiles of Heat and Power Demand for Single Shift Operation

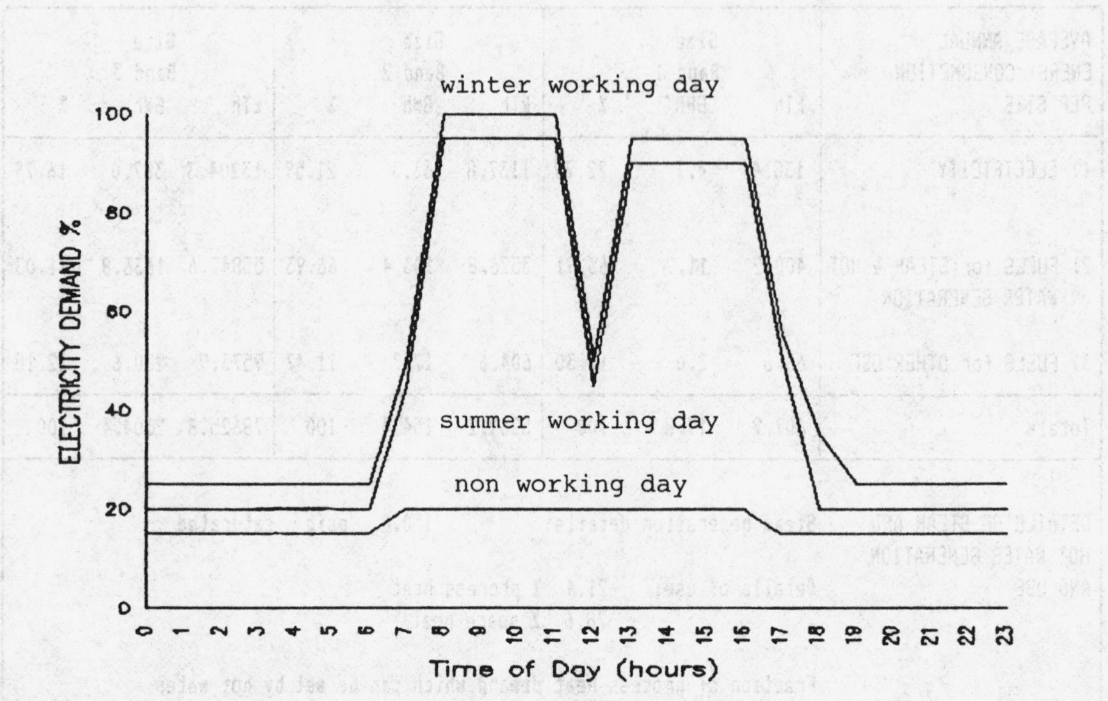

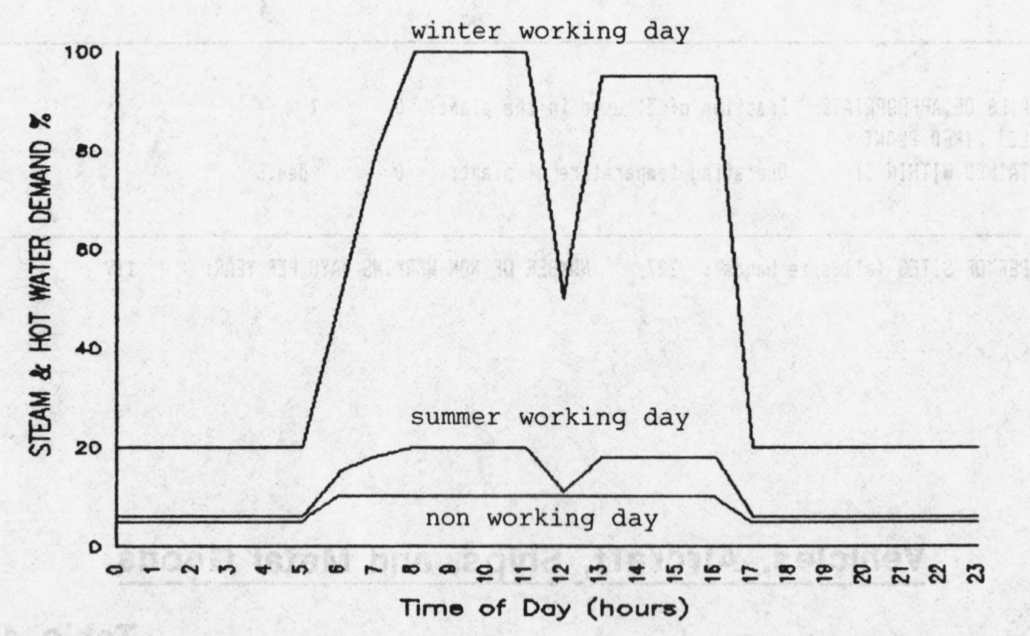

Vehicles, Aircraft, Ships, and Metal Goods

Fig. 41a

Typical Daily Profiles of Heat and Power Demand for Double Shift Operation

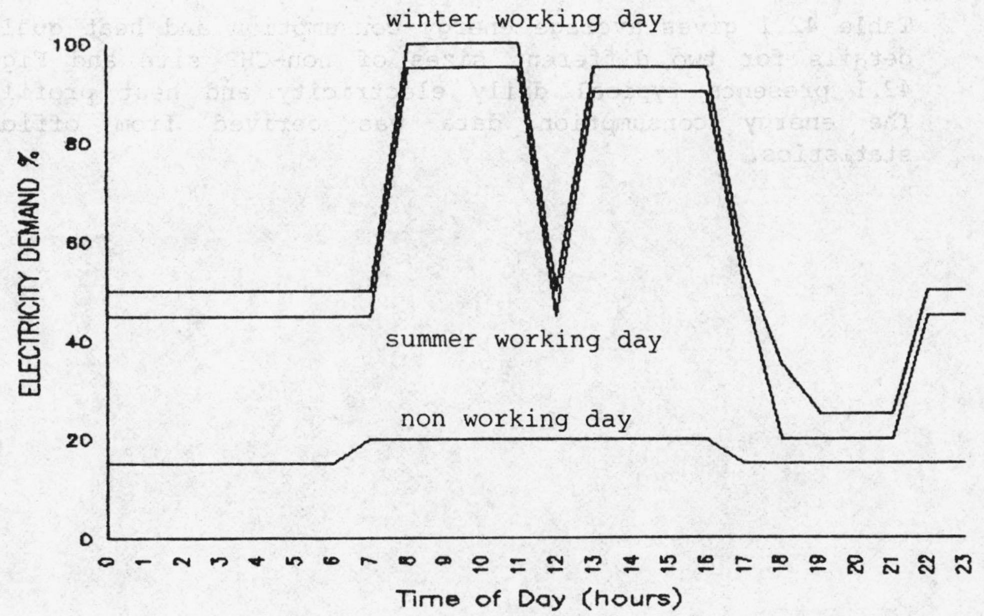

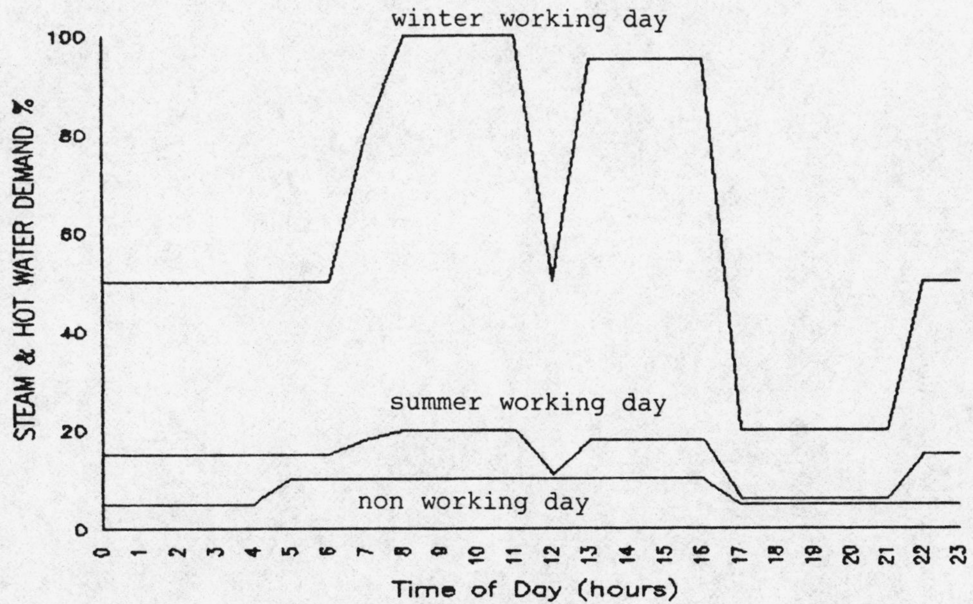

Vehicles, Aircraft, Ships, and Metal Goods

Fig. 41b

42 GENERAL TEXTILES (excluding Textile finishing) [SIC 43 excluding 4370]

Profile of the Industry

The textile industry includes : the spinning and weaving of woollen, worsted, cotton and blended fibres; knitted goods; flax; rope; twine; carpets; jute and textiles. Textile finishing is considered separately in Section 43.

Table 42.1 gives average energy consumption and heat quality details for two different sizes of non-CHP site and Figure 42.1 presents typical daily electricity and heat profiles. The energy consumption data was derived from official statistics.

Energy Consumption and Heat Quality Details for the General Textiles Industry

AVERAGE ANNUAL ENERGY CONSUMPTION PER SITE	Size Band 1 kTh	GWh	%	Size Band 2 kTh	GWh	%	Size Band 3 kTh	GWh	%
1) ELECTRICITY	136.1	4.0	20.87	1184.5	34.7	25.95			
2) FUELS for STEAM & HOT WATER GENERATION	516.0	15.1	79.13	3380.5	99.1	74.05			
3) FUELS for OTHER USE	0	0	0	0	0	0			
Totals	652.1	19.1	100	4565.0	133.8	100			

DETAILS OF STEAM AND HOT WATER GENERATION AND USE	Steam generation details: 200.0 psig saturated Details of use: 37.0 % process heat 63.0 % space heat Fraction of process heat demand which can be met by hot water at < or = 85 deg.C: 0 % Boiler efficiency: 77.5 % Condensate return: 80.0 %
DETAILS OF APPROPRIATE DIRECT FIRED PLANT CONTAINED WITHIN 3)	Fraction of 3) used in the plant: 0 % Operating temperature of plant: 0 deg.C

NUMBER OF SITES (all size bands): 293   NUMBER OF NON-WORKING DAYS PER YEAR: 119

General Textiles

Table 42

Typical Daily Profiles of Heat and Power Demand for the General Textiles Industry

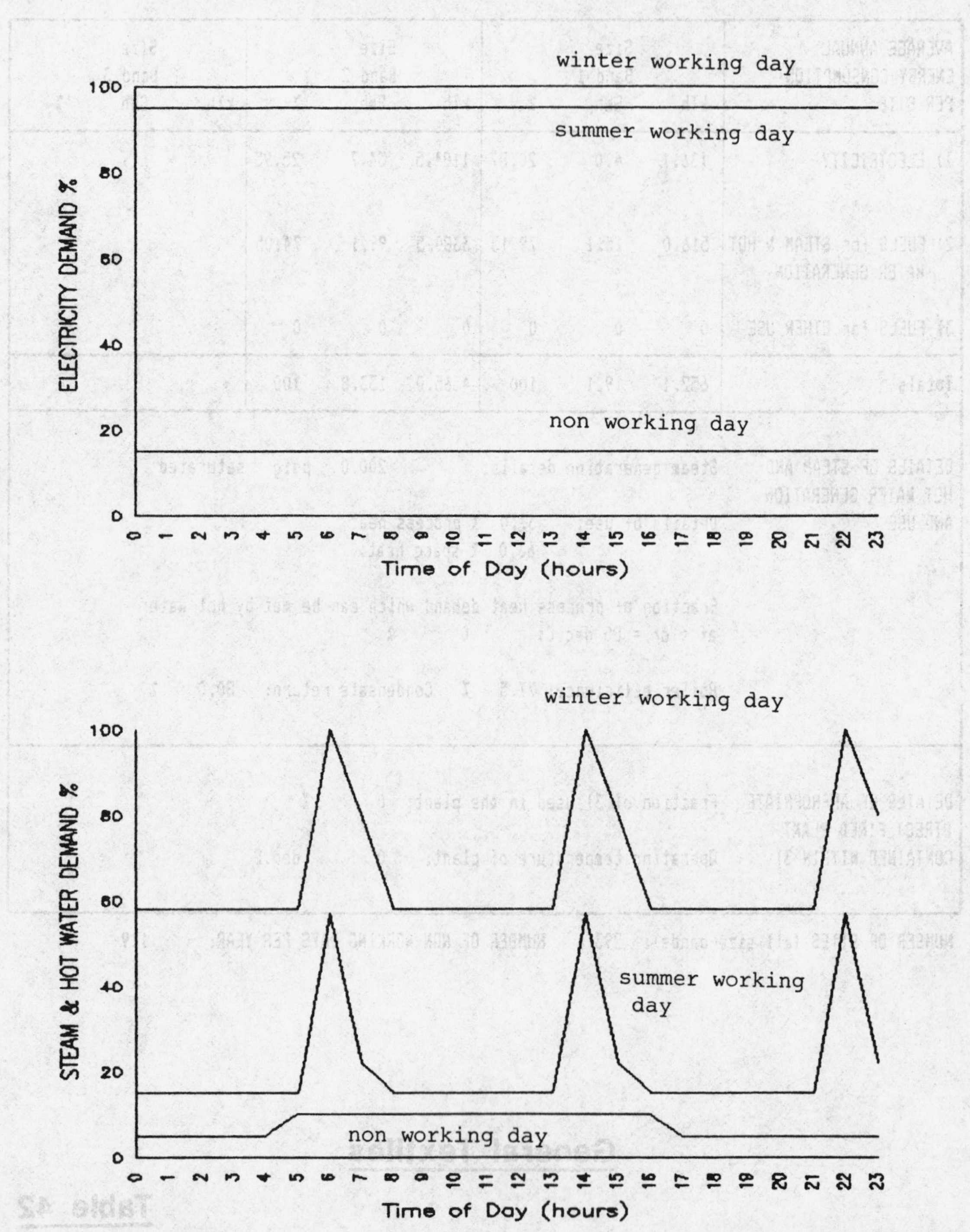

General Textiles

Fig. 42

43 TEXTILE FINISHING [SIC 4370]

Profile

Textile finishing involves a series of operations in which the fibres are wetted, treated and then dried. The purpose of the treatment may be cleaning, dyeing, surface treatment or the application of surface coatings.

Typical energy consumption and heat quality details for two different sizes of non-CHP site are given in Table 43.1 and representative electricity and heat demand profiles are presented in Figure 43.1

Energy Consumption and Heat Quality Details for the Textile Finishing Industry

AVERAGE ANNUAL ENERGY CONSUMPTION PER SITE	Size Band 1 kTh	GWh	%	Size Band 2 kTh	GWh	%	Size Band 3 kTh	GWh	%
1) ELECTRICITY	32.3	.9	8.84	134.0	3.9	8.84			
2) FUELS for STEAM & HOT WATER GENERATION	308.1	9.0	84.23	1276.9	37.4	84.23			
3) FUELS for OTHER USE	25.4	.7	6.94	105.2	3.1	6.94			
Totals	365.8	10.7	100	1516.0	44.4	100			

DETAILS OF STEAM AND HOT WATER GENERATION AND USE

Steam generation details: 200.0 psig saturated

Details of use: 82.4 % process heat
17.6 % space heat

Fraction of process heat demand which can be met by hot water at < or = 85 deg.C: 0 %

Boiler efficiency: 77.5 % Condensate return: 50.0 %

DETAILS OF APPROPRIATE DIRECT FIRED PLANT CONTAINED WITHIN 3)

Fraction of 3) used in the plant: 50.0 %

Operating temperature of plant: 220.0 deg.C

NUMBER OF SITES (all size bands): 112 NUMBER OF NON-WORKING DAYS PER YEAR: 119

Textile Finishing

Table 43

Typical Daily Profiles of Heat and Power Demand for the Textile Finishing Industry

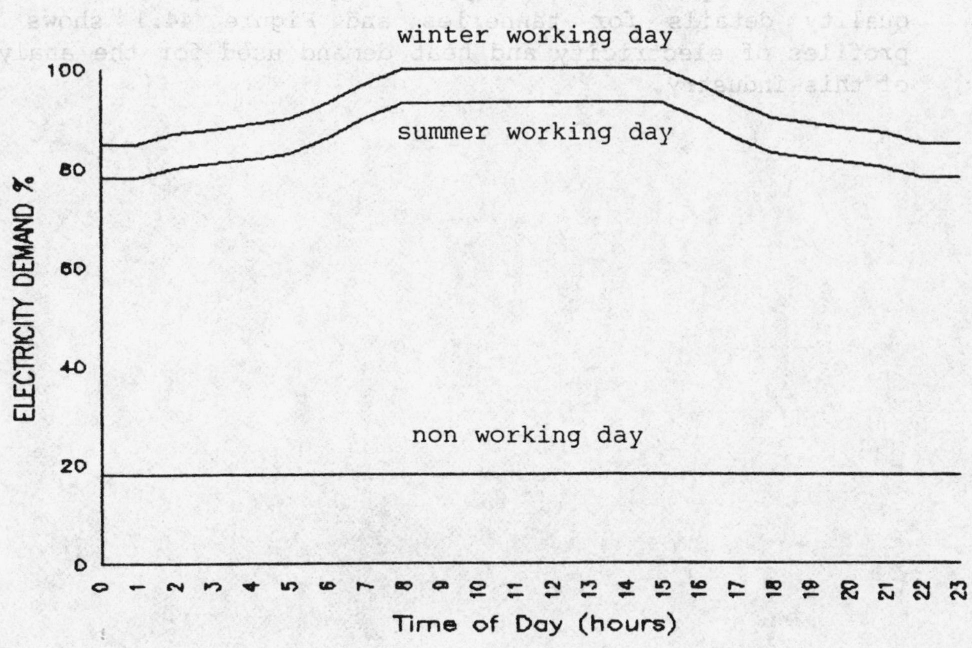

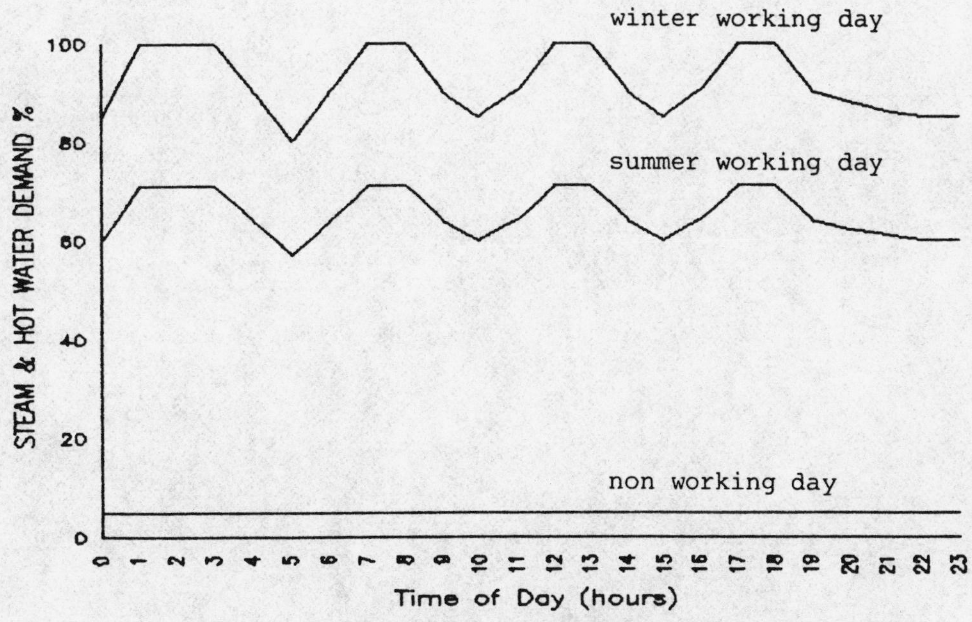

Textile Finishing

Fig. 43

44. LEATHER (TANNERIES) [part of SIC 44 and 45]

Profile of the Industry

Process heat represents a high proportion of the total energy requirements of tanneries and the heat is supplied in the form of steam or hot water.

Table 44.1 presents average energy consumption and heat quality details for tanneries and Figure 44.1 shows the profiles of electricity and heat demand used for the analysis of this industry.

Energy Consumption and Heat Quality Details for the Tanneries

AVERAGE ANNUAL ENERGY CONSUMPTION PER SITE	Size Band 1 kTh	GWh	%	Size Band 2 kTh	GWh	%	Size Band 3 kTh	GWh	%
1) ELECTRICITY	76.5	2.2	7.76	903.5	26.5	10.38			
2) FUELS for STEAM & HOT WATER GENERATION	782.7	22.9	79.35	6709.5	196.6	77.09			
3) FUELS for OTHER USE	127.2	3.7	12.89	1090.3	32.0	12.53			
Totals	986.4	28.9	100	8703.3	255.1	100			

DETAILS OF STEAM AND HOT WATER GENERATION AND USE

Steam generation details: 100.0 psig saturated

Details of use: 69.0 % process heat
31.0 % space heat

Fraction of process heat demand which can be met by hot water at < or = 85 deg.C: 50.0 %

Boiler efficiency: 77.5 % Condensate return: 80.0 %

DETAILS OF APPROPRIATE DIRECT FIRED PLANT CONTAINED WITHIN 3)

Fraction of 3) used in the plant: 0 %

Operating temperature of plant: 0 deg.C

NUMBER OF SITES (all size bands): 27 NUMBER OF NON-WORKING DAYS PER YEAR: 119

Tanneries

Table 44

Typical Daily Profiles of Heat and Power Demand for the Tanneries

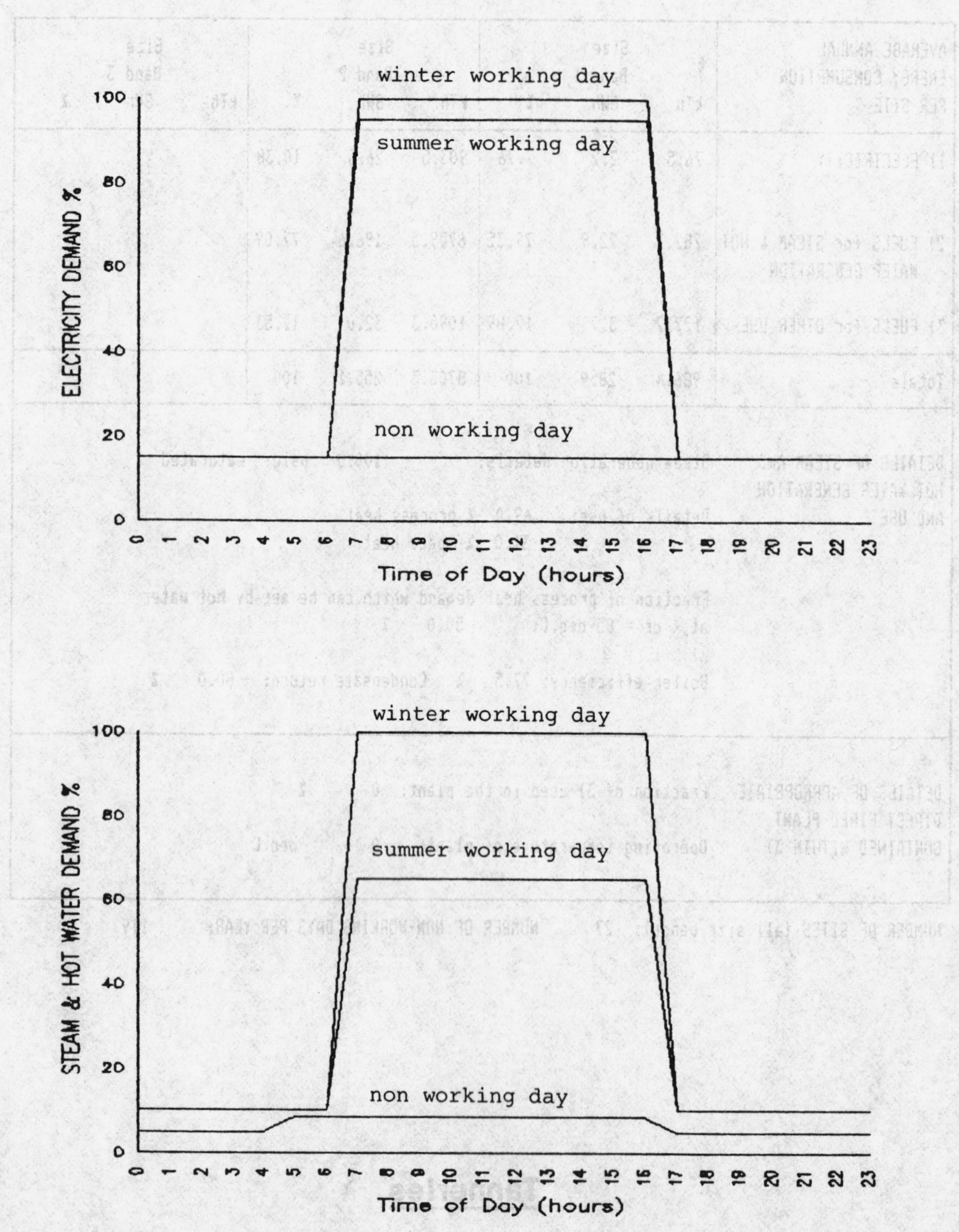

Tanneries

Fig. 44

45. LEATHER GOODS [part of SIC 44 and 45]

Profile of the Industry

This section primarily concerns the manufacture of leather clothing and leather footwear and, in general, space heating is the dominant energy use. The activities of the industry largely comprise cutting, sewing, cleaning and polishing.

Table 45.1 presents average energy consumption and heat quality details and Figure 45.1 shows the profiles of electricity and heat demand used for the analysis of this industry.

Energy Consumption and Heat Quality Details for the Leather Goods Industry

AVERAGE ANNUAL ENERGY CONSUMPTION PER SITE	Size Band 1 kTh	GWh	%	Size Band 2 kTh	GWh	%	Size Band 3 kTh	GWh	%
1) ELECTRICITY	76.5	2.2	21.84	903.5	26.5	27.78			
2) FUELS for STEAM & HOT WATER GENERATION	239.7	7.0	68.39	2054.8	60.2	63.19			
3) FUELS for OTHER USE	34.2	1.0	9.77	293.5	8.6	9.03			
Totals	350.5	10.3	100	3251.8	95.3	100			

DETAILS OF STEAM AND HOT WATER GENERATION AND USE

Steam generation details: 100.0 psig saturated

Details of use: 7.0 % process heat
93.0 % space heat

Fraction of process heat demand which can be met by hot water at < or = 85 deg.C: 0 %

Boiler efficiency: 77.5 % Condensate return: 80.0 %

DETAILS OF APPROPRIATE DIRECT FIRED PLANT CONTAINED WITHIN 3)

Fraction of 3) used in the plant: 0 %

Operating temperature of plant: 0 deg.C

NUMBER OF SITES (all size bands): 81 NUMBER OF NON-WORKING DAYS PER YEAR: 119

Leather Goods

Table 45

Typical Daily Profiles of Heat and Power Demand for the Leather Goods Industry

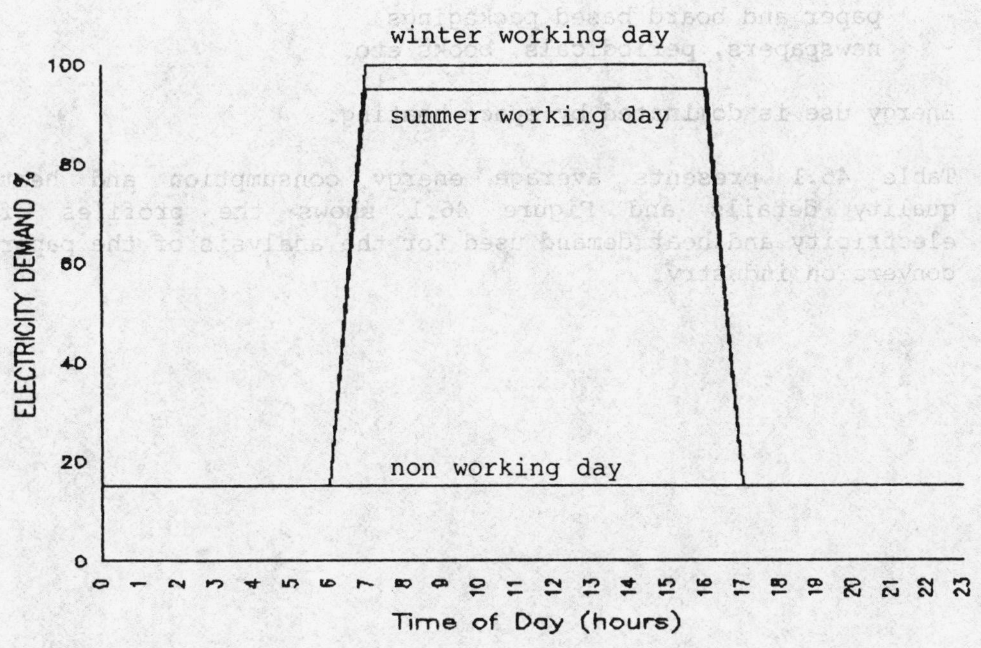

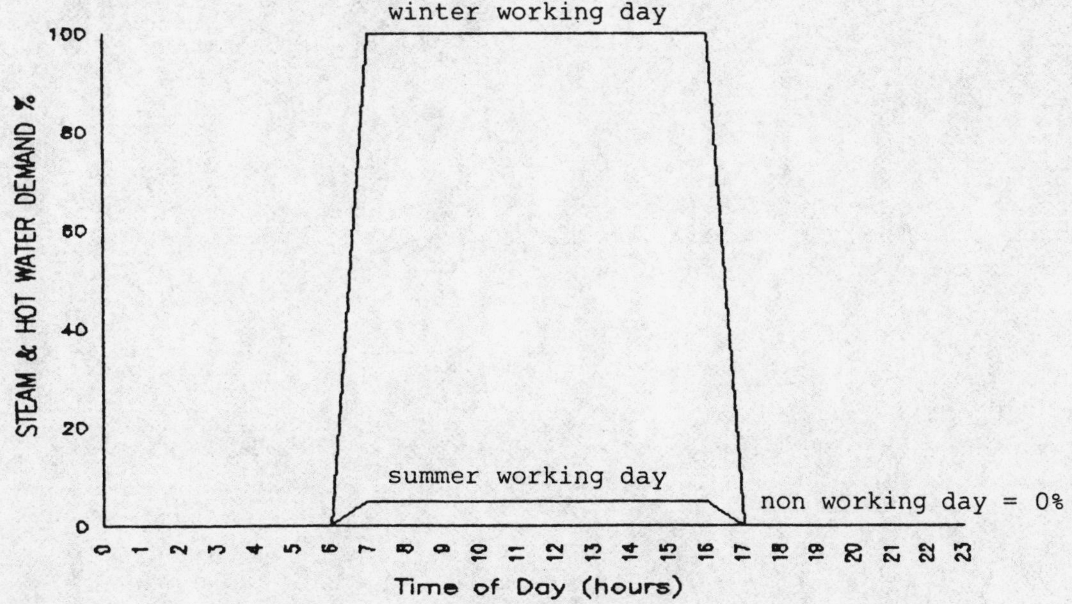

Leather Goods

Fig. 45

46 PAPER CONVERSION [SIC 47 excluding 4710 Paper and Board]

Profile of the Industry

The industry grouping concerns industrial sites which use paper or board as their main raw material and inlcudes the manufacture of the following products:

- stationery and wall coverings
- paper and board based packagings
- newspapers, periodicals, books etc.

Energy use is dominated by space heating.

Table 46.1 presents average energy consumption and heat quality details and Figure 46.1 shows the profiles of electricity and heat demand used for the analysis of the paper conversion industry.

Energy Consumption & Heat Quality Details for the Paper Conversion Industry

AVERAGE ANNUAL ENERGY CONSUMPTION PER SITE	Size Band 1 kTh	GWh	%	Size Band 2 kTh	GWh	%	Size Band 3 kTh	GWh	%
1) ELECTRICITY	111.5	3.3	21.85	628.4	18.4	16.12			
2) FUELS for STEAM & HOT WATER GENERATION	384.7	11.3	75.41	3156.2	92.5	80.95			
3) FUELS for OTHER USE	13.9	.4	2.73	114.4	3.4	2.93			
Totals	510.1	15.0	100	3899.1	114.3	100			

DETAILS OF STEAM AND HOT WATER GENERATION AND USE

Generation details: Band 1 hot water

Band 2 steam at 100 psig

Details of use: 6.0 % process heat
94.0 % space heat

Fraction of process heat demand which can be met by hot water at < or = 85 deg.C: 6.0 %

Boiler efficiency: 75.0 % Condensate return: 95.0 %

DETAILS OF APPROPRIATE DIRECT FIRED PLANT CONTAINED WITHIN 3)

Fraction of 3) used in the plant: 0 %

Operating temperature of plant: 0 deg.C

NUMBER OF SITES (all size bands): 19 NUMBER OF NON-WORKING DAYS PER YEAR: 67

Paper Conversion

Table 46

Typical Daily Profiles of Heat & Power Demand for the Paper Conversion Industry

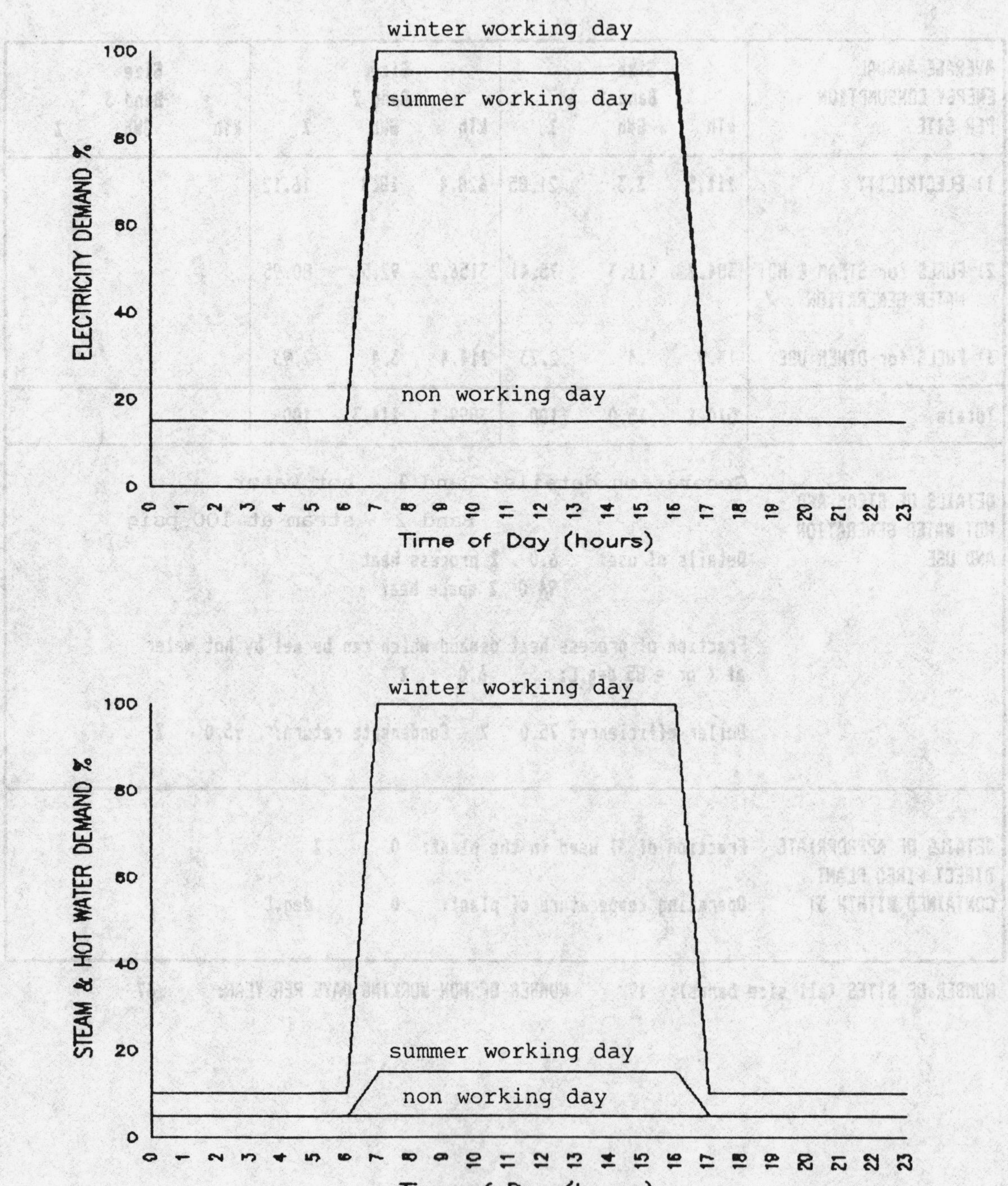

Paper Conversion

Fig. 46

47 PAPER AND BOARD MANUFACTURERS [SIC 4710]

Profile of the Industry

Paper making is now a continuous process and the principal raw material input is wood. The processing of the wood by pulping and bleaching is almost exclusively carried out abroad and bales of dried pulp are imported to this country to supply the needs of the industry. Recycled paper and other sources of cellulose fibres are also used.

The paper-making process consists firstly of dispersion of the cellulose fibres (wood pulp, recycled paper etc) in water. The slurry is then cleaned, beaten or refined and is passed onto a continuously moving wire mesh. The water drains through leaving a wet layer of paper on top of the wire. The wet paper web is then taken onto a felt and passed through roll presses to squeeze out more water. The web is next passed over a series of steam heated drying cylinders and the amount of finishing depends on the end use of the product. Refining is the single most important electricity using step and steam heating of dryers is the single most important heat use.

Typical energy consumption and heat quality of details for three different sizes of non-CHP paper making site are presented in Table 47.1 and profiles of electricity and heat demand are presented in Figure 47.1

Energy Consumption & Heat Quality Details for the Paper & Board Industry

AVERAGE ANNUAL ENERGY CONSUMPTION PER SITE	Size Band 1 kTh	GWh	%	Size Band 2 kTh	GWh	%	Size Band 3 kTh	GWh	%
1) ELECTRICITY	113.4	3.3	19.00	824.4	24.2	19.00	6286.5	184.2	19.00
2) FUELS for STEAM & HOT WATER GENERATION	483.4	14.2	81.00	3514.5	103.0	81.00	26800.3	785.5	81.00
3) FUELS for OTHER USE	0	0	0	0	0	0	0	0	0
Totals	596.8	17.5	100	4338.9	127.2	100	33086.8	969.7	100

DETAILS OF STEAM AND HOT WATER GENERATION AND USE

Steam generation details: 120.0 psig saturated

Details of use: 95.0 % process heat
5.0 % space heat

Fraction of process heat demand which can be met by hot water at < or = 85 deg.C: 0 %

Boiler efficiency: 80.0 % Condensate return: 85.0 %

DETAILS OF APPROPRIATE DIRECT FIRED PLANT CONTAINED WITHIN 3)

Fraction of 3) used in the plant: 0 %

Operating temperature of plant: 0 deg.C

NUMBER OF SITES (all size bands): 37 NUMBER OF NON-WORKING DAYS PER YEAR: 28

Paper and Board

Table 47

Typical Daily Profiles of Heat & Power Demand for the Paper & Board Industry

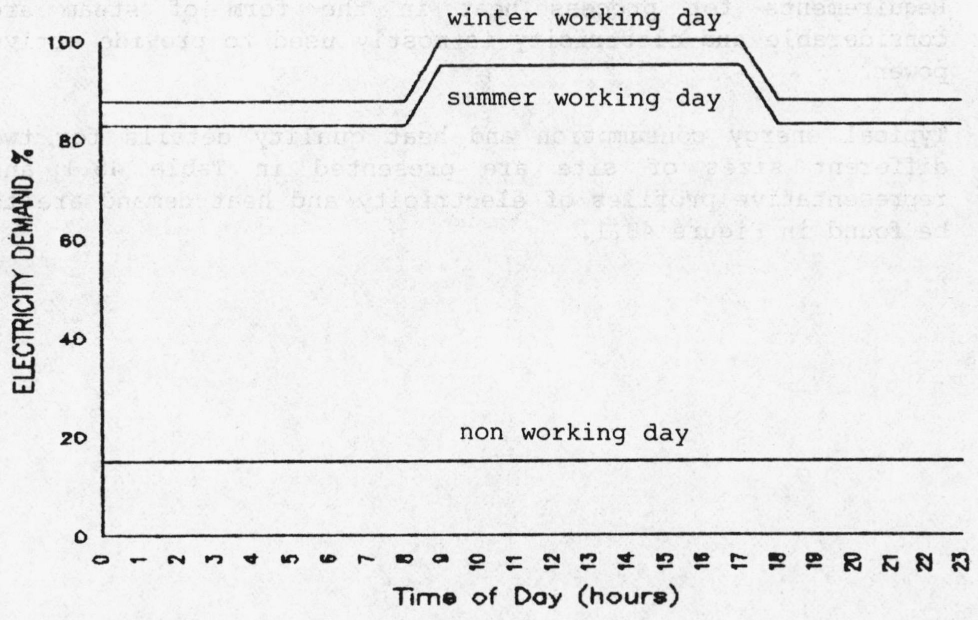

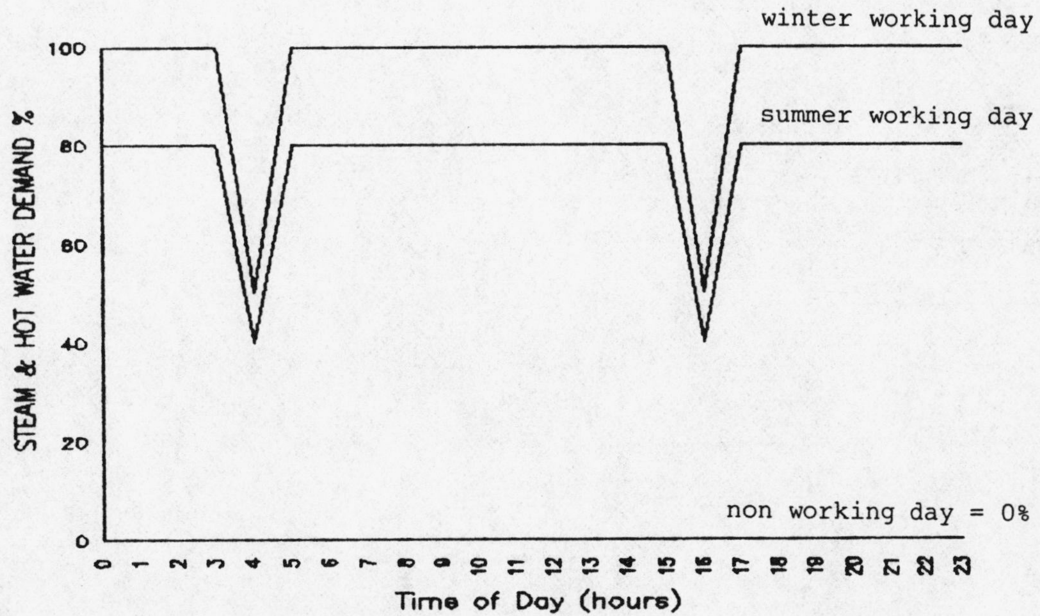

Paper and Board

Fig. 47

48. RUBBER TYRES [part of SIC 481/482]

Profile of the Industry

Rubber tyres are generally manufactured by the overlaying of plies of rubber-coated fabrics (which are produced by calendering) and extruded tread onto collapsible drums. The resulting uncured tyres are removed and vulcanized. Requirements for process heat in the form of steam are considerable and electricity is mostly used to provide motive power.

Typical energy consumption and heat quality details for two different sizes of site are presented in Table 48.1 and representative profiles of electricity and heat demand are to be found in Figure 48.1.

Energy Consumption & Heat Quality Details for the Rubber Tyre Industry

AVERAGE ANNUAL ENERGY CONSUMPTION PER SITE	Size Band 1 kTh	Size Band 1 GWh	Size Band 1 %	Size Band 2 kTh	Size Band 2 GWh	Size Band 2 %	Size Band 3 kTh	Size Band 3 GWh	Size Band 3 %
1) ELECTRICITY				359.7	10.5	16.89	3745.5	109.8	19.87
2) FUELS for STEAM & HOT WATER GENERATION				1674.3	49.1	78.60	14281.5	418.6	75.77
3) FUELS for OTHER USE				96.3	2.8	4.52	821.2	24.1	4.36
Totals				2130.3	62.4	100	18848.2	552.4	100

DETAILS OF STEAM AND HOT WATER GENERATION AND USE

Steam generation details: 250.0 psig saturated

Details of use: 69.0 % process heat
31.0 % space heat

Fraction of process heat demand which can be met by hot water at < or = 85 deg.C: 0 %

Boiler efficiency: 77.5 % Condensate return: 50.0 %

DETAILS OF APPROPRIATE DIRECT FIRED PLANT CONTAINED WITHIN 3)

Fraction of 3) used in the plant: 0 %

Operating temperature of plant: 0 deg.C

NUMBER OF SITES (all size bands): 5 NUMBER OF NON-WORKING DAYS PER YEAR: 114

Rubber Tyres

Table 48

Typical Daily Profiles of Heat & Power Demand for the Rubber Tyre Industry

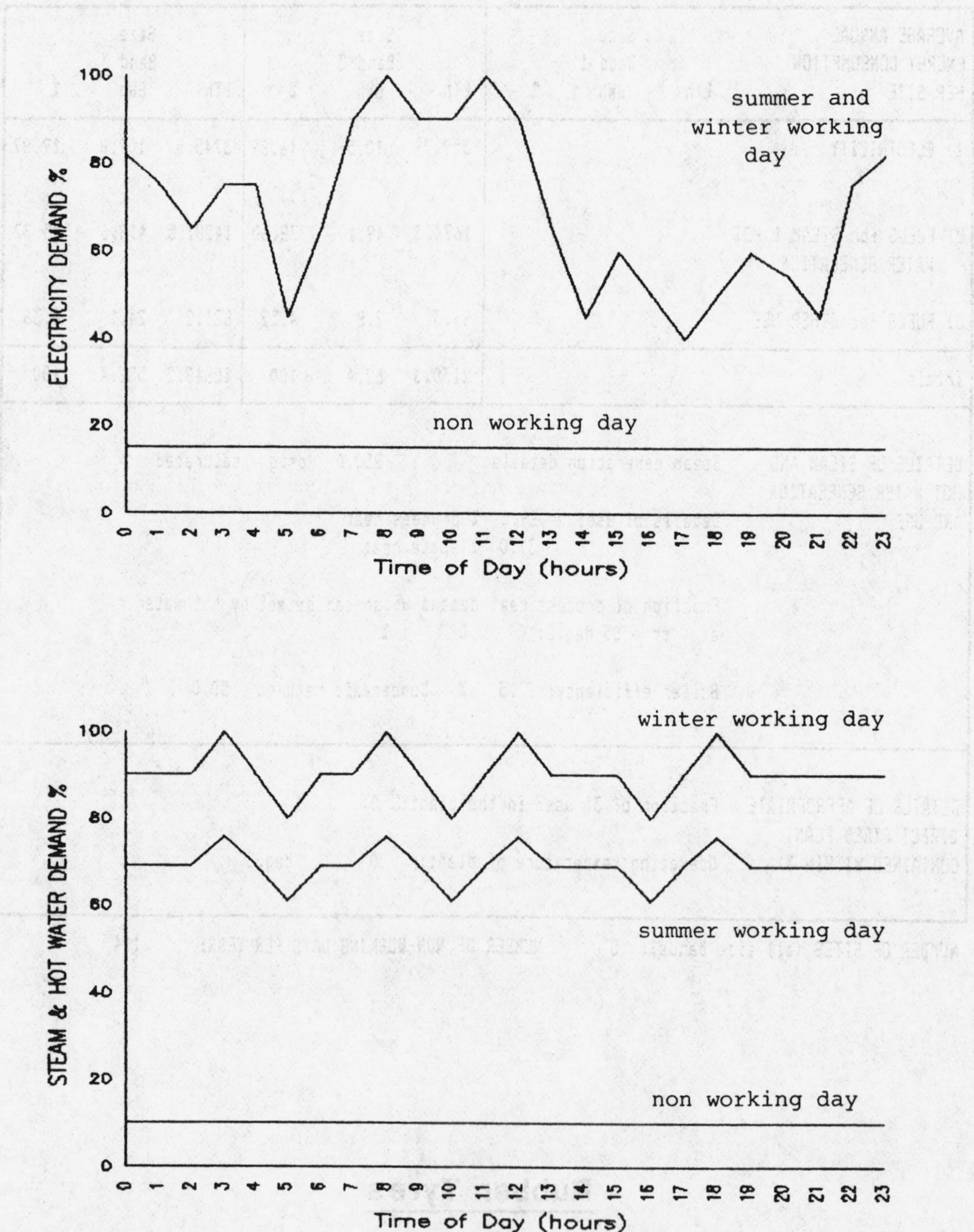

Rubber Tyres

Fig. 48

49. RUBBER PRODUCTS (OTHER THAN TYRES)
[part of SIC 481/482]

Profile of the Industry

The manufacture of rubber products generally involves the compounding of the rubber with fillers, softeners and vulcanizing agents and the resulting mix is then shaped and vulcanized. The main shaping processes are extrusion and calendering. Vulcanization uses pressure and heat (steam or hot air). Electricity is used principally to provide motive power.

Typical energy consumption and heat quality details for two different sizes of sites are to be found in Table 49.1 and representative profiles of electricity and heat demand are presented in Figure 49.1.

Energy Consumption & Heat Quality Details for the Rubber Products Industry

AVERAGE ANNUAL ENERGY CONSUMPTION PER SITE	Size Band 1 kTh	GWh	%	Size Band 2 kTh	GWh	%	Size Band 3 kTh	GWh	%
1) ELECTRICITY	147.7	4.3	28.05	359.7	10.5	16.89			
2) FUELS for STEAM & HOT WATER GENERATION	358.3	10.5	68.04	1674.3	49.1	78.60			
3) FUELS for OTHER USE	20.6	.6	3.91	96.3	2.8	4.52			
Totals	526.6	15.4	100	2130.3	62.4	100			

DETAILS OF STEAM AND HOT WATER GENERATION AND USE	Steam generation details: 250.0 psig saturated Details of use: 69.0 % process heat 31.0 % space heat Fraction of process heat demand which can be met by hot water at < or = 85 deg.C: 0 % Boiler efficiency: 77.5 % Condensate return: 50.0 %
DETAILS OF APPROPRIATE DIRECT FIRED PLANT CONTAINED WITHIN 3)	Fraction of 3) used in the plant: 0 % Operating temperature of plant: 0 deg.C

NUMBER OF SITES (all size bands): 5 NUMBER OF NON-WORKING DAYS PER YEAR: 114

Rubber Products

Table 49

Typical Daily Profiles of Heat & Power Demand for the Rubber Products Industry

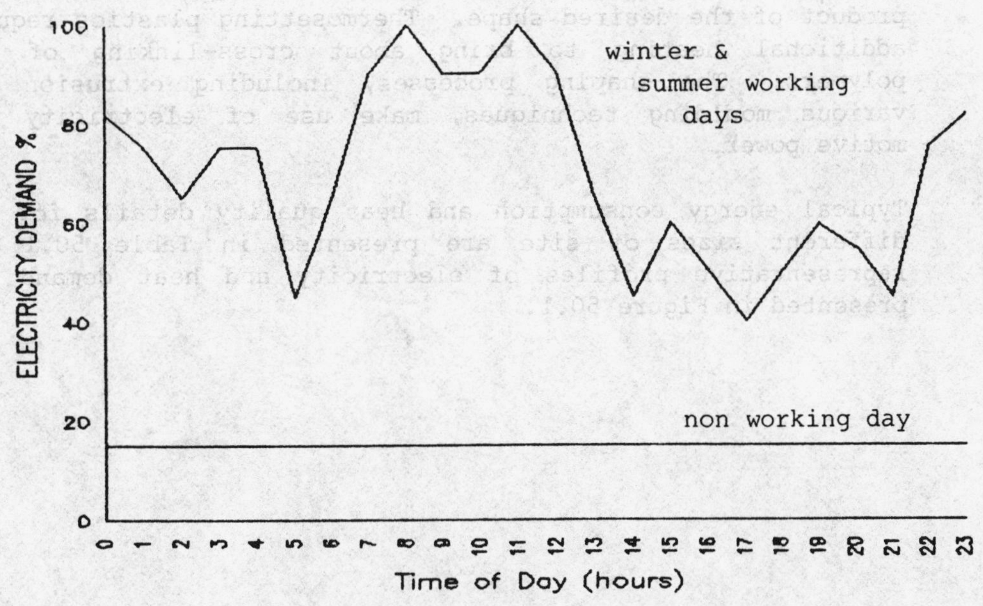

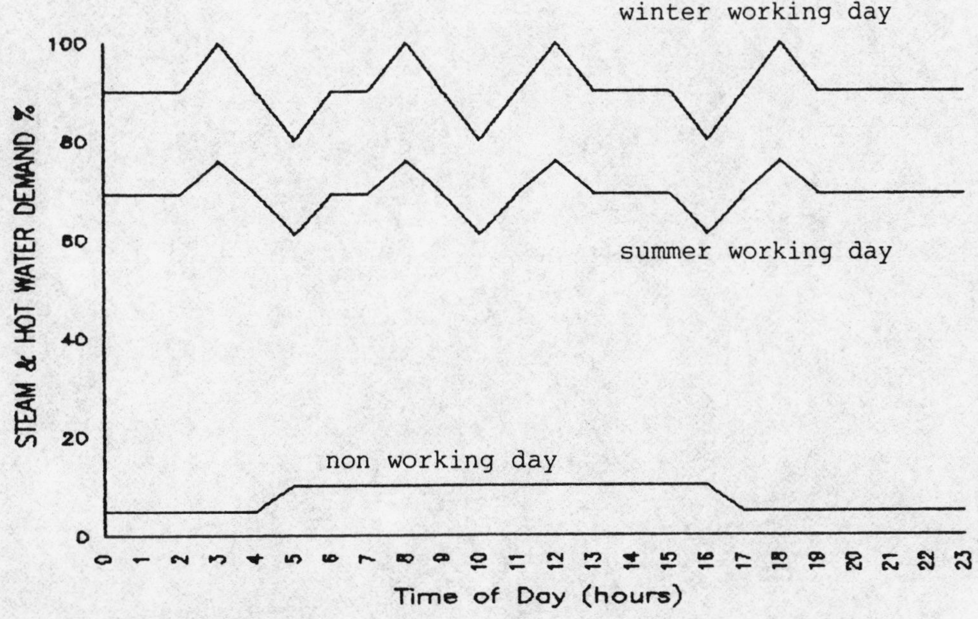

Rubber Products

Fig. 49

50. PLASTICS CONVERSION [SIC 483]

Profile of the Industry

The industry is concerned with the manufacture of products from thermoplastic and thermosetting plastics which are generally purchased in the form of pellets or powders. In the case of thermoplastics, the raw material is melted by the application of heat, shaped and cooled to produce the solid product of the desired shape. Thermosetting plastics require additional heating to bring about cross-linking of the polymers. The shaping processes, including extrusion and various moulding techniques, make use of electricity for motive power.

Typical energy consumption and heat quality details for two different sizes of site are presented in Table 50.1 and representative profiles of electricity and heat demand are presented in Figure 50.1.

Energy Consumption & Heat Quality Details for the Plastics Conversion Industry

AVERAGE ANNUAL ENERGY CONSUMPTION PER SITE	Size Band 1 kTh	GWh	%	Size Band 2 kTh	GWh	%	Size Band 3 kTh	GWh	%
1) ELECTRICITY	203.8	6.0	43.92	481.4	14.1	18.42			
2) FUELS for STEAM & HOT WATER GENERATION	247.1	7.2	53.26	2024.5	59.3	77.47			
3) FUELS for OTHER USE	13.1	.4	2.82	107.3	3.1	4.11			
Totals	464.0	13.6	100	2613.2	76.6	100			

DETAILS OF STEAM AND HOT WATER GENERATION AND USE

Generation details: Band 1 hot water

Band 2 steam at 150 psig

Details of use: 50.0 % process heat
50.0 % space heat

Fraction of process heat demand which can be met by hot water at < or = 85 deg.C: 0 %

Boiler efficiency: 77.5 % Condensate return: 80.0 %

DETAILS OF APPROPRIATE DIRECT FIRED PLANT CONTAINED WITHIN 3)

Fraction of 3) used in the plant: 0 %

Operating temperature of plant: 0 deg.C

NUMBER OF SITES (all size bands): 31 NUMBER OF NON-WORKING DAYS PER YEAR: 114

Plastics Conversion

Table 50

Typical Daily Profiles of Heat & Power Demand for the Plastics Conversion Industry

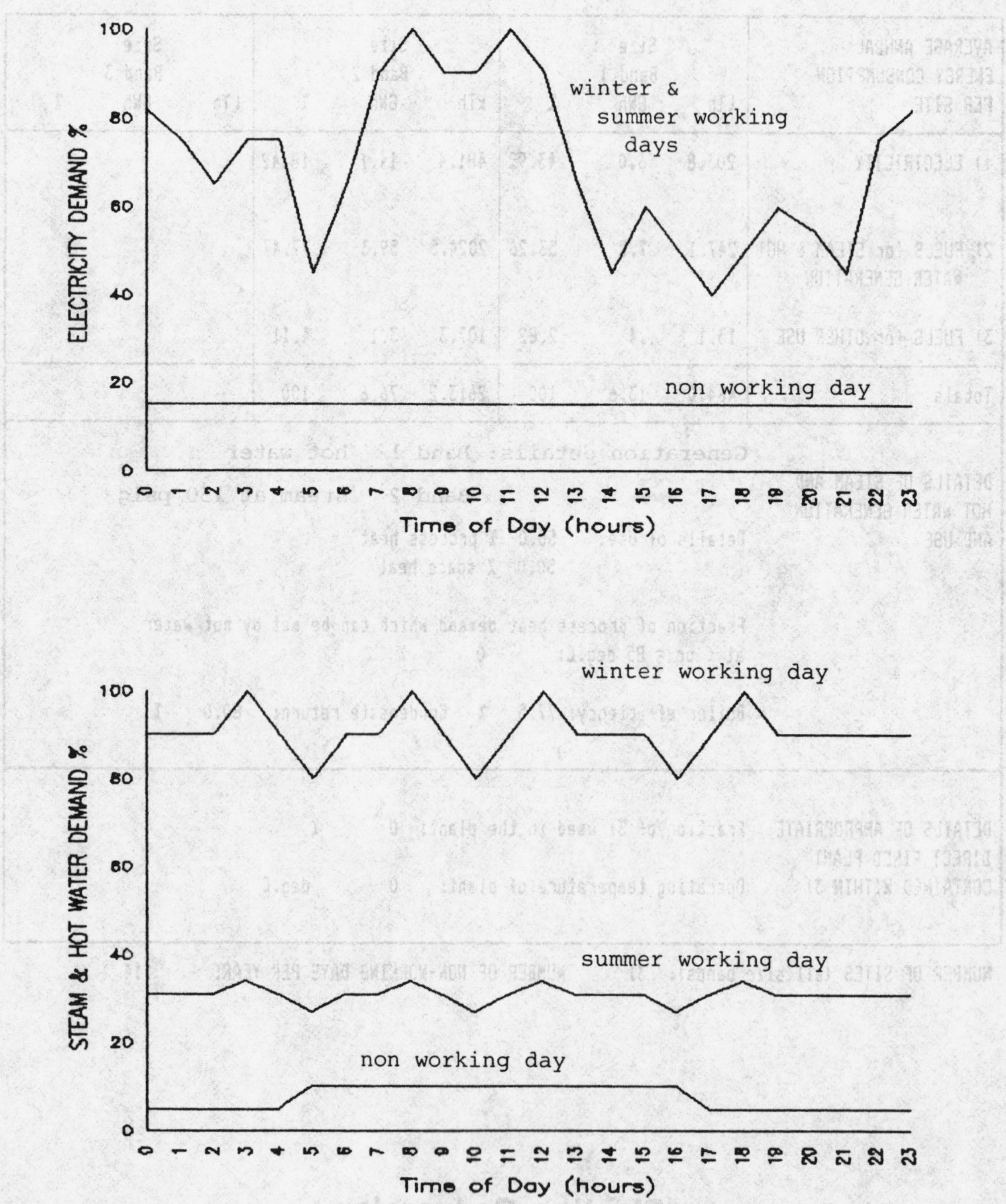

Plastics Conversion

Fig. 50

INDUSTRIAL COMBINED HEAT AND POWER
THE POTENTIAL FOR NEW USERS

APPENDIX 4

DIRECT DRIVE SYSTEMS & RETRO-FIT STEAM TURBINE CHP PLANT

CONTENTS

DIRECT DRIVE SYSTEMS AND RETRO-FIT STEAM TURBINE CHP PLANT

1 This appendix examines the scope for small retro-fit CHP systems of under 1 MW power output. Two concepts are considered:

- direct drive systems, where a prime mover, instead of driving an electricity generator, is applied to a suitable mechanical load directly; and

- small steam turbine based electricity generating CHP systems retrofitted to an existing (possibly upgraded low pressure) boiler plant to supply a relatively small part of the total site electricity demand.

Throughout this appendix to following fuel prices apply:

Gas	28.3p/therm	£2.68/GJ
Heavy Fuel Oil	11.0p/litre	£2.68/GJ
Coal	£62.5/tonne	£2.20/GJ

1.0 Direct Drive Systems

2 Two systems have been considered where a CHP plant can be used to drive an item of plant directly. These are:

(1) Back pressure steam turbine drives operating between a medium pressure distribution main and a low pressure demand, driving an air compressor, pump, refrigeration plant, or other suitable mechanical load. Steam would be provided from existing boiler plant.

(2) Spark ignition engine driving an air compressor, with heat recovery from the engine as well as the air compressor, inter- and after coolers.

1.1 Back Pressure Steam Turbine Plant

3 Small single stage steam turbines with inlet steam conditions of 250 psig (13.2 bar abs) saturated and a back pressure of 80 psig (6.5 bar abs) were considered. The performance of these turbines are described in section 2.8 of Appendix 2.

4 The turbine can be used to drive an air compressor or refrigeration machine directly through a gearbox and the exhaust steam can be used for other processes.

5 As the plant is driven directly from the steam turbine there are no electrical losses that would occur if a turbine generating set had been used. Although the net electrical efficiency of a small single stage steam turbine is low, the overall efficiency of the system is high, if the turbine exhaust steam is utilised, as the only losses are the mechanical losses in the turbine and gearbox. The mechanical power to drive the plant is thus obtained at the expense of only a marginal increase in fuel for the boiler. Even taking into consideration the boiler efficiency the cost of boiler fuel is much cheaper than electricity and so the running cost savings can be quite large.

6 The disadvantage of these systems is the low mechanical output of the turbine for a given steam flow. The heat to power ratio for the turbine considered is over 21:1, indicating that a steam flow of over 69000 lbs/hr is required to provide 1 MW of shaft power.

Savings

7 The net savings resulting from the installation of a direct drive steam turbine CHP plant are derived from the saving in electricity otherwise required to run the item of plant. This is offset against the extra fuel for the boiler and the maintenance cost.

8 Two systems have been analysed for steam turbines of 250 and 750 kW shaft output. The net cost savings are shown below in Table 1 assuming:

(a) An efficiency of 92% for an electric motor that would otherwise be used to drive the plant.

(b) Operation for 8000 hours/year.

(c) Maintenance cost of 0.14p/kWh of shaft power.

(d) A boiler efficiency of 75% based on higher calorific value.

Table 1
Annual Savings for Steam Turbine Direct Drive CHP Plant

Turbine Size	250	750	(kW)
Electricity Saving	2150	6500	(MWh/year)
Extra Boiler Fuel	10430	31290	(GJ/year)
Net National Energy Saving	580	1760	(TCE/year)
Net Annual Cost Savings:			
Gas	42900	130000	(£/year)
Heavy Fuel Oil	42900	130000	(£/year)
Coal	47900	144500	(£/year)

The fuel and electricity costs are those used for the rest of the study.

Capital costs

9 The capital costs for steam turbine systems for direct drive applications are given in section 2.4 of Appendix 1. The costs do not include the driven machine as it is assumed that a steam turbine would be retrofitted onto existing equipment.

The capital costs for the two systems considered were

(a) £ 85000 for 250 kW steam turbine.
(b) £160000 for 750 kW steam turbine.

The full scope of the capital costs is given in Appendix 2.

Payback Period

10 The payback periods for the two systems are given below in Table 2.

Table 2

Payback Periods for Steam Turbine Direct Drive CHP Plant

Turbine Size (kW)	Payback Period (for various boiler fuels)		
	Gas (years)	Heavy Fuel Oil (years)	Coal (years)
250	2.0	2.0	1.8
750	1.2	1.2	1.1

11 The payback period for the larger size of turbine is short, but a large steady site steam demand is needed. For a 250 MW turbine; a steady steam demand of 5.3 MW (17000 lbs/hr) is required and for the 750 kW turbine; 16.0 MW of steam (52000 lbs/hr). If the turbine is not used at full load for 8000 hours per year then the savings will be reduced in a roughly linear proportion and hence the payback periods will increase proportionately.

12 In situations where it is not practical to couple a steam turbine to an existing compressor the payback period would increase, perhaps by a factor of two.

1.2 Industrial Spark Ignition Engines

13 Four-stroke spark ignition internal combustion engines operating on natural gas in the size range 100 kW - 1 MW were considered. The performance of these engines is described in section 2.5 of Appendix 2.

14 The engine was used to drive an air compressor directly. Heat was recovered as hot water from the jacket cooling water and engine exhaust. In addition hot water was also recovered from the intercoolers and aftercooler of the air compressor.

15 In theory nearly all of the shaft power used to drive an air compressor can be recovered as heat from the air compressor. In practice the amount of heat recovered depends on the temperature required for the heat and the temperature of the cold feed. The heat may be recovered as hot air or water depending on the application. Hot air can be used for space heating however this limits the heat recovery to the heating season only.

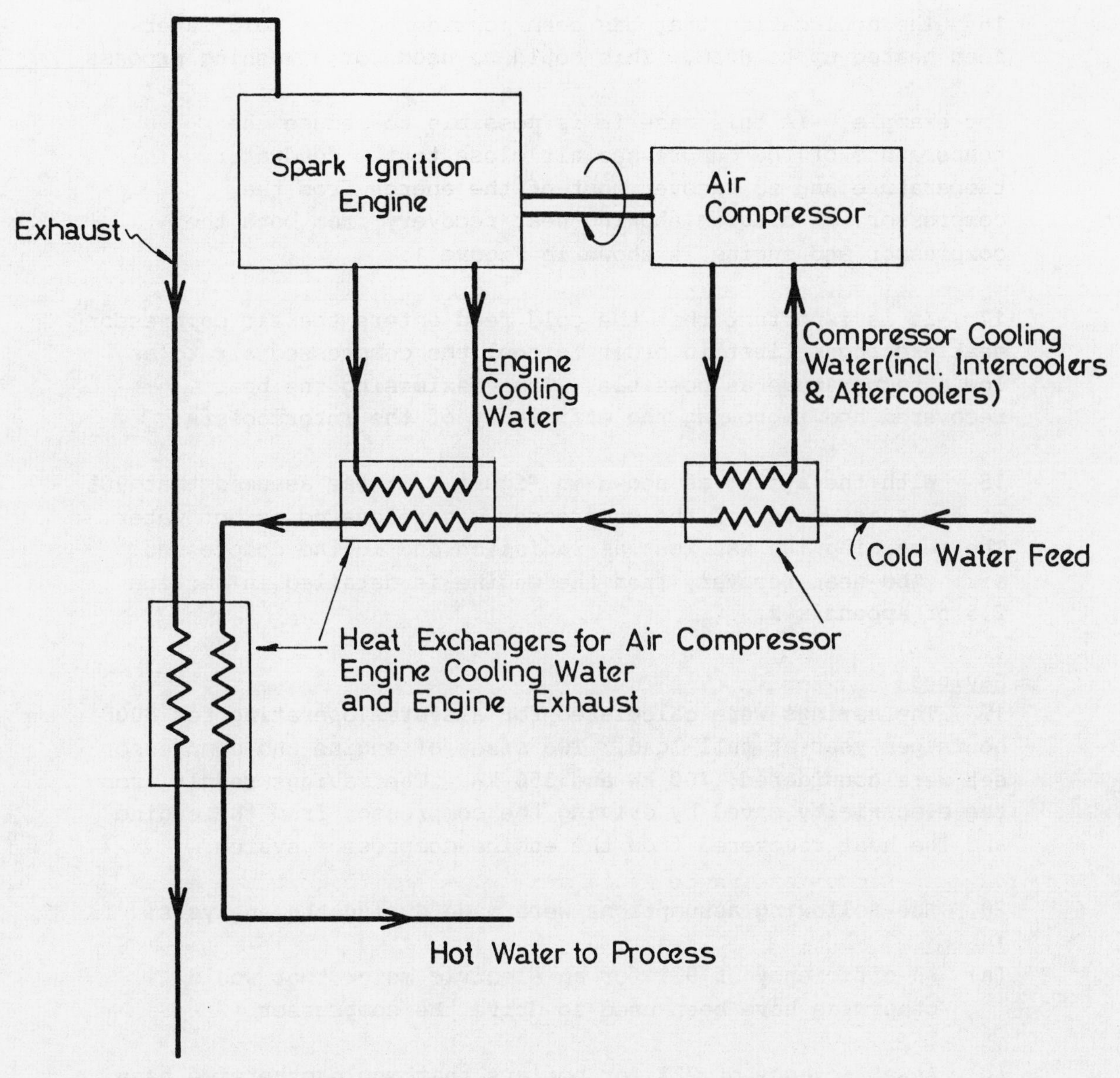

HEAT RECOVERY FROM A SPARK IGNITION ENGINE AND AIR COMPRESSOR **Fig.1**

16 The application that has been considered is a cold water feed heated up to 80°C. This could be used for a washing process for example. In this case it is possible to reduce the temperature of the compressed air close to the feedwater temperature and so recover most of the energy from the compressor. A diagram showing heat recovery from both the compressor and engine is shown in Figure 1.

17 It is important that the cold feed enters the air compressor heat exchanger first in order to cool the compressed air to as low a temperature as possible. Thus maximising the heat recovered and improving the efficiency of the intercoolers.

18 With the system as shown in Figure 1 it was assumed that 90% of the shaft power of the compressor was recovered as hot water. The remaining 10% was lost as radiation and in the compressed air. The heat recovery from the engine is detailed in section 2.5 of Appendix 2.

<u>Savings</u>

19 The savings were calculated for a system operating for 8000 hours per year at full load. Two sizes of engine and compressor set were considered: 100 kW and 550 kW. The savings result from the electricity saved by driving the compressor from the engine and the heat recovered from the engine-compressor system.

20 The following assumptions were made during the analysis:

(a) An efficiency of 92% for an electric motor that would otherwise have been used to drive the compressor.

(b) An efficiency of 75% for boilers that would otherwise have been used to provide the hot water.

(c) A maintenance cost of 0.47 p/kWh of shaft power from the engine.

(d) A temperature of 150°C for the exhaust gas from the engine after heat recovery.

(e) In calculating the savings it was assumed that there would be no heat recovery from an electrically driven compressor. Thus the savings resulting from the recovery of heat from the compressor could be added to those resulting from the installation of the spark ignition engine.

The net annual energy savings are shown below in Table 3.

Table 3
Annual Energy Savings for Spark Ignition Engine Driven Air Compressor with Heat Recovery

Engine Size (kW shaft)	Saving in Electricity (MWh/year)	Boiler Fuel Saved (GJ/year)	Fuel for Engine (GJ/year)	National Saving (TCE/year)
100	870	9675	9550	400
550	4800	53210	52525	690

Annual cost savings were calculated from the energy costs used in the rest of the study. These gave the savings shown in Table 4.

Table 4
Annual Cost Savings

Engine Size	Net Annual Cost Savings (for various boiler fuels)		
(kW shaft)	Gas	Heavy Fuel Oil	Coal
100	£25280	£25280	£20640
550	£139560	£139560	£114010

Capital Costs

21 The capital costs for spark ignition engine systems for direct drive applications are given in Section 3.1 of Appendix 2. These costs do not include an air compressor or the heat recovery equipment for the compressor.

22 It was assumed that a spark ignition engine drive could be retrofitted onto an existing air compressor but heat recovery equipment would have to be provided for the compressor.

23 The capital costs for spark ignition engine installations including the costs for heat recovery from the air compressor are as follows:

(a) £ 75000 for 100 kW engine
(b) £195000 for 550 kW engine

Payback Period

24 The payback periods for the two systems described are given below in Table 5.

Table 5
Payback Periods for Spark Ignition
Engine Direct Drive CHP Plant

Turbine Size (kW)	Payback Period (years) (for various boiler fuels)		
	Gas	Heavy Fuel Oil	Coal
100	3.0	3.0	3.6
550	1.4	1.4	1.7

25 It can be seen that short payback periods can be obtained especially for the larger engine. However these paybacks assume operation for 8000 hours per year. A reduction in the operating hours would reduce the annual savings in roughly linear proportion and this would increase the paybacks. Furthermore, it is assumed that all of the heat from the compressor and engine is recovered. In some cases it may not be practical to recover the heat from the air compressor and this would decrease the annual savings by over 30%, increasing the payback correspondingly. The savings would also decrease by a similar amount if the compressor already had heat recovery.

26 If it is not practical to couple the spark ignition engine to an existing compressor the payback period would increase, perhaps by a factor of two.

1.3 Potential for Direct Drive CHP Systems

27 The gross and investment potentials could not be calculated for direct drive CHP plant to the same degree of precision as the potential for whole site CHP, as detailed information on the applications suitable for direct drive CHP was not available. The potential estimated here has therefore been derived using broad assumptions about the applications suitable for direct drive CHP.

28 The foregoing analysis shows that direct drive CHP systems in the size range of 100 to 750 kW, and operating 8000 hrs per year, offer simple paybacks from 1 to 4 years depending on the size of the installation and the fuel used.

29 The analysis is based on the assumption that the drives can be applied to an existing load, be it compressor, pump, fan, or other mechanical equipment. The assumption is likely to be wrong in many cases, particularly where a reciprocating engine is considered. If the driven equipment must be replaced this will

increase the payback period, perhaps by a factor of two. However, if replacement is due in any case, the assumption remains valid.

Applicability

30 The application of direct drive CHP systems examined here is circumscribed by:

the availability of a suitable single load capable of absorbing the full output of the driver;

and

in the case of steam turbine based system, the availability of a suitable steam supply and steam demand on a sufficient scale to allow the interposition of a turbine;

or

in the case of spark ignition driven compressors, a suitable low temperature heat load to absorb the heat rejected by the engine and compressor intercoolers.

and

attainment of a sufficiently attractive payback to justify the investment.

Availability of a Suitable Load

31 From the data supplied on the assessment forms completed by 25 individuals in industry, information was collected on the main electrical drives installed at a range of sites promising for CHP application. The drives served refrigeration plant, air compressors, and other mechanical requirements (granulator for example).

32 At 17 out of the 18 sites where data were made available on individual drive requirements, opportunities exist to install direct drive systems of the size range considered in this study (100 to 750 kW).

33 At all but two of the 17 sites, there is scope for several direct drive systems in the 100 to 750 kW range.

34 At three out of the 17 sites it would be necessary to replace several small air or refrigeration compressors by a few larger ones to bring the load up to the 100 kW minimum assumed in this analysis.

35 In view of the wide range of activities and establishment sizes covered by the study, it is concluded that within the overall catchment area for CHP plant, the availability of a

suitable shaft power load is not a major limitation on the applicability of direct drive CHP systems.

Steam Supply and Heat Load Limitations

36 Single stage low pressure turbines require around 70 lb/hr of steam per kW net output. Assuming a steam supply efficiency of 70% at the turbine stop valve, the following is the minimum site fuel demand required to support a steam based direct CHP system, assuming 8000 hr/yr operation.

System output	Minimum site fuel demand:
250 kW	2000 kTherms per year
750 kW	6000 kTherms per year

37 Spark ignition engines driving an air compressor with heat recovery produce low pressure hot water equivalent to around 8 lb/hr of steam per kW net output. The minimum site fuel demand required to support a spark ignition engine driven compressor CHP system is shown below, assuming 8000 hr/yr operation:

System output	Minimum site fuel demand:
100 kW	100 kTherms per year
500 kW	500 kTherms per year

38 Where the system operates less than 8000 hours, the minimum annual site fuel demand needed to support a direct CHP system diminishes pro rata.

39 It is concluded that steam based direct drive systems are limited to medium and large sites (depending on the number of operating hours per year) by the available steam demand. Spark ignition engine driven compressors are not so limited but require a demand for low pressure hot water.

Attainable Payback

40 The payback attainable with direct drive systems operating 8000 hrs/yr ranges from 1 to 4 years, depending on the fuel and assuming that the existing driven machinery can be retained without costly modification.

41 The need to modify or replace the driven machinery would increase the payback, possibly doubling it. In some cases this may still result in an acceptable payback, in other cases it may

be assumed that direct drive CHP plant would only be installed when the driven machinery needs to be replaced for other reasons.

42 Operating fewer than 8000 hr/yr would increase the payback but reduce the minimum site steam demand needed to allow direct drive CHP plant, pro rota.

Assessment of Investment Potential

43 For the present purposes it was assumed that the average payback is 3 years, the average power 200 kW, the average capital commitment £100,000, and the corresponding propensity to invest 40% (using the data in Appendix 1).

44 During the course of preliminary target selection for the CHP study, data on the pattern of energy consumption at UK manufacturing sites was examined. These were classified into 3 size bands to represent small, medium and large sites to reflect the scope for CHP plant installations.

45 Size bands 2 & 3 representing medium and large sites provide a sufficient annual steam demand to permit the installation of direct drive CHP systems of 250 kW or over. There are a total of 934 such sites recorded in the data base.

46 At a fraction of these sites, the installation of a large CHP plant is a possibility. The present assessment assumes that this will actually happen at perhaps 30 to 40 sites. Such sites are removed from the catchment area for direct drive CHP plant.

47 For the present purposes, it was assumed that the total catchment area for direct CHP systems is 900 sites. At an average installed size of 200 kW the gross potential is thus 72 MW (0.4 x 900 x 0.2).

48 Not all the sites included in the catchment area in fact provide a suitable load for direct drive, or where such load does exists, the payback may not be acceptable. For the present purposes, the arbitrary assumption was made that half the sites in the catchment area can accommodate a direct driven CHP plant giving an acceptable payback.

Energy Saving

49 Savings calculated for direct drive systems amount to 2.3 ktce per MW in the case of steam driven units, and 4 ktce per MW in the case of spark ignition engine driven compressors with heat recovery, both operating 8000 hrs/yr.

50 The mix of installations, and the average yearly operation cannot be estimated with confidence. For the present purposes,

it was assumed that on average, the savings amount to 2 ktce per MW.

51 The results of the analysis and the assumptions stated above are summarised below in terms of estimated installed capacity, level of investment, and energy saving implications.

Investment Potential for Direct Drive CHP Plant

Manufacturing Industry	Estimated Installed Capacity (MWe)	Estimated Level of Investment (£Million)	Estimated Primary Energy Saving (ktce)
Total	35	18	70

2.0 Retro-fit Steam Turbines for Low Pressure Steam Boilers

52 The main part of the study is concerned with the installation of new CHP plant for the generation electricity and supply of heat for industrial processes. There is however an example where the retro-fitting of a steam turbine to an existing boiler installation will provide attractive paybacks.

53 Many industrial sites generate saturated steam at low pressures up to 250 psig (18.2 bar abs) using shell and tube type boilers. Such boilers cannot operate at pressures much above 250 psig and are not appropriate for high pressure CHP steam turbine installations. In many industries the steam raising boilers are capable of operating up to 250 psig and yet the steam distribution pressures may be below 100 psig. It can be feasible in such circumstances to generate steam at a high pressure and reduce the pressure in a steam turbine generating set to the site distribution pressure. Single stage steam turbines are most appropriate for these applications on account of their relatively low cost (for small sizes) and the small pressure ratio required.

54 An example is given below of a typical application with details of the savings and payback that could be achieved.

2.1 Analysis of the Systems

55 In a typical system saturated steam at 250 psig (18.2 bar abs) would be generated by the existing boiler installation. This could be passed through a back pressure single stage steam turbine generating set exhausting at 80 psig (6.5 bar abs). The performance of a steam turbine of this type is given in section 2.8 of Appendix 2. The shaft output given should be multiplied by 0.95 to allow for losses in the electricity generator.

56 A feature of single stage steam turbines operating over low pressure ratios is a high heat to power ratio. Thus a high steam flow rate is required per kW of electrical output. Allowing for the electrical efficiency of a generator a steam flow of 70000 lbs/hr is required to generate 1 MWe.

57 The effect of installing the steam turbine would be to increase slightly the steam output required from the boilers. At the turbine exhaust the steam would be slightly wet and a proportion of the boiler steam flow (about 5%) would be removed as condensate. Thus a slight increase in boiler output is required to supply the existing steam demand. The extra boiler fuel is offset by the electricity generated and the marginal efficiency is favourable. For the system considered an increase in boiler fuel consumption of 10 kW would produce 6.6 kW of electricity while maintaining the same process steam demand.

Savings

58 The net savings from the installation of the steam turbine generator result from the savings in electricity after accounting for the extra boiler fuel and maintenance costs.

59 Two systems have been analysed for installations of 250 kW and 750 kW electrical output. The net cost savings are shown in Table 6 and involve the following assumptions:

(a) The steam turbine operates at full load for 8000 hours per year.

(b) The maintenance cost for the turbine is 0.15 p/kWh of electricity generated.

(c) The efficiency of the boiler is 75% based on higher calorific value.

(d) The electricity generated displaces imported electricity and the savings are therefore costed at that rate.

Table 6

Annual Energy Savings From Steam Turbine Generating Set

Generating Set Size (kWe)	Electricity Generated (MWh/year)	Extra Boiler Fuel required (GJ/year)	Net. National Energy Saving (TCE/year)
250	2000	10880	490
750	6000	32650	1470

60 The annual cost savings resulting from the energy savings above and including maintenance costs are shown below in Table 7. They are based on the fuel prices used in the main part of the study.

Table 7
Annual Cost Savings

Generating Set Size (kWe)	Net Annual Cost Savings (for various boiler fuels)		
	Gas	Heavy Fuel Oil	Coal
250	£33840	£33840	£39060
750	£101500	£101500	£117200

Capital Costs

61 The capital costs for complete backpressure steam turbine installations operating from low pressure boilers are given in section 3.4 of Appendix 2. For the sizes considered they are:

(a) £110000 for 250 kWe turbine
(b) £190000 for 750 kWe turbine

Payback Period

62 The simple payback period for the two systems considered is shown in Table 8 for various boiler fuels.

Table 8
Payback Period for Low Pressure Steam Turbine Generating Sets

Turbine Size (kWe)	Payback Period (Years) (for various boiler fuels)		
	Gas	Heavy Fuel Oil	Coal
250	3.3	3.3	2.8
750	1.9	1.9	1.6

63 Table 8 shows that the simple paybacks for the larger turbine are quite attractive. The payback is based on operation for 8000 hours per year and therefore assumes a steady steam demand. The high heat to power ratio of these turbines means that a steady steam flow of 53000 lbs/hour is required for the 750 kWe turbine.

2.2 Potential for Retro-fit Steam Turbines

64 The potential could not be estimated to the same degree of precision as the potential for whole site CHP, as detailed information on the applications suitable for retrofit steam turbines was not available. The potential estimated here has therefore been established using broad assumptions about suitable applications.

65 The foregoing analysis shows that retrofit steam turbine CHP systems in the size range of 250 to 750 kW, and operating 8000 hrs per year, offer simple paybacks from 1.6 to 5.1 years depending on the size of the installation and the fuel used.

Applicability

66 The application of retrofit steam turbine CHP systems examined here is circumscribed by:

- the availability of a suitable steam supply and steam demand on a sufficient scale to allow the interposition of a turbine;
- the availability of sufficient electrical base load to absorb the output of the generator; and
- attainment of a sufficiently attractive payback to justify the investment.

Steam Supply and Heat Load Limitations

67 Single stage low pressure turbines require around 70 lb/hr of steam per kW net output. Assuming a steam supply efficiency of 70% at the turbine stop valve, the following is the minimum site fuel demand required to support a steam based direct CHP system, assuming 8000hrs/yr operation.

System output	Minimum site fuel demand:
250 kW	2,000 kTherms per year
750 kW	6,000 kTherms per year

68 It is concluded that the application area for retrofit steam turbine based CHP systems overlaps that of direct drive CHP systems in medium and large sites.

Attainable Payback

69 The payback attainable with direct drive systems operating 8000 hrs/yr ranges from 1.6 to 5.1 years, depending on the fuel and size of the installation. In general, this is inferior to

the payback obtainable from direct drive CHP systems if there is no need to replace the driven machinery, but probably similar to that obtained if the driven machinery needs replacement.

70 Operating fewer than 8000 hr/yr would increase the payback but reduce the minimum site steam demand needed to allow direct drive CHP plant, pro rota.

Assessment of Investment Potential

71 It is concluded that retrofit steam turbine electricity generating CHP systems, being steam demand limited, represent an interesting alternative to direct drive systems, but make no addition to the investment potential for CHP.

72 The concept of small retrofit electricity generating CHP systems increases the range of options available, and removes some of the limitations imposed by the possible unavailability of suitable loads on the use of direct drive systems. For the present purposes, the arbitrary assumption was made that such concepts could be applied in the case of half the sites previously excluded from the scope for direct drive systems, that is 90 sites (900 x 0.4 x 0.25).

Energy Saving

73 Savings calculated for small electricity generating CHP systems amount to 2 ktce per MW, operating 8000 hrs/yr.

74 For the present purposes, it will be assumed that operating hours are below 8000, and that the annual savings are similar to those realised in conventional CHP plant, or around 1.6 ktce per MW.

75 An average installed cost of £150,000, corresponding to an average unit size of around 300 kW, was assumed. This suggests savings of 0.48 ktce per installation.

76 The results of the analysis and assumptions stated above are summarised below in terms of installed capacity, likely level of investment, and energy saving implications.

Investment Potential for Retro-fit Steam Turbine CHP Plant

Manufacturing Industry	Estimated Installed Capacity (MWe)	Estimated Level of Investment (£Million)	Estimated Primary Energy Saving (ktce)
Total	27	13	43

2.3 Estimated Total Investment Potential

77 Summing the estimated investment potentials for both direct drive and retrofit steam turbine CHP gives the following result:

Investment Potential	Estimated Installed Capacity (MWe)	Estimated Capital Investment (£Million)	Estimated Primary Energy Saving (ktce)
Direct Drive CHP Plant	35	18	70
Retrofit Steam Turbine Sets	27	13	43
Total(rounded)	60	30	110

78 The estimate of realistic potential is based upon broad assumptions about the number of applications that would be suitable for direct drives and retrofit steam turbines.

79 The results indicate that the realistic potential for these types of plant is in the order 60 MWe.

INDUSTRIAL COMBINED HEAT AND POWER
THE POTENTIAL FOR NEW USERS

APPENDIX 5
ASSESSMENT OF GROUP CHP SCHEMES

CONTENTS

1.0 Introduction

1 There are instances where the heat and power demands of an individual site are such that a CHP plant installation will not offer an attractive payback.

2 In such circumstances the summation of the heat and power demands of two or more nearby sites may produce the opportunity for a more economic CHP scheme. In such a scheme one CHP plant would be installed to provide heat and power for a group of industrial sites. The CHP plant may be owned by one of the industrial sites and heat and power sold to the other sites.Alternatively a utility company may install a plant and sell heat and power to the nearby industrial sites.

3 Assessing the potential within the United Kingdom for such group CHP schemes is beyond the scope of this study, on account of the large effort needed to assemble all of the geographical information. However, a method is described here to enable a preliminary assessment of group schemes to be made wherever the possibility is identified.

2.0 Assessment of a Group CHP Scheme

4 The simple graphical method, described in Appendix 1 (section 3), for estimating the most cost effective payback period for a CHP plant can also be used to make an initial estimate of the payback period for group CHP schemes. The heat and electricity demands for each site in the group can be added together and the values of the five parameters established, as described in Appendix 1, for the group as a whole. The payback period for the group can then be estimated from the assessment charts.

Example

5 An example is given below showing the benefit of a group CHP scheme for the two sites A and B.

	Site A	Site B
Total Annual Heat Demand (kTherms)	1750 (185000 GJ)	2532 (267100 GJ)
Total Annual Electrical Consumption (MWh)	10790 (38840 GJ)	30790 (110800 GJ)
Typical Maximum Heat Steady Demand (Therms/hr)	250 (7.33 MW)	455 (13.3 MW)
Price of fuel for boilers (p/Therm)	28.2 (£2.68/GJ)	28.2 (£2.68/GJ)

6 The values of the five parameters needed to calculate the payback for a gas turbine CHP installation on each site are:

	Site A	Site B
Average Site Heat Demand	5.85 MW	8.47 MW
Average Heat/Power ratio	4.75	2.41
Average Load Factor	0.8	0.634
CHP/Site Fuel Price Ratio	1.0	1.0
CHP Fuel Price	28p/therm	28p/therm

7 The payback periods for the most cost effective gas turbine installations are calculated from the chart (Appendix 1, Fig. 2) to be:

5.9 years for Site A
5.3 years for Site B

8 Combining the heat and power demands for each site to produce a group scheme gives the following annual heat demands:

Group annual heat demand (Therms)	4282 (452100 GJ)
Group annual electrical consumption (MWh)	41580 (149640 GJ)
Group maximum heat demand (Therms/hour)	705 (20.63 MW)

9 The values of the five parameters needed to calculate the payback period for a gas turbine CHP installation for the group are:

Average group heat demand	14.32 MW
Average group heat/power ratio	3.02
Average group load factor	0.69
CHP/Site fuel price ratio	1.0
CHP Fuel Price	28p/therm

10 The payback period for the group scheme is 4.5 years which is lower than the payback period for either site individually. This gives an indication of the benefits of a group scheme.

11 The charts in Figs. 5 - 7 of Appendix 4 should only be used to give an initial estimate of the payback for a group scheme. If a favourable payback is found then a more detailed analysis should be undertaken to investigate the effect of combining the heat and power demands. Taking into account factors such as the phasing of the demands on each site.

3.0 Demonstration of Payback Periods for Group Schemes

12 The graphical method established in the previous section was used to estimate the payback periods resulting from group CHP schemes. Four schemes were investigated representing typical combinations of industrial site. The schemes are fictitious in that they do not refer to any known industrial estates or combinations of site. However they do represent typical groups of industrial site that may be found.

13 The graphical charts can only be used to make a preliminary estimate of the payback period for group schemes. They are accurate to within ±20% when used on individual sites. Group schemes are likely to contain more elements of uncertainty and variation from site to site. Consequently the charts should be used with more care for the estimation of payback period for group schemes.

Example I

14 Example I comprises four sites, each of approximately the same size, typical of small factories in the mechanical engineering and manufactured goods industries. Sites in these industries are characterised by having fairly low load factors for the heat demand as the heat load is predominantly for space heating.

15 The heat and power demands are shown in Table 1 (page 505). The sites all burn gas in their boilers and conversion to gas turbine CHP plant is investigated. The payback periods for the best gas turbine CHP plant at each of the four sites and the group scheme are estimated from the graphical charts as follows:

Site	Best Payback Period for Gas Turbine CHP
(a)	8.6 years
(b)	7.5 years
(c)	9.0 years
(d)	9.0 years
Group Scheme	5.7 years

16 It can be seen that the CHP payback period for the group scheme is considerably better than for CHP at any individual site.

Example II

17 Example II comprises two sites in the food industry; a

distillery and a frozen food factory. Each has a similar load factor. But the distillery has a high heat to power ratio whereas the frozen food factory had a much lower heat to power ratio on account of the high electrical demand for refrigeration. The heat and power demands are shown in Table 2 (page 505).

18 The sites both burn gas in their boilers and conversion to gas turbine CHP plant is investigated. The payback periods for the best CHP plant at each site and the group scheme are estimated from the graphical charts and the results are as follows:

Site	Best Payback Period for Gas Turbine CHP
(a)	5.1
(b)	4.7
Group Scheme	3.8

19 Each site has, individually, a similar payback period for gas turbine CHP plant in spite of the different heat loads and heat to power ratios. Combining the sites results in a much improved payback.

Example III

20 Example III comprises three sites in the paper industry. One medium sized papermill and two sites in the paper conversion industry: a site making packages and a printing works. The heat and power demands are shown in Table 3 (page 506). The sites all burn gas in their existing boilers and conversion to CHP plant is investigated. The payback periods for the best gas turbine CHP plant at each site and the group scheme are estimated from the graphical charts as follows:

Site	Best Payback Period for Gas Turbine CHP
(a)	4.8
(b)	6.5
(c)	9.8
Group Scheme	4.7

21 The payback period for the group CHP scheme is not much lower than the payback period for CHP at site (a), although the improvement for the other sites is quite considerable. Site (a)

is the largest site and has a very high load factor, consequently it has a reasonable payback for CHP as an individual site. The effect of adding a number of smaller sites with lower load factor is not as great therefore as if all the sites had been of a similar size.

Example IV

22 Example IV comprises four sites; one large process site in the soap and detergents industry and three smaller manufacturing sites comprising a rubber products site and two light engineering products sites. The heat and power demands are shown in Table 4 (page 506).

23 The sites all burn heavy fuel oil in their boilers and conversion to coal fired steam turbine CHP plant was investigated. The payback periods for the best CHP plant at each site and the group scheme were estimated from the graphical charts and the results are as follows:

Site	Best Payback Period for Steam Turbine CHP
(a)	7.1
(b)	13.8
(c)	11.0
(d)	11.0
Group Scheme	5.4

24 Site (a) has a high heat to power ratio and high load factor. Consequently when converting from heavy fuel oil to coal firing and steam turbine CHP the payback period, although not particulary attractive, is substantially better than those for the smaller sites, which have much lower heat demands and load factors. The effects of bringing the smaller sites into a group scheme however is to improve significantly on the payback achievable at site (a) alone.

Table 1
Group Scheme - Example I
Heat and Power Demands

Site	Mean Heat Demand [MW]	Heat to Power Ratio	Load Factor	Fuel for Existing Boilers
(a) Engineering	3.0	2.25	0.24	Heavy Fuel Oil
(b) Engineering	4.0	2.1	0.28	Heavy Fuel Oil
(c) Engineering	2.5	2.1	0.28	Heavy Fuel Oil
(d) Engineering	2.5	2.25	0.24	Heavy Fuel Oil
Group Scheme	12.0	2.2	0.26	Gas Fired CHP

Table 2
Group Scheme - Example II
Heat and Power Demands

Site	Mean Heat Demand [MW]	Heat to Power Ratio	Load Factor	Fuel for Existing Boilers
(a) Distillery	27.4	11.3	0.50	Gas
(b) Frozen Food Factory	6.8	0.9	0.64	Gas
Group Scheme	34.2	3.4	0.53	Gas Fired CHP

Table 3
Group Scheme - Example III
Heat and Power Demands

Site	Mean Heat Demand [MW]	Heat to Power Ratio	Load Factor	Fuel for Existing Boilers
(a) Papermill	9.4	3.4	0.80	Gas
(b) Paper Packaging	7.9	3.8	0.31	Gas
(c) Printing	1.0	3.6	0.25	Gas
Group Scheme	18.3	3.5	0.45	Gas Fired CHP

Table 4
Group Scheme - Example IV
Heat and Power Demands

Site	Mean Heat Demand [MW]	Heat to Power Ratio	Load Factor	Fuel for Existing Boilers
(a) Soap Factory	18.2	5.0	0.83	Oil
(b) Rubber Products	4.4	3.6	0.57	Oil
(c) Engineering	9.2	2.4	0.28	Oil
(d) Engineering	9.2	2.4	0.28	Oil
Group Scheme	41.3	3.3	0.43	Coal Fired CHP

4.0 Conclusions on the Choice of Group Schemes

25 The examples of group CHP schemes shown in the previous section clearly demonstrate how the grouping together of several sites can produce the opportunity for a more attractive CHP installation. Regarding the choice of sites that are suitable for group schemes, several general comments can be made.

(i) In general the payback period for a CHP installation decreases as the size of the heat demands and load factor increase. Below 10 MW average heat demand the payback period decreases rapidly with increasing heat demand. Whereas above 20 MW average heat demand the payback decreases much more slowly. The reason for this is mainly attributable to the higher capital cost of CHP plant per unit of output at the lower unit sizes. It follows therefore that the benefit, in terms of reduction in payback, is much greater when small sites are combined than when larger sites are combined in a group scheme. This is illustrated in the cases shown in Examples I to IV.

(ii) There are two main types of group CHP scheme that can be envisaged: those where the sites of the group are all of similar size and those where a large site is grouped with one or more smaller sites.

Where several sites of the same size are grouped together then, following on from the previous comments, a much greater improvement in payback period will occur where smaller sites are involved rather than larger sites. Furthermore the distances involved in linking large sites together may increase the capital cost so much that it may be better to operate individual CHP plants on each site.

Where a large plant is grouped with several smaller sites the payback period that can be achieved from the group scheme will only be a marginal improvement on the payback period achievable at the larger site, but a substantial improvement on the payback periods achievable at the smaller sites. This is illustrated in Examples III and IV.

26 In conclusion, the payback for a group scheme will depend upon a complex relationship between the heat and power demands of the sites in the group. The method described in this appendix can enable a preliminary assessment of this payback to be made. Following this a more detailed analysis must be undertaken to establish a more accurate payback, this would also identify the optimum size and type of plant and the magnitude of the cost savings and capital cost involved.

5.0 Use of the Electricity Board's System to Transfer Electricity

27 Since the Energy Act 1983 the electricity boards have been obliged to publish tariffs for private generators who wish to transfer electricity from one site to another via the electricity board's system. This tariff is applicable to group CHP schemes where a generator on one site may wish to transfer electricity to another site in the group.

The "use of the system" tariffs are complicated and involve several parts including:-

- A standing charge
- A use of the system charge, which includes allowances for losses dependant on the voltage at which the electricity is transferred.

28 In addition there are top up charges for extra electricity drawn by the neighbouring site on top of the amount transferred from the generating site.

29 For group CHP schemes the sites must be adjacent or very close to one another otherwise the cost of transferring the heat would be excessive. It is possible therefore that a private "across the fence" electricity connection can be made to transfer electricity. These are thus three ways in which electricity can be transferred from between sites involved in a group scheme:

- via the electricity board's system utilising a "use of the system" tariff.

- via a private 'across the fence' connection.

- by the generating site exporting all surplus electricity to the electricity board and the neighbouring site purchasing all of its electricity on a conventional maximum demand tariff.

30 The most economical way of transferring electricity will depend on the particular circumstances of the sites in a group scheme and each case must be considered individually. There are some general points which should be considered however.

31 Firstly, the top-up charges which apply in conjunction with a use of the system tariff need to be judged against the costs of installing a private electrical connection, and can be high during times of peak demand in the winter. If the generating site in a group cannot supply the bulk of the electrical demand

of the neighbouring sites then the 'use of the system' tariff may be a more expensive way of transferring the electricity.

32 Secondly, a private electrical connection between two neighbouring sites only involves a once and for all capital expenditure. It also provides a certain amount of independence from the electricity board. Where the sites are close together this option may well be the most economical option.

33 The electricity board's system can be used to transfer electricity between the sites involved in a group CHP scheme however the other ways of transferring the electricity should be considered carefully to establish the most economical means.

INDUSTRIAL COMBINED HEAT AND POWER
THE POTENTIAL FOR NEW USERS

APPENDIX 6
OTHER WORK IN THIS FIELD

CONTENTS

1.0 Work in the United Kingdom

1 Little significant work has been undertaken in the UK to quantify the potential for industrial CHP although much work has been done for city-wide CHP combined with district heating schemes, as there is currently a considerable interest in this area.

2 Energy Paper No. 35 (published by HMSO 1979) is a report on Combined Heat and Power in the UK. It is concerned primarily with CHP district heating but industrial CHP is also discussed on a qualitative level. In this report case studies were examined and it was concluded that industrial CHP schemes were technically feasible and could make substantial savings in primary energy but the economics were often unattractive to industrial companies.

3 A report on the potential for industrial CHP in the UK was presented as evidence to the Sizewell Inquiry. 'Industrial Combined Heat and Power Generation', - Sizewell B Inquiry, Proof of Evidence, by B. Wilkins for the Council for the Protection of Rural England. CPRE/P/5. June 1983. The potential is estimated to be within the range 5-10 GWe by the year 2000 and 7 to 14 GWe by 2010. This is substantially greater than that estimated in this study. The estimate is based upon general published information on the number and size of the consumers supplied with electricity and gas in the UK. The economic analysis identifies payback periods significantly lower than those identified in this study, and based upon capital costs for CHP installations typically less than a third of those used in this study. The estimates of installed capacity suggest that CHP is installed almost as a matter of course. The report is based upon far more generalised information than this study. It is felt therefore that the estimates of potential are significantly in excess of what is likely to be installed.

4 Most of the other publications concerned with industrial CHP have been descriptions of case studies or general discussions of the technical and economic factors affecting CHP. Some of these are listed in the bibliography.

2.0 Work Overseas

5 In the United States of America a number of studies have been undertaken to quantify a potential for industrial CHP. Some of these have involved identifying the potential for industrial CHP in certain areas of the country, in certain industries or for particular types of CHP plant. Two studies in particular have been undertaken for the US Department of Energy to estimate the national potential for industrial CHP in all industries. These are:

"Industrial Cogeneration Potential (1980 - 2000) Targeting of Opportunities at the Plant Site (TOPS)", by General Energy Associates, Published in 1983 for the US Department of Energy.

"Industrial Cogeneration Potential (1980 - 2000) for application of four commercially available prime movers at the plant site", by Dun and Bradstreet Technical Economic Services, Published in August 1984 for the US Department of Energy.

6 Both of these studies were of a similar nature and the second was an update of the first. They both utilised the Industrial Plant Energy Profile Data Base of General Energy Associates. This data base contains industrial plant energy profile data for the top 10000 industrial sites in the USA. A data base of CHP plant was then used in conjunction with this industrial site data base to determine the most cost effective CHP plant at each of the 10000 sites. The results were then summed to obtain the total potential for all of industry.

7 In some ways the studies were similar to this one but they had the advantage of being able to use an existing data base of industrial plant energy profiles.

8 The studies only quantified the 'gross potential' for CHP based on one economic criterion (that the internal rate of return was greater than 7%) and they did not assess the quantity of CHP that may be actually installed ('investment potential').

9 Based on the above economic criterion it was estimated that the potential for industrial CHP in the USA was in the order of 40000MW. United States industry is much larger than UK industry and has a different mix. It is therefore difficult to make any realistic comparisons of the potential for CHP between the two countries.

10 There are also many published case studies of industrial CHP installations in the USA and related technical and economic studies of industrial CHP. Some of these are listed in the bibliography.

INDUSTRIAL COMBINED HEAT AND POWER
THE POTENTIAL FOR NEW USERS

BIBLIOGRAPHY

The bibliography contains the references of some other work relevant to industrial CHP.

UK References

'Combined Heat and Power and Electricity Generation in British Industry 1983-1988', Energy Efficiency Series No. 5, Energy Efficiency Office, HMSO, 1986

'Technical and Economic Impact of Co-generation', Seminar at Institution of Mechanical Engineers, London, 22nd May 1986.

'Combined Heat and Power' Symposium organised jointly by the Institution of Chemical Engineers and the Institute of Energy, Sheffield University, 9th April 1986.

'Combined Heat and Electrical Power Generation in the United Kingdom': Energy Paper No. 35, HMSO, 1979.

'Study of Combined Heat and Power and Electricity Generation in Industry': Merz and McLellan, 1976.

'Report on Combined Power and Space Heating in Industry': McLellan and Partners, 1979.

'Combined Heat and Power - a Case History': J.D. Gurney, Energy Management Conference, Birmingham, October 1978.

'The Practicalities of Combined Heat and Power': Conference of the Combustion Engineering Association, Buxton, November 1984.

Combined Power and Heat Loads in Industry - a Case Study': Baille W.E. and O'Shea J.A., Conference on Local Energy Centres, Imperial College, London, July 1977.

These are also many other references of case studies of particular installations or discussions of the technical and economic aspects of industrial CHP.

Foreign References

'Industrial Cogeneration Potential (1980-2000) for Application of Four Commercially Available Prime Movers at the Plant Site': Dun and Bradstreet Technical Economic Services, Philadelphia, PA, USA, August 1984.

'Industrial Cogeneration Potential (1980-2000) Targeting of Opportunities at the Plant Site (TOPS): General Energy Associates Philadelphia, PA, USA, May 1983.

'Cogeneration Technology Alternatives Study CTAS' NASA, Cleveland, Ohio, USA, August 1984.

'Energy Conservation Potential of Industrial Cogeneration': Jody B.J., Fieldhouse I., Snow R.H., Energy Technology, Washington DC., USA, June 1982.

'The Potential for Industrial Cogeneration Development by 1990': Resource Planning Associates, Cambridge, MA, (USA), July 1981.

'Study of the Potential for Cogeneration in Canada: Industrial Steam Turbines': Acres Shawinigan Ltd. Toronto, Ontario, Canada, December 1979.

'Industrial Cogeneration Case Studies': Faucett (Jack) Associates, Chevy Chase, MD, USA, May 1980.

'Institutional Obstacles to Industrial Cogeneration': Dean, N.L., Selected Studies on Energy: Background Papers for Energy: The Next Twenty Years; Ballinger Pub Co., Cambridge, MA, USA, 1980.

'A.G.A. Manual Cogeneration Feasibility Analysis': Orlando J.A., American Gas Association, Energy Systems Committee and Boston Gas Company, April 1982.

'Cogeneration Technology Handbook Orlando, J.A., Government Institutes Inc., Rockville, Maryland, USA, April 1984.

'Cogeneration Case Studies Using the COGEN2 Model'; Mathtech inc Princeton Junction, NJ, USA, June 1982.

'Cogeneration Source Book': Payne F.W., Fairmont Press, 1985.

Periodicals

There are many journals which report on industrial CHP from time to time, two which regularly report on such systems are:

'Modern Power Systems', United Trade Press, London.

'Cogeneration', Pequot Publishing, Southport, CT, USA.

Other Information

The Combined Heat and Power Association represents a number of organisations and individuals who have an interest in CHP. Further information may be obtained from: CHPA, Bedford House, Stafford Road, Cateham, Surrey CR3 6JA.

Printed in the United Kingdom for Her Majesty's Stationery Office
Dd289032 12/88 C10 G3379 10170